CDX Learning Systems™

FUNDAMENTALS OF

Mobile Heavy Equipment

TASKSHEET MANUAL | AED FOUNDATION TECHNICAL STANDARDS

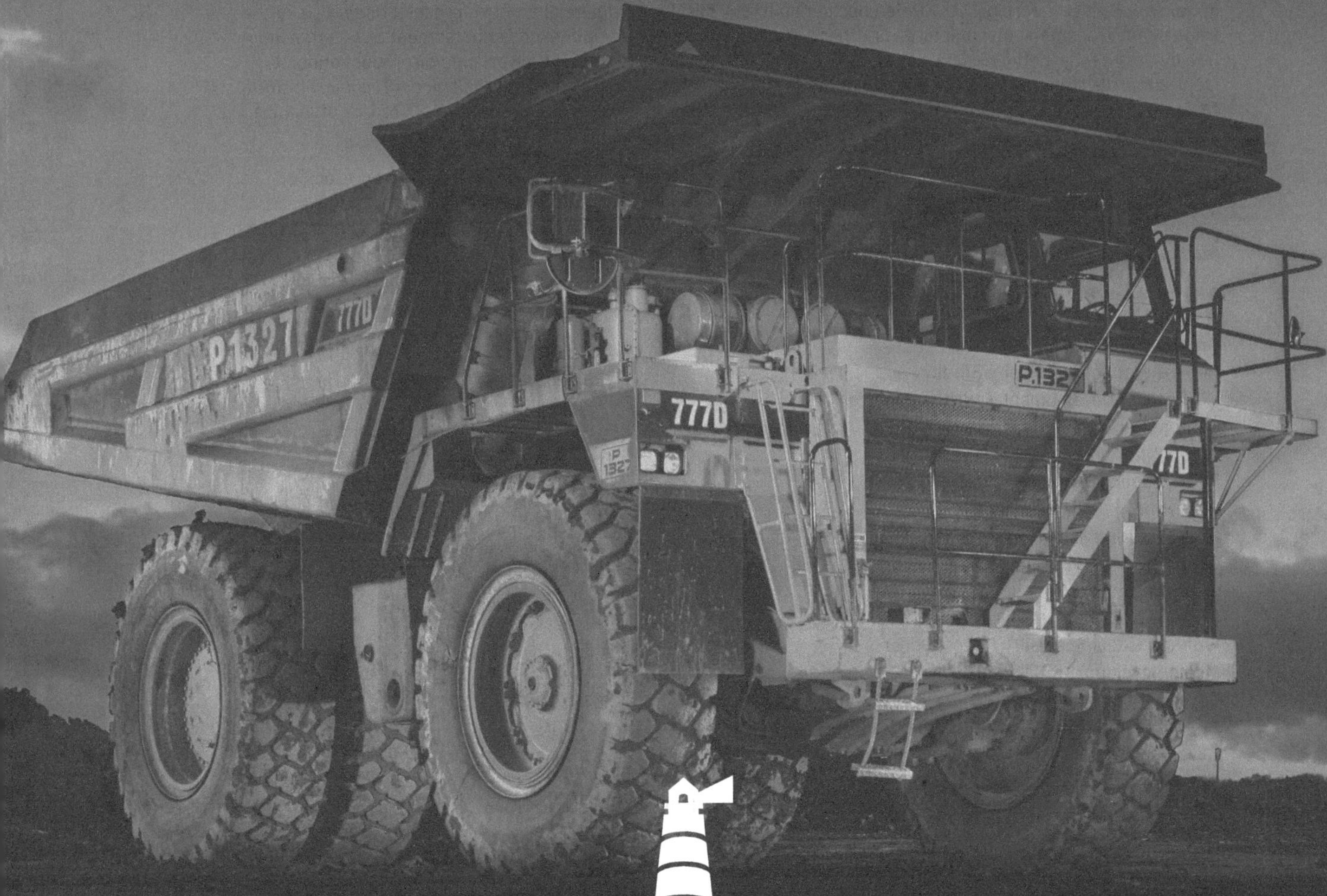

JONES & BARTLETT
LEARNING

World Headquarters
Jones & Bartlett Learning
25 Mall Road
Burlington, MA 01803
978-443-5000
info@jblearning.com
www.jblearning.com

Jones & Bartlett Learning books and products are available through most bookstores and online booksellers. To contact Jones & Bartlett Learning directly, call 800-832-0034, fax 978-443-8000, or visit our website, www.jblearning.com.

Substantial discounts on bulk quantities of Jones & Bartlett Learning publications are available to corporations, professional associations, and other qualified organizations. For details and specific discount information, contact the special sales department at Jones & Bartlett Learning via the above contact information or send an email to specialsales@jblearning.com.

978-1-284-17832-6

Production Credits
General Manager: Kimberly Brophy
VP, Product Development: Christine Emerton
Content Services Manager: Kevin Murphy
Product Manager: Jesse Mitchell
Project Specialist: Molly Hogue
Marketing Manager: Amanda Banner
Manufacturing and Inventory Control Supervisor: Amy Bacus
Composition and Project Management: Integra Software Services Pvt. Ltd.
Cover Design: Scott Moden
Director of Rights & Media: Joanna Gallant
Rights & Media Specialist: Robert Boder
Media Development Editor: Shannon Sheehan
Cover Image (Title Page): © Chris Henderson/Getty Images
Printing and Binding: Gasch Printing
Cover Printing: Gasch Printing

Library of Congress Cataloging-in-Publication Data unavailable at time of printing.

6048

Printed in the United States of America

26   10 9 8 7 6 5 4

Contents

Section A1: Safety-Administrative

CONTENTS

Use of Hand Tools

Student/Intern information:

Name _________________________________ Date ___________ Class ___________________

Vehicle used for this activity:

Year _______________ Make _____________________________ Model _____________________

Odometer ____________ Hour meter ___________ VIN _____________________________

Learning Objective/Task	CDX Tasksheet Number	2014 Edition Rev2/12/16 AED Standard
• Identify and correctly name the basic hand tools.	E0001	1a.1
• Demonstrate the proper use of the designed application and safe operating procedure for each.	E0002	1a.1
• Demonstrate a proper source for calibration of precision hand tools.	E0003	1a.1

Materials Required
- Standard toolkit and other tools as required
- Precision measuring tools: micrometer, dial-indicator, dial-caliper, etc.

Some Safety Issues to Consider
- Comply with personal and environmental safety practices associated with clothing; eye protection; hand tools; power equipment; proper ventilation; and the handling, storage, and disposal of chemicals/materials in accordance with local, state, and federal safety and environmental regulations.
- Tools allow us to increase our productivity and effectiveness. However, they must be used according to the manufacturer's procedures. Failure to follow those procedures can result in serious injury or death.

Performance Standard
0–No exposure: No information or practice provided during the program; complete training required

1–Exposure only: General information provided with no practice time; close supervision needed; additional training required

2–Limited practice: Has practiced job during training program; additional training required to develop skill

3–Moderately skilled: Has performed job independently during training program; limited additional training may be required

4–Skilled: Can perform job independently with no additional training

▶ **TASK** Identify and correctly name the basic hand tools.

AED
1a.1

Time off_______________

Time on_______________

CDX Tasksheet Number: E0001

1. **Identify instructor-designated tools and record their proper usage.**

    a. **Tool and usage:**

    ____

    ____

    ____

Total time_______________

    b. **Tool and usage:**

    ____

    ____

    ____

    c. **Tool and usage:**

    ____

    ____

    ____

    d. **Tool and usage:**

    ____

    ____

    ____

2. **Instructor Comments:**

    ____

    ____

    ____

Performance Rating

CDX Tasksheet Number: E0001

0	1	2	3	4

Supervisor/instructor signature ____ Date _____________

Student/Intern information:

Name _________________________________ Date ____________ Class _______________________

Vehicle used for this activity:

Year _______________ Make _________________________ Model _______________________

Odometer ____________ Hour meter ___________ VIN _________________________________

▶ TASK Demonstrate the proper use of the designed application and safe operating procedure for each.

AED
1a.1

Time off_______________

Time on_______________

CDX Tasksheet Number: E0002

Total time_______________

1. **Demonstrate safe handling and use of instructor-designated tools on an instructor-designated vehicle.**

    a. **Tool to be used:** _________________________________
       Appropriate usage: Yes: ___________ **No:** ___________
    b. **Tool to be used:** _________________________________
       Appropriate usage: Yes: ___________ **No:** ___________
    c. **Tool to be used:** _________________________________
       Appropriate usage: Yes: ___________ **No:** ___________
    d. **Tool to be used:** _________________________________
       Appropriate usage: Yes: ___________ **No:** ___________

2. **Instructor Comments:**

    ____
    ____
    ____

Performance Rating

CDX Tasksheet Number: E0002

0	1	2	3	4

Supervisor/instructor signature ____ Date ____________

Student/Intern information:

Name ___________________________ Date ___________ Class ___________________

Vehicle used for this activity:

Year _____________ Make ___________________________ Model ___________________

Odometer ___________ Hour meter ___________ VIN ___________________________

▶ TASK Demonstrate a proper source for calibration of precision
hand tools.

AED
1a.1

CDX Tasksheet Number: E0003

1. **The student will read and list each of the measurements shown in the
   following pages. Once those are complete, the student will need to
   demonstrate actual measurements with the various measuring tools and
   components. These tasks will require observation of the student over a
   prolonged period after the initial check. Ask your instructor to give you a
   date for your evaluation.**

   a. **Write that date here:** ___________________________

2. **Continue with your other projects, demonstrating safe handling, proper
   cleaning, maintenance, and storage of the tools until the date of your
   evaluation.**

3. **On or after that date, have your instructor verify satisfactory completion of
   each task.**

Interpreting Metric Micrometer Readings

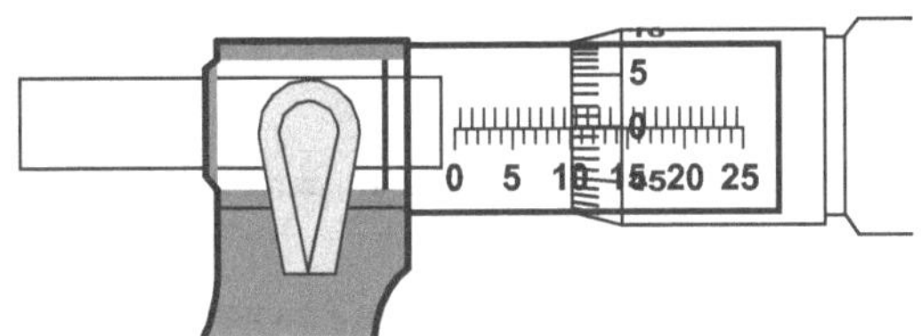

1. Actual Reading ___________________________

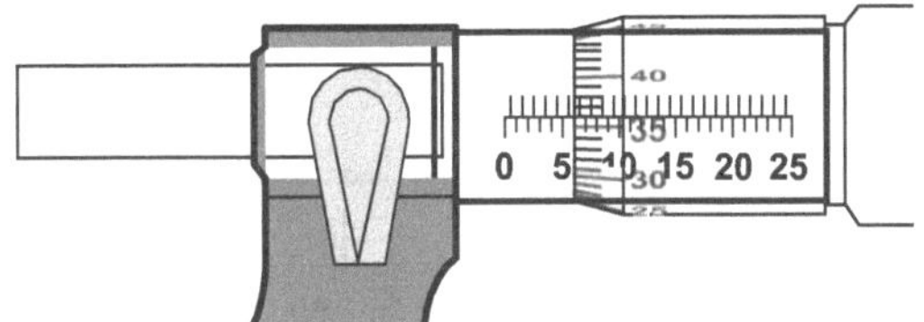

2. Actual Reading ___________________________

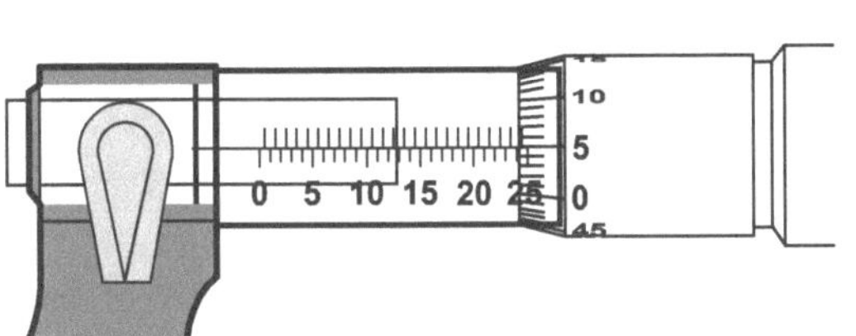

3. Actual Reading ___________________________

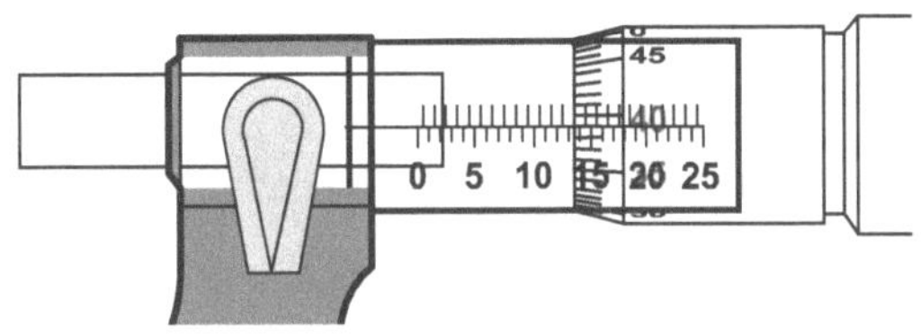

4. Actual Reading ___________________________

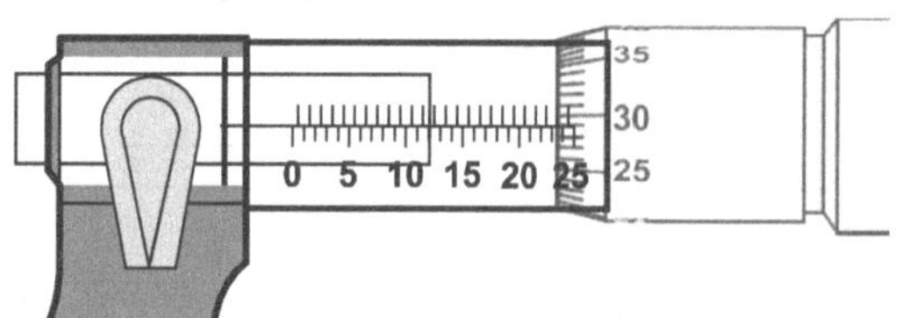

5. Actual Reading ____________

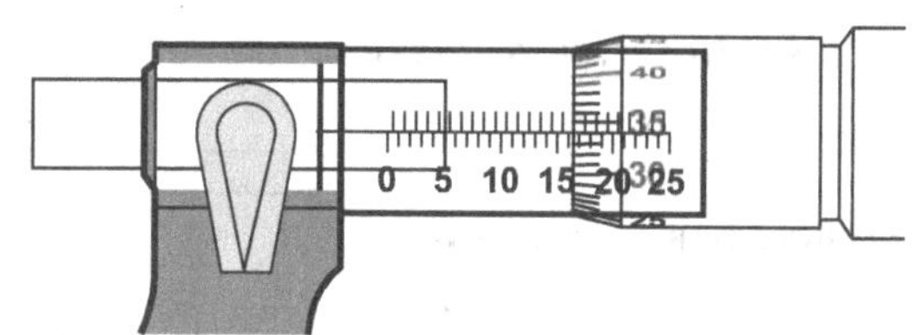

6. Actual Reading ____________

Interpreting Standard Micrometer Readings

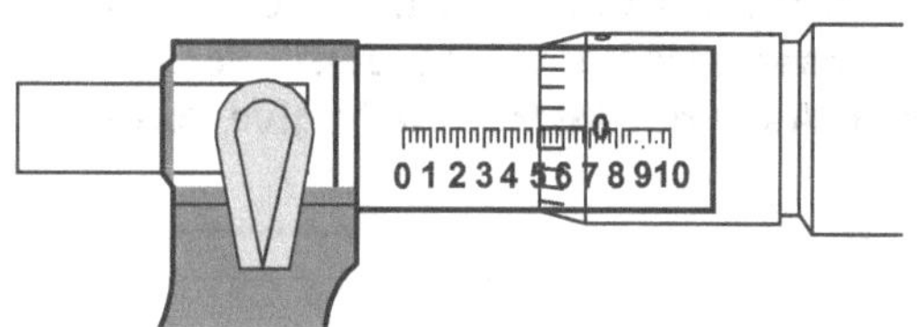

1. Actual Reading ____________

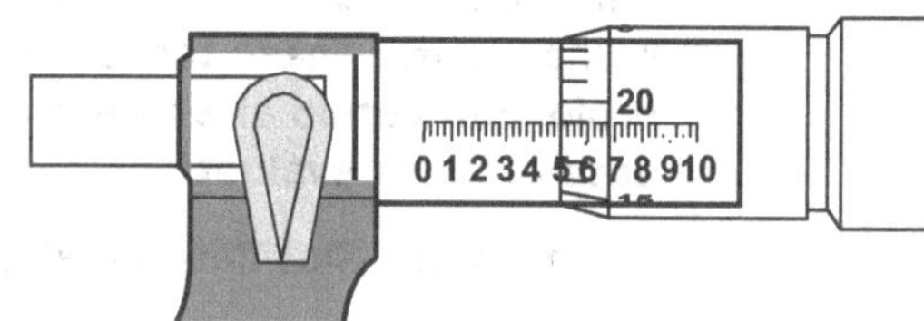

2. Actual Reading ____________

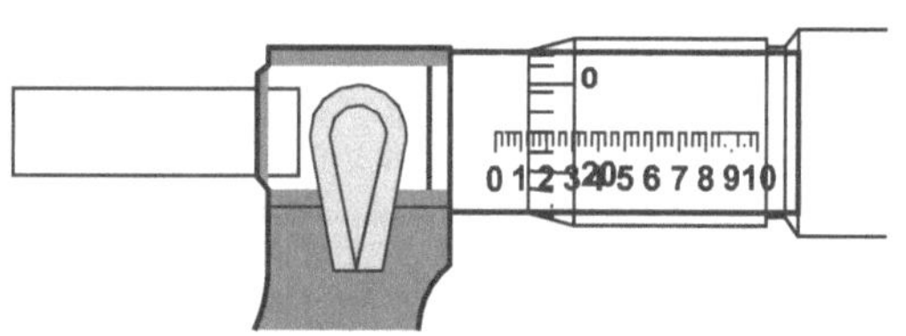

3. Actual Reading ____________

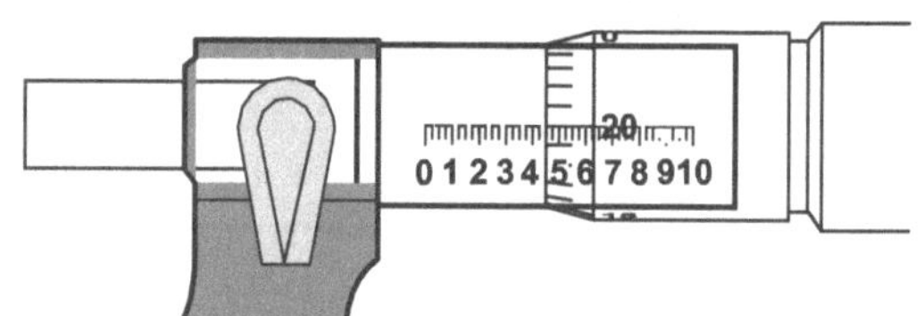

4. Actual Reading ____________

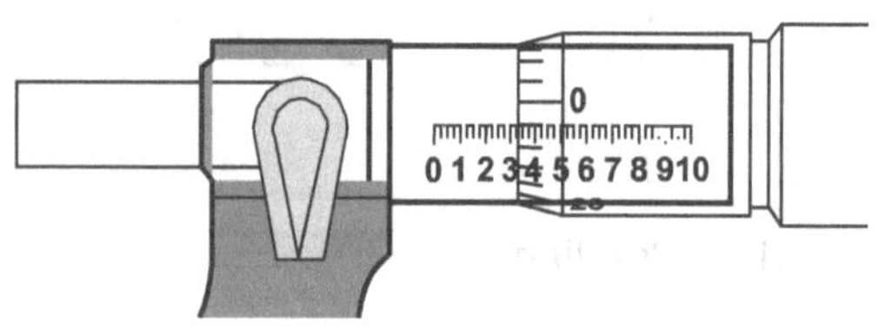

5. Actual Reading ____________

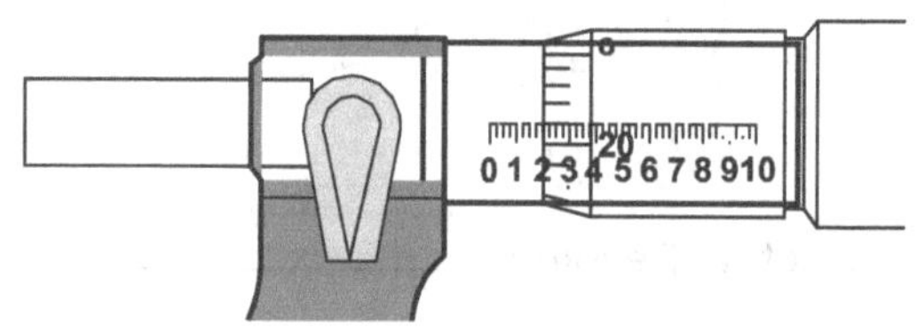

6. Actual Reading ____________

Interpreting Metric Dial-Indicator Readings

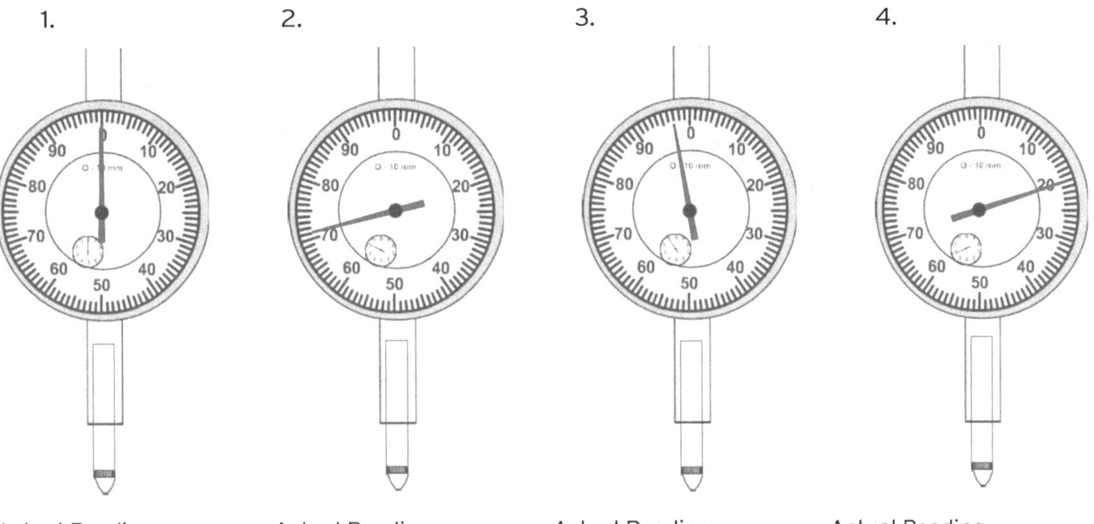

1.   2.   3.   4.

Actual Reading _______   Actual Reading _______   Actual Reading _______   Actual Reading _______

Interpreting Dial-Indicator Readings

1.   2.   3.

4.   5.   6.

Actual Reading _______   Actual Reading _______   Actual Reading _______

Interpreting Standard Dial-Caliper Readings

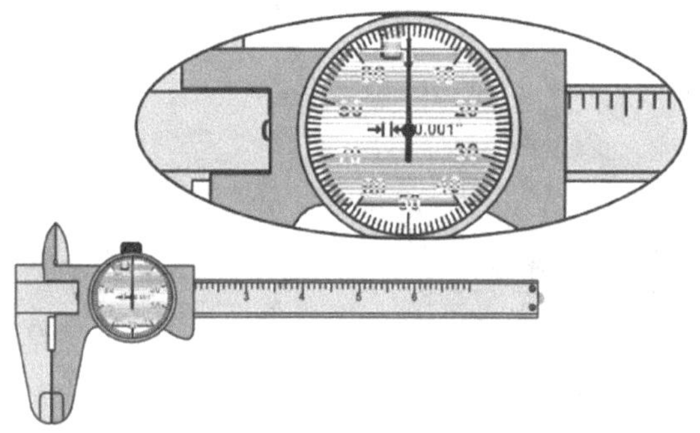

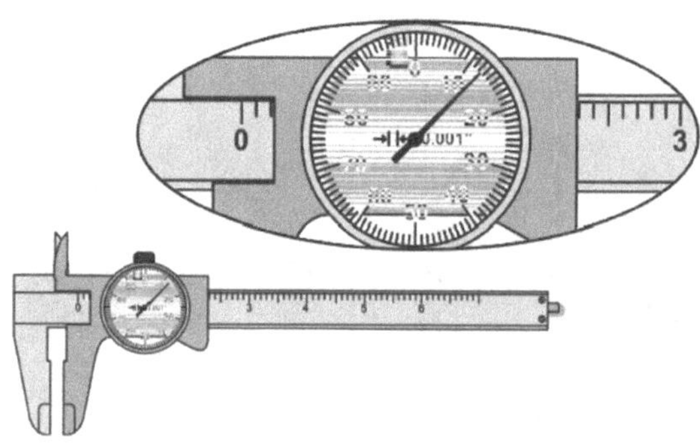

Actual Reading _______________   Actual Reading _______________

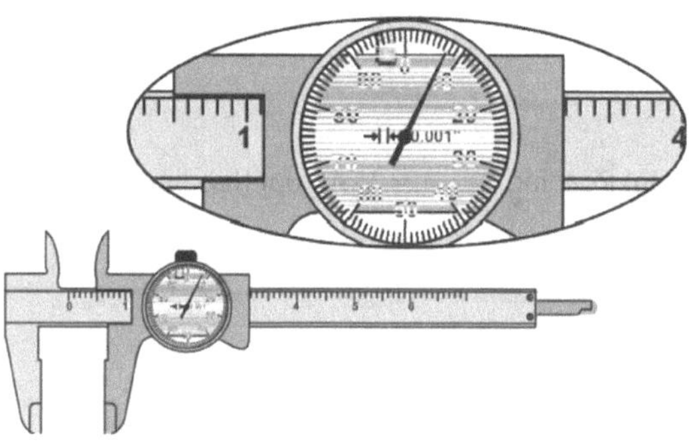

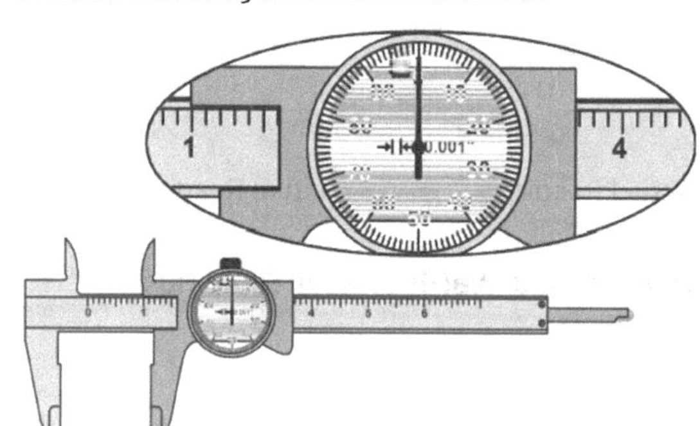

Actual Reading _______________   Actual Reading _______________

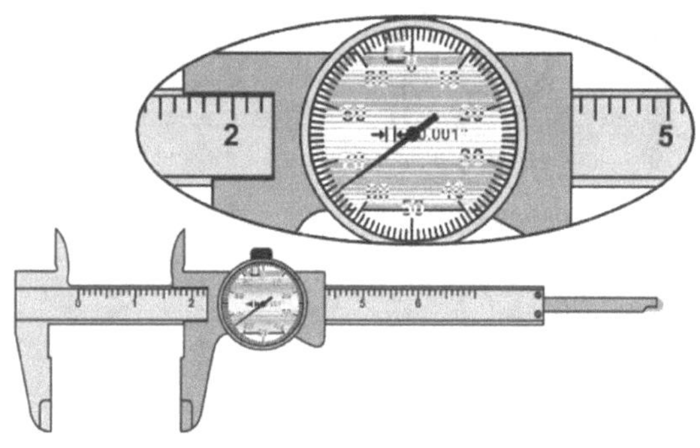

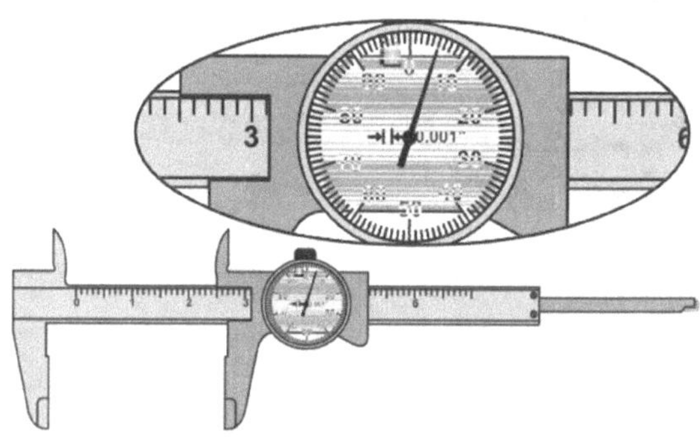

Actual Reading _______________   Actual Reading _______________

Interpreting Vernier Caliper Readings

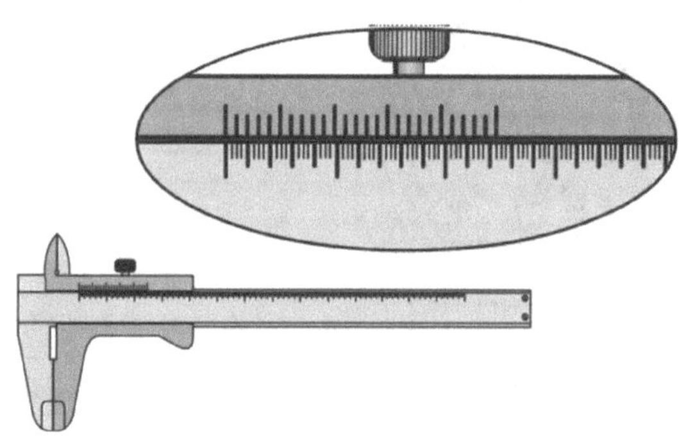

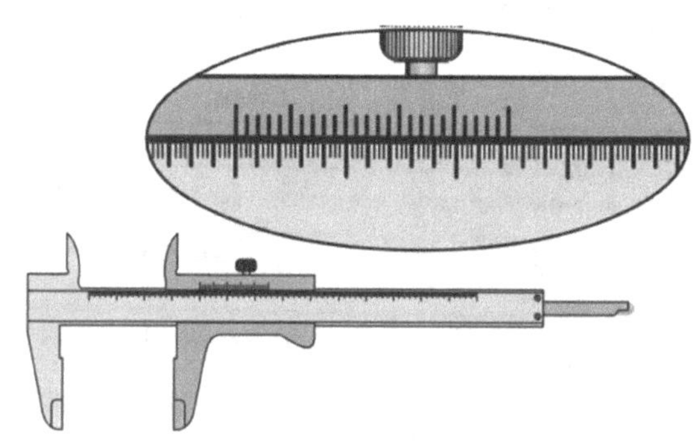

Actual Reading _______________   Actual Reading _______________

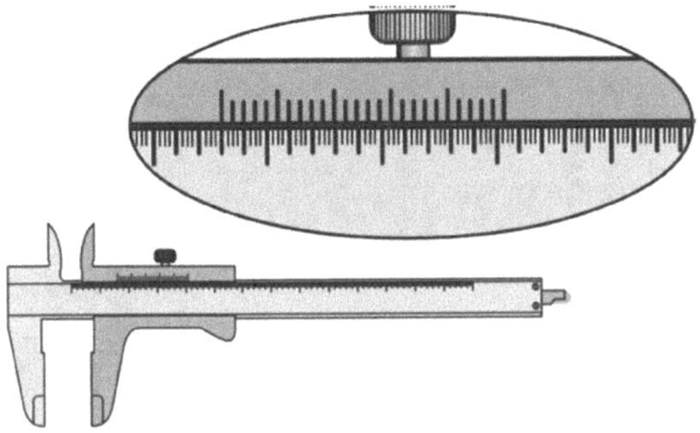

Actual Reading ________________

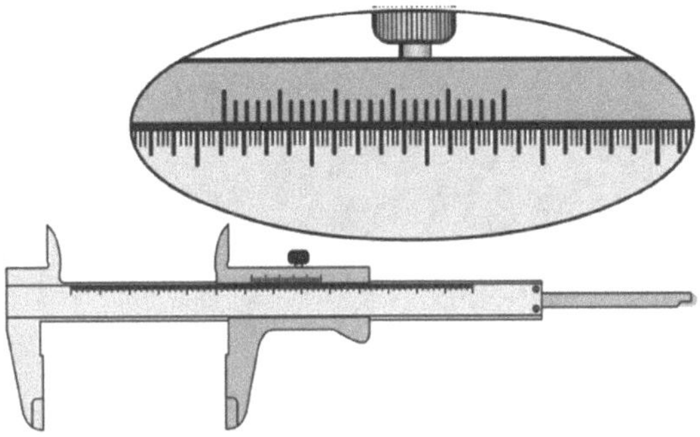

Actual Reading ________________

Actual Reading ________________

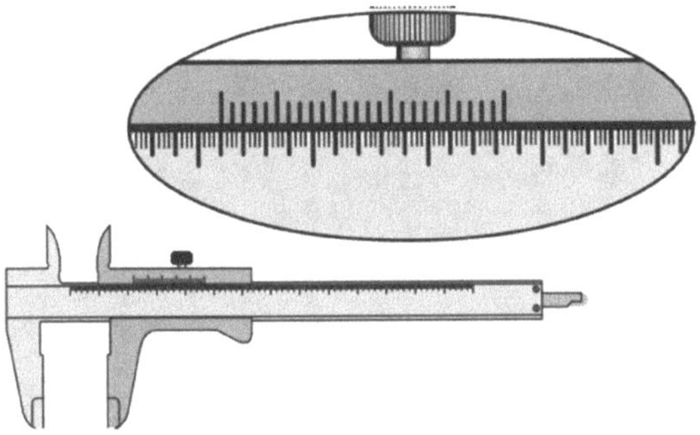

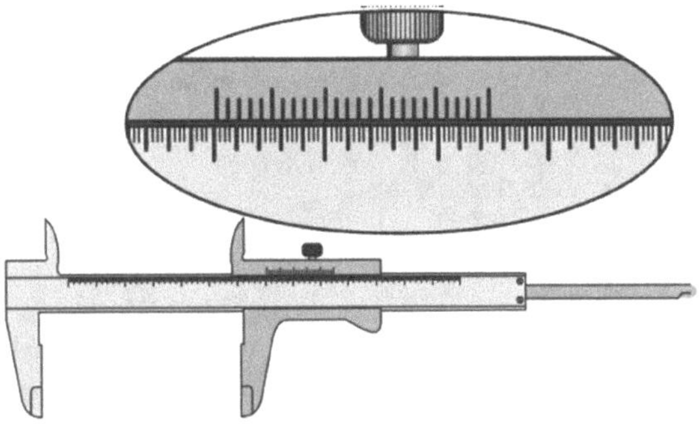

Actual Reading ________________

4. **Discuss your performance with your instructor to identify any problem areas.**

Instructor Comments:

____

____

____

Performance Rating

CDX Tasksheet Number: E0003

☐	☐	☐	☐	☐
0	1	2	3	4

Supervisor/instructor signature ________________________________ Date ____________

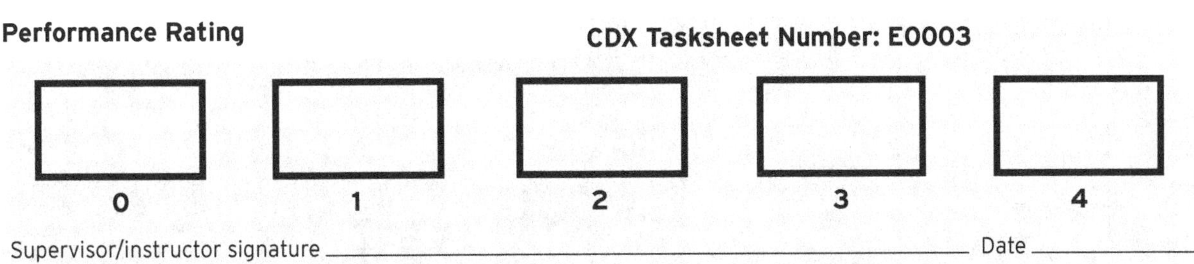

Use of Electric Tools

Student/Intern information:

Name _________________________ Date ___________ Class _________________________

Vehicle used for this activity:

Year _______________ Make _________________________ Model _________________________

Odometer ____________ Hour meter ___________ VIN _________________________

Learning Objective/Task	CDX Tasksheet Number	2014 Edition Rev2/12/16 AED Standard
• Identify and correctly name the electrical tool.	E0004	1a.2
• Demonstrate the proper use of the designed application and safe operating procedure for each.	E0005	1a.2

Time off_______________

Time on_______________

Total time_______________

Materials Required
- Standard electric hand tools and other tools as required

Some Safety Issues to Consider
- Comply with personal and environmental safety practices associated with clothing; eye protection; hand tools; power equipment; proper ventilation; and the handling, storage, and disposal of chemicals/materials in accordance with local, state, and federal safety and environmental regulations.
- Electric tools allow us to increase our productivity and effectiveness. However, they must be used according to the manufacturer's procedures. Failure to follow those procedures can result in serious injury or death.

Performance Standard
0—No exposure: No information or practice provided during the program; complete training required

1—Exposure only: General information provided with no practice time; close supervision needed; additional training required

2—Limited practice: Has practiced job during training program; additional training required to develop skill

3—Moderately skilled: Has performed job independently during training program; limited additional training may be required

4—Skilled: Can perform job independently with no additional training

Student/Intern information:

Name _________________________________ Date ____________ Class _________________________________

Vehicle used for this activity:

Year ________________ Make _________________________________ Model _________________________________

Odometer ____________ Hour meter ____________ VIN _________________________________

▶ TASK Identify and correctly name the electrical tool.

AED
1a.2

Time off_________________

Time on_________________

Total time_________________

CDX Tasksheet Number: E0004

1. **Identify instructor-designated electrical tools and record their proper usage.**

    a. **Tool and usage:**

    ____
    ____
    ____

    b. **Tool and usage:**

    ____
    ____
    ____

    c. **Tool and usage:**

    ____
    ____
    ____

    d. **Tool and usage:**

    ____
    ____
    ____

2. **Instructor Comments:**

    ____
    ____
    ____

Performance Rating

CDX Tasksheet Number: E0004

0	1	2	3	4

Supervisor/instructor signature _________________________________ Date ____________

Student/Intern information:

Name _________________________________ Date ____________ Class ______________________________

Vehicle used for this activity:

Year ________________ Make ________________________________ Model ________________________________

Odometer _____________ Hour meter ____________ VIN ____________________________________

▶ TASK Demonstrate the proper use of the designed application and safe-operating procedure for each.

AED
1a.2

Time off______________

Time on______________

CDX Tasksheet Number: E0005

1. **Demonstrate safe handling and use of instructor-designated electrical tools.**

    a. **Tool to be used:** ____
       Appropriate usage: Yes: ____________ **No:** ____________
    b. **Tool to be used:** ____
       Appropriate usage: Yes: ____________ **No:** ____________
    c. **Tool to be used:** ____
       Appropriate usage: Yes: ____________ **No:** ____________
    d. **Tool to be used:** ____
       Appropriate usage: Yes: ____________ **No:** ____________

Total time______________

2. **Discuss your performance with your instructor to identify any problem areas.**

    Instructor Comments:

    ____
    ____
    ____

Performance Rating

CDX Tasksheet Number: E0005

0	1	2	3	4

Supervisor/instructor signature ___________________________________ Date ____________

Use of Air Tools

Student/Intern information:

Name _________________________ Date __________ Class _________________________

Vehicle used for this activity:

Year ____________ Make _____________________ Model _________________________

Odometer __________ Hour meter __________ VIN _________________________

Learning Objective/Task	CDX Tasksheet Number	2014 Edition Rev2/12/16 AED Standard
• Identify and correctly name the basic air tool.	E0006	1a.3
• Demonstrate the proper use of the designed application and safe operating procedure for each.	E0007	1a.3
• Demonstrate the proper use of the designed application and safe operating procedure for each.	E0008	1a.3

Time off_____________

Time on_____________

Total time_____________

Materials Required
- Standard air tools and other tools as required.

Some Safety Issues to Consider
- Comply with personal and environmental safety practices associated with clothing; eye protection; hand tools; power equipment; proper ventilation; and the handling, storage, and disposal of chemicals/materials in accordance with local, state, and federal safety and environmental regulations.
- Air tools allow us to increase our productivity and effectiveness. However, they must be used according to the manufacturer's procedures. Failure to follow those procedures can result in serious injury or death.

Performance Standard
0–No exposure: No information or practice provided during the program; complete training required

1–Exposure only: General information provided with no practice time; close supervision needed; additional training required

2–Limited practice: Has practiced job during training program; additional training required to develop skill

3–Moderately skilled: Has performed job independently during training program; limited additional training may be required

4–Skilled: Can perform job independently with no additional training

Student/Intern information:

Name _______________________ Date ___________ Class _______________________

Vehicle used for this activity:

Year _____________ Make _____________________ Model _______________________

Odometer ___________ Hour meter ___________ VIN _______________________

▶ TASK Identify and correctly name the basic air tool.

AED
1a.3

Time off_______________

CDX Tasksheet Number: E0006

Time on_______________

1. **Identify instructor-designated basic air tools and record their proper usage.**

    a. **Tool and usage:**

        ____

        ____

        ____

    Total time_______________

    b. **Tool and usage:**

        ____

        ____

        ____

    c. **Tool and usage:**

        ____

        ____

        ____

    d. **Tool and usage:**

        ____

        ____

        ____

2. **Instructor Comments:**

    ____

    ____

    ____

Performance Rating   **CDX Tasksheet Number: E0006**

0	1	2	3	4

Supervisor/instructor signature _______________________________ Date ___________

Student/Intern information:

Name _________________________ Date _________ Class _________________________

Vehicle used for this activity:

Year _____________ Make _____________________ Model _____________________

Odometer __________ Hour meter __________ VIN _________________________

▶ TASK Demonstrate the proper use of the designed application
and safe-operating procedure for each.

AED
1a.3

Time off __________

Time on __________

CDX Tasksheet Number: E0007

1. **Demonstrate safe handling and use of instructor-designated basic air tools.**

    a. **Tool to be used:** _________________________
       Appropriate usage: Yes: __________ **No:** __________
    b. **Tool to be used:** _________________________
       Appropriate usage: Yes: __________ **No:** __________
    c. **Tool to be used:** _________________________
       Appropriate usage: Yes: __________ **No:** __________
    d. **Tool to be used:** _________________________
       Appropriate usage: Yes: __________ **No:** __________

Total time __________

2. **Discuss your performance with your instructor to identify any problem areas.**

   Instructor Comments:

   ____
   ____
   ____

Performance Rating

CDX Tasksheet Number: E0007

| 0 | 1 | 2 | 3 | 4 |

Supervisor/instructor signature _________________________________ Date __________

Student/Intern information:

Name _________________________________ Date ___________ Class ___________________________

Vehicle used for this activity:

Year ______________ Make _________________________ Model ________________________________

Odometer ____________ Hour meter ____________ VIN _____________________________________

▶ **TASK** Demonstrate the proper inspection, care, maintenance, and storage for basic air tools.

AED
1a.3

Time off________________

Time on________________

Total time________________

CDX Tasksheet Number: E0008

1. **Demonstrate the proper inspection, care, maintenance, and storage for basic air tools.**

   a. **Tool to be used:** _______________________________________
      Appropriate usage: Yes: ___________ **No:** ___________
   b. **Tool to be used:** _______________________________________
      Appropriate usage: Yes: ___________ **No:** ___________
   c. **Tool to be used:** _______________________________________
      Appropriate usage: Yes: ___________ **No:** ___________
   d. **Tool to be used:** _______________________________________
      Appropriate usage: Yes: ___________ **No:** ___________

2. **Discuss your performance with your instructor to identify any problem areas.**

   Instructor Comments:

   ____
   ____
   ____

Performance Rating

CDX Tasksheet Number: E0008

0	1	2	3	4

Supervisor/instructor signature ____ Date ___________

Use of Lifting Equipment

Student/Intern information:

Name _______________________________ Date ____________ Class ______________________

Vehicle used for this activity:

Year ______________ Make __________________________ Model __________________________

Odometer ____________ Hour meter ____________ VIN _________________________________

Learning Objective/Task	CDX Tasksheet Number	2014 Edition Rev2/12/16 AED Standard
• Identify and correctly name the various types of lifting equipment.	E0009	1a.5
• Demonstrate the proper inspection, care, maintenance, and storage for lifting equipment.	E0010	1a.5
• Demonstrate the proper use of the designed application and safe operating procedure for each.	E0011	1a.5
• Understand current regulations and standards for use, inspection, and certification of lifting equipment.	E0012	1a.5

Time off____________

Time on____________

Total time____________

Materials Required

• Standard shop lifting equipment

  a. Jack stands
  b. Hoists (overhead and floor type)
  c. Hydraulic jacks
  d. Blocking and cribbing
  e. Come-alongs (chain and cable type)
  f. Lifting chains–lifting eyes, links, spreader bars, etc.
  g. Slings
  h. Securing chains.

• Various equipment to be lifted

Some Safety Issues to Consider

• Comply with personal and environmental safety practices associated with clothing; eye protection; hand tools; power equipment; proper ventilation; and the handling, storage, and disposal of chemicals/materials in accordance with local, state, and federal safety and environmental regulations.

• Tools allow us to increase our productivity and effectiveness. However, they must be used according to the manufacturer's procedures. Failure to follow those procedures can result in serious injury or death.

Performance Standard

0—No exposure: No information or practice provided during the program; complete training required

1—Exposure only: General information provided with no practice time; close supervision needed; additional training required

2—Limited practice: Has practiced job during training program; additional training required to develop skill

3—Moderately skilled: Has performed job independently during training program; limited additional training may be required

4—Skilled: Can perform job independently with no additional training

Student/Intern information:

Name _________________________________ Date _____________ Class _________________________

Vehicle used for this activity:

Year ______________ Make _________________________________ Model _________________________

Odometer _____________ Hour meter _____________ VIN _________________________________

▶ TASK Identify and correctly name the various types of lifting equipment.

AED
1a.5

Time off________________

Time on________________

Total time________________

CDX Tasksheet Number: E0009

1. **Identify and correctly name the various types of lifting equipment including:**
   a. **Jack stands**
   b. **Hoists (overhead and floor type)**
   c. **Hydraulic jacks**
   d. **Blocking and cribbing**
   e. **Come-alongs (chain and cable type)**
   f. **Lifting chains—lifting eyes, links, spreader bars, etc.**
   g. **Slings**
   h. **Securing chains.**

2. **Identify instructor-designated lifting equipment.**
   a. **Lifting equipment to be used:** _________________________________
      Appropriate identification: Yes: ____________ **No:** ____________
   b. **Lifting equipment to be used:** _________________________________
      Appropriate identification: Yes: ____________ **No:** ____________
   c. **Lifting equipment to be used:** _________________________________
      Appropriate identification: Yes: ____________ **No:** ____________
   d. **Lifting equipment to be used:** _________________________________
      Appropriate identification: Yes: ____________ **No:** ____________
   e. **Lifting equipment to be used:** _________________________________
      Appropriate identification: Yes: ____________ **No:** ____________
   f. **Lifting equipment to be used:** _________________________________
      Appropriate identification: Yes: ____________ **No:** ____________
   g. **Lifting equipment to be used:** _________________________________
      Appropriate identification: Yes: ____________ **No:** ____________

3. **Instructor Comments:**

____

____

____

Performance Rating

CDX Tasksheet Number: E0009

| 0 | 1 | 2 | 3 | 4 |

Supervisor/instructor signature _______________________________________ Date ____________

Student/Intern information:

Name _________________________________ Date _____________ Class _________________________

Vehicle used for this activity:

Year _______________ Make _________________________________ Model _________________________

Odometer _____________ Hour meter _____________ VIN _________________________________

▶ TASK Demonstrate the proper inspection, care, maintenance, and storage for lifting equipment.

AED
1a.5

CDX Tasksheet Number: E0010

1. **Using Chapter 8 research the procedures to properly inspect, care, maintain, and store various types of lifting equipment.**

2. **Demonstrate the proper inspection, care, maintenance, and storage for lifting equipment.**

   a. **Lifting equipment to be used:** _________________________________
      Demonstrated properly?: Yes: _____________ **No:** _____________
   b. **Lifting equipment to be used:** _________________________________
      Demonstrated properly?: Yes: _____________ **No:** _____________
   c. **Lifting equipment to be used:** _________________________________
      Demonstrated properly?: Yes: _____________ **No:** _____________
   d. **Lifting equipment to be used:** _________________________________
      Demonstrated properly?: Yes: _____________ **No:** _____________
   e. **Lifting equipment to be used:** _________________________________
      Demonstrated properly?: Yes: _____________ **No:** _____________
   f. **Lifting equipment to be used:** _________________________________
      Demonstrated properly?: Yes: _____________ **No:** _____________
   g. **Lifting equipment to be used:** _________________________________
      Demonstrated properly?: Yes: _____________ **No:** _____________

3. **Instructor Comments:**

   ____
   ____
   ____

Performance Rating

CDX Tasksheet Number: E0010

0	1	2	3	4

Supervisor/instructor signature _________________________________ Date _____________

Student/Intern information:

Name _________________________________ Date ____________ Class _____________________________

Vehicle used for this activity:

Year _______________ Make _________________________________ Model ______________________________

Odometer _____________ Hour meter ____________ VIN ____

▶ TASK Demonstrate the proper use of the designed application and safe-operating procedure for each.

AED 1a.5

Time off_____________

Time on_____________

CDX Tasksheet Number: EOO11

1. **Demonstrate safe handling and use of instructor-designated lifting equipment by lifting, blocking, cribbing, rigging, etc.**

Total time_____________

   a. **Lifting equipment to be used:** ___________________________________

     Appropriate usage: Yes: ____________ **No:** ____________

   b. **Lifting equipment to be used:** ___________________________________

     Appropriate usage: Yes: ____________ **No:** ____________

   c. **Lifting equipment to be used:** ___________________________________

     Appropriate usage: Yes: ____________ **No:** ____________

   d. **Lifting equipment to be used:** ___________________________________

     Appropriate usage: Yes: ____________ **No:** ____________

   e. **Lifting equipment to be used:** ___________________________________

     Appropriate usage: Yes: ____________ **No:** ____________

   f. **Lifting equipment to be used:** ___________________________________

     Appropriate usage: Yes: ____________ **No:** ____________

2. **Discuss your performance with your instructor to identify any problem areas.**

Instructor Comments:

____

____

____

Performance Rating

CDX Tasksheet Number: EOO11

0	1	2	3	4

Supervisor/instructor signature _________________________________ Date ____________

Student/Intern information:

Name _________________________________ Date _____________ Class _________________________________

Vehicle used for this activity:

Year _______________ Make _________________________________ Model _________________________________

Odometer _____________ Hour meter _____________ VIN _________________________________

▶ TASK Understand current regulations and standards for use, inspection, and certification of lifting equipment.

AED
1a.5

Time off_________________

Time on_________________

CDX Tasksheet Number: E0012

1. **Using Chapter 8 research the current regulations and standards for use, inspection, and certification of lifting equipment. Answer the questions below.**

Total time_________________

    a. **Name the governing bodies that set the rules and regulations regarding lifting and lifting equipment.**

    ____

    ____

    ____

    ____

    b. **List the rules and regulations for overhead and gantry cranes below:**

    ____

    ____

    ____

    ____

    c. **List the rules and regulations for hoisting and lifting equipment below:**

    ____

    ____

    ____

    ____

2. **Instructor Comments:**

____

____

____

Performance Rating

CDX Tasksheet Number: E0012

0	1	2	3	4

Supervisor/instructor signature ____ Date _____________

Use of Various Cleaning Equipment

Student/Intern information:

Name _________________________ Date ___________ Class _______________________

Vehicle used for this activity:

Year ______________ Make _______________________ Model ___________________

Odometer ___________ Hour meter __________ VIN ______________________________

Learning Objective/Task	CDX Tasksheet Number	2014 Edition Rev2/12/16 AED Standard
• Identify the risks, hazards, and precautions for cleaning materials, both personal and environmental.	EOO13	1a.6
• Demonstrate an understanding of Safety Data Sheets (SDS) and requirements to meet OSHA standards.	EOO14	1a.6

Time off__________

Time on__________

Total time__________

Materials Required
- Program's shop policy and other safety information
- Shop cleaning materials
- Safety Data Sheets (SDS)

Some Safety Issues to Consider
- Comply with personal and environmental safety practices associated with clothing; eye protection; hand tools; power equipment; proper ventilation; and the handling, storage, and disposal of chemicals/materials in accordance with local, state, and federal safety and environmental regulations.
- Tools allow us to increase our productivity and effectiveness. However, they must be used according to the manufacturer's procedures. Failure to follow those procedures can result in serious injury or death.
- Shop rules and procedures are critical to your safety. Please give this your utmost attention.

Performance Standard
0—No exposure: No information or practice provided during the program; complete training required
1—Exposure only: General information provided with no practice time; close supervision needed; additional training required
2—Limited practice: Has practiced job during training program; additional training required to develop skill
3—Moderately skilled: Has performed job independently during training program; limited additional training may be required
4—Skilled: Can perform job independently with no additional training

Student/Intern information:

Name _______________________________ Date ____________ Class _______________________________

Vehicle used for this activity:

Year ________________ Make _______________________________ Model _______________________________

Odometer ____________ Hour meter ____________ VIN _______________________________

▶ **TASK** Identify the risks, hazards, and precautions for cleaning materials, both personal and environmental.

AED
1a.6

CDX Tasksheet Number: E0013

1. **Locate the various cleaning materials used in the school's shop. In the table below list all of the cleaning materials that you found.**

2. **Ask your instructor where the safety data sheets are kept at in your shop, then look up the physical hazards, health hazards, and environmental hazards, and list in the table below.**

3. **Read the SDS hazard statement for each of the cleaning materials in your list.**

Cleaning Material	Physical Hazard	Health Hazard	Environmental Hazard
Brakleen	Gases under pressure	Skin corrosion/ irritation Carcinogenicity	Aquatic environment Long term

4. **In your own words describe precautions required for eliminating injury or death from harmful exposure.**

____

____

____

____

____

5. Instructor Comments:

____

____

____

Performance Rating　　　　**CDX Tasksheet Number: E0013**

0	1	2	3	4

Supervisor/instructor signature ____________________________________ Date ____________

▶ TASK Demonstrate an understanding of Safety Data Sheets (SDS) and requirements to meet OSHA standards.

AED
1a.8

Time off_________________

Time on_________________

CDX Tasksheet Number: E0014

1. **Research safety data sheets (SDS) and answer the questions below.**

   a. **List the safety precautions when handling motor oil:**

   ____
   ____
   ____
   ____

Total time_________________

   b. **List the flash point of gasoline/petroleum:** _________________°F/°C

   c. **Which fire extinguisher should be used to put out a gasoline/ petroleum fire?**

   ____
   ____
   ____
   ____

   d. **List the first aid treatment for battery acid in the eyes:**

   ____
   ____
   ____
   ____

   e. **List the first aid treatment for ingestion of anti-freeze (ethylene glycol):**

   ____
   ____
   ____
   ____

2. **Instructor Comments:**

Performance Rating

CDX Tasksheet Number: E0014

0	1	2	3	4

Supervisor/instructor signature ____________________ Date ________

Use of Fluid Pressure Testing Equipment

Student/Intern information:

Name ___________________________ Date __________ Class ___________________

Vehicle used for this activity:

Year ______________ Make _______________________ Model ___________________

Odometer ___________ Hour meter __________ VIN ___________________________

Learning Objective/Task	CDX Tasksheet Number	2014 Edition Rev2/12/16 AED Standard
• Explain at least three dangers of working with fluids under pressure.	E0015	1a.7

Time off__________

Time on__________

Total time__________

Materials Required
- Program's shop policy and other safety information
- Shop cleaning materials
- Safety Data Sheets (SDS)

Some Safety Issues to Consider
- Comply with personal and environmental safety practices associated with clothing; eye protection; hand tools; power equipment; proper ventilation; and the handling, storage, and disposal of chemicals/materials in accordance with local, state, and federal safety and environmental regulations.
- Tools allow us to increase our productivity and effectiveness. However, they must be used according to the manufacturer's procedures. Failure to follow those procedures can result in serious injury or death.
- Shop rules and procedures are critical to your safety. Please give this your utmost attention.

Performance Standard
0–No exposure: No information or practice provided during the program; complete training required
1–Exposure only: General information provided with no practice time; close supervision needed; additional training required
2–Limited practice: Has practiced job during training program; additional training required to develop skill
3–Moderately skilled: Has performed job independently during training program; limited additional training may be required
4–Skilled: Can perform job independently with no additional training

Name _____________________ Date __________ Class _____________________

Vehicle used for this activity:

Year ____________ Make _________________ Model _____________________

Odometer __________ Hour meter __________ VIN _____________________

▶ TASK Explain at least three dangers of working with fluids under pressure.

AED 1a.7

Time off___________

Time on___________

Total time___________

CDX Tasksheet Number: E0015

1. **Research the dangers of working with fluids under pressure using Chapter 3, and the Internet.**

2. **List below at least three dangers of working with fluids under pressure and describe the health risks as well as the steps to avoid them:**

    i. ____
    ____
    ____
    ____
    ____
    ____
    ____
    ____

    ii. ____
    ____
    ____
    ____
    ____
    ____
    ____
    ____

    iii. ____
    ____
    ____
    ____
    ____
    ____
    ____
    ____

3. **List below the machine systems that you think will store fluid under pressure:**

    i. _______________________
    ii. _______________________
    iii. _______________________
    iv. _______________________
    v. _______________________
    vi. _______________________

4. **List the steps below you would take to avoid injury when working around each of the machine systems listed above.**

____
____
____
____
____

5. Instructor Comments:

____
____
____
____
____

Performance Rating   **CDX Tasksheet Number: E0015**

| 0 | 1 | 2 | 3 | 4 |

Supervisor/instructor signature ________________________________ Date __________

Environment of Service Facility

Student/Intern information:

Name _________________________________ Date ____________ Class _________________________

Vehicle used for this activity:

Year _______________ Make ___________________________ Model _____________________

Odometer ______________ Hour meter ____________ VIN ___________________________

Learning Objective/Task	CDX Tasksheet Number	2014 Edition Rev2/12/16 AED Standard
• Explain why carbon monoxide and diesel smoke can be hazardous to your health and the precautions required for eliminating injury or death.	E0016	1a.8
• Recognize symptoms of exposure to carbon monoxide, diesel smoke, and other hazardous materials.	E0017	1a.8

Materials Required

- Program's shop policy and other safety information
- Shop exhaust system equipment

Some Safety Issues to Consider

- Comply with personal and environmental safety practices associated with clothing; eye protection; hand tools; power equipment; proper ventilation; and the handling, storage, and disposal of chemicals/materials in accordance with local, state, and federal safety and environmental regulations.
- Tools allow us to increase our productivity and effectiveness. However, they must be used according to the manufacturer's procedures. Failure to follow those procedures can result in serious injury or death.
- Shop rules and procedures are critical to your safety. Please give this your utmost attention.

Performance Standard

0–No exposure: No information or practice provided during the program; complete training required

1–Exposure only: General information provided with no practice time; close supervision needed; additional training required

2–Limited practice: Has practiced job during training program; additional training required to develop skill

3–Moderately skilled: Has performed job independently during training program; limited additional training may be required

4–Skilled: Can perform job independently with no additional training

Name _________________________________ Date ____________ Class _____________________________

Vehicle used for this activity:

Year _______________ Make _________________________________ Model _____________________________

Odometer _____________ Hour meter _____________ VIN _________________________________

▶ **TASK** Explain why carbon monoxide and diesel smoke can be hazardous to your health and the precautions required for eliminating injury or death.

AED
1a.8

Time off_________________

Time on_________________

Total time_________________

CDX Tasksheet Number: EO016

1. OSHA personal exposure limits for exhaust gases found in diesel emissions including carbon monoxide, nitric oxide, nitrogen dioxide, and sulfur dioxide can be dangerous to human health. Always make sure to attach some kind of OSHA-approved exhaust gas extraction equipment before running the machine to prevent harmful exposure.

2. In your own words describe how carbon monoxide and diesel smoke can be hazardous to your health.

____

____

____

____

____

3. In your own words describe precautions required for eliminating injury or death from harmful exposure.

____

____

____

____

____

4. Properly position a machine in a work stall.

5. Properly connect the exhaust extraction system to the machine exhaust.

6. With your instructor's permission, start the vehicle and verify that the extraction equipment is secure and operating properly.

7. Turn off the vehicle, return the exhaust hoses to their proper storage places, and shut off the extraction system if it isn't being used anymore.

8. Have your instructor verify satisfactory completion of this procedure, any observations found, and any necessary action(s) recommended.

9. **Instructor Comments:**

____

____

____

Performance Rating

CDX Tasksheet Number: E0016

0	1	2	3	4

Supervisor/instructor signature _____________________________ Date __________

© 2019 Jones & Bartlett Learning, LLC, an Ascend Learning Company

Student/Intern information:

Name _____________________________ Date ___________ Class _____________________

Vehicle used for this activity:

Year _______________ Make _____________________________ Model _____________________

Odometer ____________ Hour meter ____________ VIN _____________________________

▶ **TASK** Recognize symptoms of exposure to carbon monoxide, diesel smoke, and other hazardous materials.

AED
1a.8

Time off_____________

Time on_____________

CDX Tasksheet Number: EOO17

Total time_____________

1. **Using Chapter 3 research the symptoms of exposure to carbon monoxide, diesel smoke, and other hazardous materials.**

2. **In your own words describe how to recognize symptoms of exposure to carbon monoxide, diesel smoke, and other hazardous materials.**

   ____

   ____

   ____

   ____

3. **Instructor Comments:**

   ____

   ____

   ____

Performance Rating

CDX Tasksheet Number: EOO17

| 0 | 1 | 2 | 3 | 4 |

Supervisor/instructor signature ____ Date _____________

Machine Identification and Operation

Student/Intern information:

Name _________________________________ Date ____________ Class _______________________

Vehicle used for this activity:

Year ______________ Make _____________________________ Model ___________________________

Odometer ____________ Hour meter ____________ VIN _______________________________________

Learning Objective/Task	CDX Tasksheet Number	2014 Edition Rev2/12/16 AED Standard
• Identify the various types of construction equipment and forklifts, using the standard industry names accepted by equipment manufacturers.	E0018	1a.9
• Demonstrate and explain the proper, safe, and fundamental operation of the various types of machinery.	E0019	1a.9
• Understand from a user's perspective the importance of and reasons for caution/warning lights, backup alarms, seat belts, safety instructions, decals, and other customer-related safety information.	E0020	1a.9

Time off_____________

Time on_____________

Total time_____________

Materials Required

- Machinery the technicians will be involved with:
  Examples: a. Excavator, b. Skid steers, c. Backhoes, d. Compaction equipment, e. Paving equipment, f. Crawler and track type loader, g. Scraper, h. Crane, i. Scissor lift, j. Fork lift and material handler, k. Wheel loader, l. Haul truck, m. Motor grader, n. Trencher, o. Horizontal directional drill, Hybrid drive
- Equipment manufacturer's operators manual
- Personal protective equipment (PPE)

Some Safety Issues to Consider

- Activities require test driving the machine on the school grounds, which carry severe risks. Attempt this task only with full permission from your supervisor/instructor and follow all the guidelines exactly.
- Comply with personal and environmental safety practices associated with clothing; eye protection; hand tools; power equipment; proper ventilation; and the handling, storage, and disposal of chemicals/materials in accordance with federal, state, and local regulations.
- Always wear the correct protective eyewear and clothing and use the appropriate safety equipment, as well as fender covers, seat protectors, and floor mat protectors.
- Make sure you understand and observe all legislative and personal safety procedures when carrying out practical assignments. If you are unsure of what these are, ask your supervisor/instructor.

Performance Standard

0–No exposure: No information or practice provided during the program; complete training required

1–Exposure only: General information provided with no practice time; close supervision needed; additional training required

2–Limited practice: Has practiced job during training program; additional training required to develop skill

3–Moderately skilled: Has performed job independently during training program; limited additional training may be required

4–Skilled: Can perform job independently with no additional training

Student/Intern information:

Name _________________________________ Date ____________ Class _________________________

Vehicle used for this activity:

Year _______________ Make _________________________ Model _________________________

Odometer ____________ Hour meter ____________ VIN _________________________

▶ **TASK** Identify the various types of construction equipment and forklifts, using the standard industry names accepted by equipment manufacturers.

AED
1a.9

CDX Tasksheet Number: E0018

1. **Locate the machines in the shop or yard labeled a–o, and then identify each using the standard industry names accepted by equipment manufacturers. Enter name in the table below:**

Machine Letter	Machine Name
a	
b	
c	
d	
e	
f	
g	
h	
i	
j	
k	
l	
m	
n	
o	

2. **Instructor Comments:**

____

____

____

Performance Rating   **CDX Tasksheet Number: E0018**

| 0 | 1 | 2 | 3 | 4 |

Supervisor/instructor signature _______________________________ Date _____________

Student/Intern information:

Name _____________________________ Date ____________ Class _________________________

Vehicle used for this activity:

Year ______________ Make ________________________ Model _______________________________

Odometer ____________ Hour meter ____________ VIN ___________________________________

▶ TASK Demonstrate and explain the proper, safe, and fundamental operation of the various types of machinery.

AED 1a.9

Time off_______________

Time on_______________

CDX Tasksheet Number: E0019

Total time_______________

1. **Ask your instructor to assign to you a machine and mark below:**

Machine	Instructor X here
a. Excavator	
b. Skid steer	
c. Backhoe	
d. Compaction equipment	
e. Paving equipment	
f. Crawler or track-type loader	
g. Scraper	
h. Crane	
i. Scissor lift	
j. Fork lift or material handler	
k. Wheel loader	
l. Haul truck	
m. Motor grader	
n. Trencher	
o. Horizontal directional drill, Hybrid drive	

2. **Locate the machine in the shop or yard, remove the operator's manual, and perform a walk around inspection. List any defects found during your inspection.**

    Machine OK to operate: yes/no

    Defects:

    ____

    ____

    ____

    ____

    ____

3. **Operate the machine in the shop yard under your instructor's supervision and operate the typical functions as directed by your instructor. Once complete, safely park the machine and stow all implements according to manufacturer's recommendations.**

4. **Instructor Comments:**

____

____

____

Student/Intern information:

Name _____________________________ Date ___________ Class _______________________

Vehicle used for this activity:

Year _______________ Make _______________________ Model _____________________

Odometer ____________ Hour meter ___________ VIN _________________________________

▶ **TASK** Understand from a user's perspective the importance of and reasons for caution/warning lights, backup alarms, seat belts, safety instructions, decals, and other customer-related safety information.

AED
1a.9

Time off_________________

Time on_________________

Total time_________________

CDX Tasksheet Number: E0020

1. **In your own words describe the importance of and reasons for the following safety items:**

    a. **Caution/warning lights:**

    ____
    ____
    ____

    b. **Backup alarms:**

    ____
    ____
    ____

    c. **Seat belts:**

    ____
    ____
    ____

    d. **Safety instructions, decals, and other customer-related safety information:**

    ____
    ____
    ____

2. **Instructor Comments:**

Performance Rating

CDX Tasksheet Number: E0020

0	1	2	3	4

Supervisor/instructor signature _______________________________ Date __________

Mandated Regulations

Student/Intern information:

Name ___________________________ Date ___________ Class ___________________

Vehicle used for this activity:

Year ___________ Make ___________________ Model ___________________

Odometer ___________ Hour meter ___________ VIN ___________________

Learning Objective/Task	CDX Tasksheet Number	2014 Edition Rev2/12/16 AED Standard
• Identify the different types of fire extinguishers and know the applications and correct use of each type.	E0021	1a.10
• Demonstrate how to find, explain, and use an SDS for a product.	E0022	1a.10
• Explain why working safely is important, and explain the procedures for reporting unsafe working conditions and practices.	E0023	1a.10

Time off___________

Time on___________

Total time___________

Materials Required
- Program's shop policy and other safety information
- Fire extinguisher(s) from the shop
- Safety Data Sheets (SDS)

Some Safety Issues to Consider
- Comply with personal and environmental safety practices associated with clothing; eye protection; hand tools; power equipment; proper ventilation; and the handling, storage, and disposal of chemicals/materials in accordance with local, state, and federal safety and environmental regulations.
- Fire extinguishers come in a variety of types and sizes. It pays to understand the differences before they are needed.
- Tools allow us to increase our productivity and effectiveness. However, they must be used according to the manufacturer's procedures. Failure to follow those procedures can result in serious injury or death.
- Shop rules and procedures are critical to your safety. Please give this your utmost attention.

Performance Standard

0—No exposure: No information or practice provided during the program; complete training required

1—Exposure only: General information provided with no practice time; close supervision needed; additional training required

2—Limited practice: Has practiced job during training program; additional training required to develop skill

3—Moderately skilled: Has performed job independently during training program; limited additional training may be required

4—Skilled: Can perform job independently with no additional training

Student/Intern information:

Name _________________________________ Date _____________ Class _________________________________

Vehicle used for this activity:

Year _______________ Make _________________________________ Model _________________________________

Odometer _____________ Hour meter ___________ VIN _________________________________

▶ **TASK** Identify the different types of fire extinguishers and know the applications and correct use of each type.

AED
1a.10

Time off_________________

Time on_________________

CDX Tasksheet Number: E0021

1. **Research the types, location, and use of fire extinguishers.**

    i. **List the different types of fire extinguishers (dry chemical, CO_2, etc.) available:**

       Total time_________________

       Type: __________; **for use on what class of fire:** ___________

       Type: __________; **for use on what class of fire:** ___________

       Type: __________; **for use on what class of fire:** ___________

       Type: __________; **for use on what class of fire:** ___________

       In the above list, put a star next to the type(s) of fire extinguishers in this shop.

2. **List the steps for proper use of a fire extinguisher:**

    ____

    ____

    ____

    ____

    ____

    ____

3. **Instructor Comments:**

    ____

    ____

    ____

Performance Rating

CDX Tasksheet Number: E0021

0	**1**	**2**	**3**	**4**

Supervisor/instructor signature _________________________________ Date ___________

Student/Intern information:

Name _________________________________ Date ____________ Class _____________________

Vehicle used for this activity:

Year _____________ Make _________________________ Model _____________________________

Odometer ____________ Hour meter ____________ VIN _________________________________

▶ TASK Demonstrate how to find, explain, and use an SDS for a product. **AED 1a.10**

Time off____________

Time on____________

CDX Tasksheet Number: E0022

1. **Research safety data sheets (SDS) and answer the questions below.**

    a. **List the safety precautions when handling hydraulic oil:**

    ____
    ____
    ____
    ____
    ____

    Total time____________

    b. **List the flash point of diesel/petroleum:** _____________________ °F/°C

    c. **Which fire extinguisher should be used to put out a diesel/petroleum fire?**

    ____
    ____
    ____
    ____
    ____

    d. **List the first aid treatment for battery acid in the eyes:**

    ____
    ____
    ____
    ____
    ____

    e. **List the first aid treatment for ingestion of diesel exhaust fluid (DEF):**

    ____
    ____
    ____
    ____
    ____

2. **Instructor Comments:**

Student/Intern information:

Name _________________________________ Date ___________ Class _____________________

Vehicle used for this activity:

Year _______________ Make _________________________ Model ________________________

Odometer ____________ Hour meter ___________ VIN _________________________________

▶ **TASK** Explain why working safely is important, and explain the procedures for reporting unsafe working conditions and practices.

AED
1a.10

Time off_________________

Time on_________________

Total time_______________

CDX Tasksheet Number: E0023

1. **Research Workman's compensation and accident prevention using OSHAs website:**

    a. **What are the costs of accidents to employers and employee affected?**

    ____
    ____
    ____
    ____
    ____

    b. **Explain lost time injuries and how they affect employers and employee.**

    ____
    ____
    ____
    ____
    ____

    c. **Explain proper accident and injury reporting procedures and how they affect employers and employee if not reported.**

    ____
    ____
    ____
    ____
    ____

    d. **Explain why working safely is important.**

    ____
    ____
    ____
    ____
    ____

e. Explain the procedures for reporting unsafe working conditions and practices.

____

____

____

____

____

2. Instructor Comments:

____

____

____

Performance Rating

CDX Tasksheet Number: E0023

0	1	2	3	4

Supervisor/instructor signature _________________________ Date __________

Shop and In-Field Practices

Student/Intern information:

Name _______________________________ Date _____________ Class _______________________

Vehicle used for this activity:

Year _______________ Make _____________________________ Model _______________________

Odometer _____________ Hour meter ____________ VIN _________________________________

Learning Objective/Task	CDX Tasksheet Number	2014 Edition Rev2/12/16 AED Standard
• Identify safe work practices in each situation.	E0024	1a.11
• Demonstrate safe work practices in the shop or in the field.	E0025	1a.11
• Identify proper lifting and pulling techniques to avoid personal injury.	E0026	1a.11
• Demonstrate proper lifting and pulling techniques.	E0027	1a.11
• Demonstrate proper shop/facility cleanliness/appearance to dealer standards.	E0028	1a.11

Time off_______________

Time on_______________

Total time_______________

Materials Required
- Program's shop policy and other safety information
- Safety Data Sheets (SDS)

Some Safety Issues to Consider
- Comply with personal and environmental safety practices associated with clothing; eye protection; hand tools; power equipment; proper ventilation; and the handling, storage, and disposal of chemicals/materials in accordance with local, state, and federal safety and environmental regulations.
- Shop rules and procedures are critical to your safety. Please give this your utmost attention.
- Marked safety areas play an important role in maintaining a safe work environment. Always understand and heed marked safety areas.
- Fire blankets are an important component of fire safety in the shop. Always know their location, purpose, and use.
- Fire extinguishers come in a variety of types and sizes. It pays to understand the differences before they are needed.
- Eyewash stations are an important component of shop safety. Always know their location, purpose, and use.
- Pre-planned evacuation routes are an important component of shop safety. Always know the location of all evacuation routes for your shop.

- Tools allow us to increase our productivity and effectiveness. However, they must be used according to the manufacturer's procedures. Failure to follow those procedures can result in serious injury or death.

Performance Standard

0—No exposure: No information or practice provided during the program; complete training required

1—Exposure only: General information provided with no practice time; close supervision needed; additional training required

2—Limited practice: Has practiced job during training program; additional training required to develop skill

3—Moderately skilled: Has performed job independently during training program; limited additional training may be required

4—Skilled: Can perform job independently with no additional training

Student/Intern information:

Name _______________________ Date _____________ Class _______________________

Vehicle used for this activity:

Year _______________ Make _______________________ Model _______________________

Odometer _____________ Hour meter _____________ VIN _______________________

▶ TASK Identify safe work practices in each situation.

AED
1a.11

Time off_______________

Time on_______________

CDX Tasksheet Number: E0024

1. **List the location(s) of the following items.**

    a. **Program's general shop safety rules and procedures:**

    ____
    ____
    ____
    ____
    ____
    ____
    ____

Total time_______________

    b. **Material Data Safety Sheets (MSDS) book:**

    ____
    ____
    ____
    ____
    ____
    ____
    ____

    c. **Procedure for operation of a fire extinguisher:**

    ____
    ____
    ____
    ____
    ____
    ____
    ____

2. **List the shop's policy for wearing of safety glasses while in the shop:**

    ____
    ____
    ____
    ____
    ____
    ____

3. **List the shop's policy for operating machines:**

    ____
    ____
    ____
    ____
    ____
    ____

4. **List the shop's policy for type of clothing in the shop:**

____

____

____

____

____

____

5. **List the shop's policy for jewelry in the shop:**

____

____

____

____

____

____

6. **Research the uniform color code system used to designate safety areas in a shop. List each color and its designation:**

    i. Color: _______________ designates: _______________

    ii. Color: _______________ designates: _______________

    iii. Color: _______________ designates: _______________

    iv. Color: _______________ designates: _______________

7. **Research the location, purpose, and use of eyewash stations in your shop.**

    a. **Describe the purpose of an eyewash station:**

____

____

____

____

____

    b. **Describe the situations when an eyewash station should be used and the proper procedures for its use:**

____

____

____

____

____

    c. **What type of eye injury would not require the use of an eyewash station?**

____

____

____

____

8. Research the location and purpose of evacuation routes.

   a. Describe the purpose of evacuation routes:

9. Describe the hazardous environments in the shop and the actions and procedures that should be taken in an emergency:

10. Describe the shops policies regarding the safe disposal of hazardous waste:

11. Pass the shop's safety test and record your score here:

12. Instructor Comments:

Performance Rating

CDX Tasksheet Number: E0024

0	1	2	3	4

Supervisor/instructor signature _______________________________ Date __________

Student/Intern information:

Name _________________________________ Date ___________ Class ____________________

Vehicle used for this activity:

Year _______________ Make _________________________ Model ___________________

Odometer ____________ Hour meter ____________ VIN _______________________________

▶ **TASK** Demonstrate safe work practices in the shop or in the field. _____________ **AED** *1a.11*

CDX Tasksheet Number: E0025

Time off___________

Time on___________

1. **Demonstrate safe personal protective equipment (PPE) work practices in various locations and working conditions in the shop and field.**

   a. **List the PPE required when performing general duties in the shop or field:**

   ____

   ____

   ____

   ____

   ____

   b. **List the PPE required when performing welding, torching, and grinding duties in the shop or field:**

   ____

   ____

   ____

   ____

   ____

   c. **List the PPE required when working in dusty conditions in the shop or field:**

   ____

   ____

   ____

   ____

   ____

   d. **List the PPE required when using bench grinder in the shop or field:**

   ____

   ____

   ____

   ____

   ____

Total time___________

2. **These tasks require observation of the student over a prolonged period. Your instructor will evaluate your performance on a continued basis.**

Instructor Comments:

____

____

____

Performance Rating

CDX Tasksheet Number: E0025

| 0 | 1 | 2 | 3 | 4 |

Supervisor/instructor signature _______________________________________ Date _____________

Student/Intern information:

Name ________________________________ Date ____________ Class ____________________

Vehicle used for this activity:

Year ______________ Make ________________________ Model ____________________________

Odometer ____________ Hour meter ____________ VIN ________________________________

▶ TASK Identify proper lifting and pulling techniques to avoid personal injury.

AED
1a.11

Time off________________

Time on________________

CDX Tasksheet Number: E0026

1. **Research the proper lifting and pulling techniques and describe below:**

Total time________________

   a. **Explain why it is important to use proper lifting and pulling techniques daily.**

2. **Instructor Comments:**

Performance Rating

CDX Tasksheet Number: E0026

| 0 | 1 | 2 | 3 | 4 |

Supervisor/instructor signature ____ Date ____________

Student/Intern information:

Name _________________________ Date __________ Class _______________

Vehicle used for this activity:

Year ______________ Make __________________ Model _______________

Odometer ___________ Hour meter ___________ VIN _______________

▶ **TASK** Demonstrate proper lifting and pulling techniques to avoid personal injury.

AED 1a.11

Time off__________

Time on__________

CDX Tasksheet Number: E0027

1. **Demonstrate the proper lifting and pulling techniques to your instructor:**

   a. **Lifting heavy item off the ground and placing on bench.** __________

   b. **Lifting heavy item off the ground with partner.** __________

   c. **Pushing heavy load.** __________

   d. **Pulling heavy load.** __________

   Total time__________

2. **These tasks require observation of the student over a prolonged period. Your instructor will evaluate your performance on a continued basis.**

   Instructor Comments:

   ____
   ____
   ____

Performance Rating

CDX Tasksheet Number: E0027

| 0 | 1 | 2 | 3 | 4 |

Supervisor/instructor signature _______________________________ Date __________

▶ **TASK** Demonstrate proper shop/facility cleanliness/appearance to dealer standards.

**AED
1a.11**

Time off_______________

Time on_______________

Total time_______________

CDX Tasksheet Number: E0028

1. **Research an equipment dealer's standards for proper shop/facility cleanliness/appearance. List below the key areas called out in the standards.**

   ____
   ____
   ____
   ____

2. **List below the procedures and tooling /equipment the dealer recommends for proper shop/facility cleanliness/appearance:**

   ____
   ____
   ____
   ____

3. **Describe in your own words why proper shop/facility cleanliness/appearance is important:**

   ____
   ____
   ____
   ____

4. **Demonstrate proper shop/facility cleanliness/appearance to dealer standards as outlined by your instructor.**

   ____
   ____
   ____
   ____

5. These tasks require observation of the student over a prolonged period. Your
   instructor will evaluate your performance on a continued basis.

   Instructor Comments:

   ____

   ____

   ____

Performance Rating

CDX Tasksheet Number: E0028

| 0 | 1 | 2 | 3 | 4 |

Supervisor/instructor signature ___________________________________ Date ____________

Hazard Identification and Prevention

Student/Intern information:

Name ________________________________ Date ____________ Class ___________________________

Vehicle used for this activity:

Year ________________ Make ________________________________ Model ______________________

Odometer ____________ Hour meter ____________ VIN ____

Learning Objective/Task	CDX Tasksheet Number	2014 Edition Rev2/12/16 AED Standard
• Demonstrate proper lockout/tagout procedures.	E0029	1a.12
• Demonstrate understanding of the HazCom standard and how to use Safety Data Sheets and Chemical Labels.	E0030	1a.12
• Demonstrate proper work procedures in handling wheel assemblies.	E0031	1a.12

Time off____________

Time on____________

Total time____________

Materials Required
- Program's shop policy and other safety information
- Safety Data Sheets (SDS)

Some Safety Issues to Consider
- Comply with personal and environmental safety practices associated with clothing; eye protection; hand tools; power equipment; proper ventilation; and the handling, storage, and disposal of chemicals/materials in accordance with local, state, and federal safety and environmental regulations.
- Shop rules and procedures are critical to your safety. Please give this your utmost attention.
- Marked safety areas play an important role in maintaining a safe work environment. Always understand and heed marked safety areas.
- Fire blankets are an important component of fire safety in the shop. Always know their location, purpose, and use.
- Fire extinguishers come in a variety of types and sizes. It pays to understand the differences before they are needed.
- Eyewash stations are an important component of shop safety. Always know their location, purpose, and use.
- Pre-planned evacuation routes are an important component of shop safety. Always know the location of all evacuation routes for your shop
- Tools allow us to increase our productivity and effectiveness. However, they must be used according to the manufacturer's procedures. Failure to follow those procedures can result in serious injury or death.

Performance Standard

0—No exposure: No information or practice provided during the program; complete training required

1—Exposure only: General information provided with no practice time; close supervision needed; additional training required

2—Limited practice: Has practiced job during training program; additional training required to develop skill

3—Moderately skilled: Has performed job independently during training program; limited additional training may be required

4—Skilled: Can perform job independently with no additional training

Student/Intern information:

Name _________________________________ Date ____________ Class _____________________

Vehicle used for this activity:

Year _______________ Make _________________________ Model _______________________

Odometer _____________ Hour meter ____________ VIN _______________________________

▶ TASK Demonstrate proper lockout/tagout procedures.

AED
1a.12

Time off_______________

Time on_______________

Total time_______________

CDX Tasksheet Number: E0029

1. **Research the shop's lockout/tagout procedures and answer the questions below.**

    i. **When should lockout/tagout procedures be used?**

    ____
    ____
    ____
    ____
    ____
    ____
    ____

    ii. **List the shop's lockout/tagout procedures:**

    ____
    ____
    ____
    ____
    ____
    ____
    ____

    iii. **Demonstrate the procedure for locking out a machines electrical system:**

    ____
    ____
    ____
    ____
    ____
    ____
    ____

    iv. **Demonstrate the procedure for locking out a machines hydraulic system:**

    ____
    ____
    ____
    ____
    ____
    ____
    ____

v. Demonstrate the procedure for locking out a machines articulating joint (steering):

____

____

____

____

____

____

vi. Demonstrate the procedure for locking out a machines hoist system:

____

____

____

____

____

____

2. These tasks require observation of the student over a prolonged period. Your instructor will evaluate your performance on a continued basis.

Instructor Comments:

____

____

____

Performance Rating

CDX Tasksheet Number: E0029

0	1	2	3	4

Supervisor/instructor signature _______________________________________ Date ___________

Student/Intern information:

Name _________________________ Date ____________ Class ___________________

Vehicle used for this activity:

Year _____________ Make _________________________ Model ___________________

Odometer ____________ Hour meter ____________ VIN _________________________

▶ TASK Demonstrate understanding of the HazCom standard and how to use Safety Data Sheets and Chemical Labels.

AED
1a.12

Time off_____________

Time on_____________

CDX Tasksheet Number: E0030

Total time_____________

1. **Research the hazard communication policy for your shop.**

   a. **Explain why it is important to understand and follow a hazard communication plan.**

   ____
   ____
   ____
   ____
   ____

2. **Gather a variety of chemicals used in your shop and identify the SDS symbols for each chemical. List below chemical name, symbols and their meaning.**

Chemical Name	SDS Symbols	Symbol Meaning

3. **Instructor Comments:**

____

____

____

Student/Intern information:

Name _________________________________ Date ____________ Class _____________________

Vehicle used for this activity:

Year _______________ Make _________________________ Model _____________________

Odometer ____________ Hour meter ____________ VIN _________________________________

▶ **TASK** Demonstrate proper work procedures in handling wheel assemblies.

AED
1a.12

Time off_______________

Time on_______________

Total time_______________

CDX Tasksheet Number: E0031

1. **Research the proper work procedures in handling wheel assemblies using the manufacturers, service information for various machines assigned to you by your instructor. List below the procedures for the various machines including tooling and equipment that must be used.**

Machine	Procedures	Tooling	Equipment

2. **Demonstrate proper work procedures in handling wheel assemblies on an instructor assigned machine.**

Instructor Comments:

____

____

____

Comprehending Basic Academic Functions

Student/Intern information:

Name _________________________________ Date ____________ Class _________________________

Vehicle used for this activity:

Year ______________ Make _________________________ Model ____________________________

Odometer ____________ Hour meter ____________ VIN ____________________________________

Learning Objective/Task	CDX Tasksheet Number	2014 Edition Rev2/12/16 AED Standard
• Exhibit the ability to use parts and service reference/technical materials, and safety materials in print or computer format.	E0032	1b.1
• Exhibit the ability to follow written instructions.	E0033	1b.1

Time off____________

Time on____________

Total time____________

Materials Required
- Machine
- Manufacturers' service information
- Manufacturers' parts information
- Repair orders

Some Safety Issues to Consider
- Comply with personal and environmental safety practices associated with clothing; eye protection; hand tools; power equipment; proper ventilation; and the handling, storage, and disposal of chemicals/materials in accordance with local, state, and federal safety and environmental regulations.
- Shop rules and procedures are critical to your safety. Please give this your utmost attention.

Performance Standard
0—No exposure: No information or practice provided during the program; complete training required

1—Exposure only: General information provided with no practice time; close supervision needed; additional training required

2—Limited practice: Has practiced job during training program; additional training required to develop skill

3—Moderately skilled: Has performed job independently during training program; limited additional training may be required

4—Skilled: Can perform job independently with no additional training

Name _________________________ Date __________ Class _________________

Vehicle used for this activity:

Year ____________ Make _____________________ Model _________________

Odometer ___________ Hour meter __________ VIN _____________________

▶ TASK Exhibit the ability to use parts and service reference/technical materials, and safety materials in print or computer format.

AED
1b.1

Time off_______________

Time on_______________

Total time_______________

CDX Tasksheet Number: E0032

1. **Exhibit the ability to use parts and service reference/technical materials, and safety materials using an instructor-assigned machine and its service and parts information.**

    a. **Locate the following part numbers for your machine and enter the information below:**

Part	Part Number
Air filter	
Fuel filter	
Hydraulic filter	
Seat belt	

    b. **Locate the following service procedures for your machine, print a copy of the procedures, and attach to the back of this document.**
       i. **Air filter removal and replacement**
       ii. **Fuel filter removal and replacement**
       iii. **Hydraulic filter removal and replacement**
       iv. **Seat belt inspection and maintenance**

2. **Instructor Comments:**

    ____

    ____

    ____

Performance Rating

CDX Tasksheet Number: E0032

0	1	2	3	4

Supervisor/instructor signature _________________________ Date __________

Student/Intern information:

Name _______________________________ Date ___________ Class _______________________

Vehicle used for this activity:

Year _______________ Make _______________________ Model _______________________

Odometer ___________ Hour meter ___________ VIN _______________________________

▶ TASK Exhibit the ability to follow written instructions.

AED
1a.12

CDX Tasksheet Number: E0033

1. **Using an instructor-assigned machine and its information:**

    a. **Locate the following service procedures for your machine, print a copy of the procedures, and attach to the back of this document.**

    i. **Air filter removal and installation**

    b. **Read the instructions for the procedure above and describe the steps in your own words below.**

    ____

    ____

    ____

    ____

    ____

2. **Instructor Comments:**

    ____

    ____

    ____

Performance Rating

CDX Tasksheet Number: E0033

0	1	2	3	4

Supervisor/instructor signature ____ Date ___________

Awareness of Dealership Goals, Objectives, and Policies

Student/Intern information:

Name ___________________________________ Date _____________ Class _____________________

Vehicle used for this activity:

Year _______________ Make _________________________________ Model _____________________

Odometer _____________ Hour meter ____________ VIN _________________________________

Learning Objective/Task	CDX Tasksheet Number	2014 Edition Rev2/12/16 AED Standard
• Maintain awareness of sexual harassment policy, safety rules, environmental regulations, disciplinary action policy, and equal opportunity policy.	E0034	1b.3

Time off______________

Time on______________

Total time______________

Materials Required

- Program's shop policy
- Schools student hand book

Some Safety Issues to Consider
- Shop rules and procedures are critical to your safety. Please give this your utmost attention.

Performance Standard
0–No exposure: No information or practice provided during the program; complete training required
1–Exposure only: General information provided with no practice time; close supervision needed; additional training required
2–Limited practice: Has practiced job during training program; additional training required to develop skill
3–Moderately skilled: Has performed job independently during training program; limited additional training may be required
4–Skilled: Can perform job independently with no additional training

Student/Intern information:

Name _________________________________ Date _____________ Class _________________________

Vehicle used for this activity:

Year _______________ Make _____________________________ Model _________________________

Odometer ____________ Hour meter ____________ VIN _________________________________

▶ **TASK** Maintain awareness of sexual harassment policy, safety rules, environmental regulations, disciplinary action policy, and equal opportunity policy.

AED
1b.1

Time off__________________

Time on__________________

Total time__________________

CDX Tasksheet Number: E0034

1. **Research the school and programs policies regarding student conduct.**

    a. **Describe why it is important to understand and maintain awareness of the following policies.**

       i. **Sexual harassment policy:**

          ____
          ____
          ____
          ____

       ii. **Safety rules:**

          ____
          ____
          ____
          ____

       iii. **Environmental regulations:**

          ____
          ____
          ____
          ____

       iv. **Disciplinary action policy:**

          ____
          ____
          ____

       v. **Equal opportunity policy:**

          ____
          ____
          ____
          ____

2. **Instructor Comments:**

____

____

____

Performance Rating

0	1	2	3	4

Supervisor/instructor signature _______________________________________ Date _____________

Section A2: Electronics/Electrical Systems

CONTENTS

Fundamental Knowledge

Student/Intern information:

Name _________________________________ Date _____________ Class _________________________

Vehicle used for this activity:

Year _______________ Make _________________________ Model _____________________

Odometer _____________ Hour meter ___________ VIN _________________________

Time off_______________

Time on_______________

Total time_______________

Learning Objective/Task	CDX Tasksheet Number	2014 Edition Rev2/12/16 AED Standard
• Know the basic structure of conductors, insulators, and semiconductors.	E0035	2.1a
• Know the reaction of like and unlike charges.	E0036	2.1a
• Describe the differences of conventional and electron theory current flow.	E0037	2.1b
• Define resistance and its effect on current flow.	E0038	2.1b
• Demonstrate the principles of operation and the correct usage of the various types of meters to measure volts, amps, and ohms.	E0039	2.1b
• Demonstrate knowledge of the laws governing permanent magnets, electromagnets, and magnetic fields.	E0040	2.1b
• Demonstrate knowledge of the effects of magnetic forces on current carrying conductors.	E0041	2.1b
• Know the basic parts and operation of the basic types of storage batteries.	E0042	2.1b

Materials Required

- Machine or simulator with electrical circuit concerns
- Equipment manufacturer's workshop manual including schematic wiring diagrams
- Digital volt-ohmmeter (DVOM), ammeter, current clamp
- Electrical spare parts, including fuses, circuit breakers, relays, and solenoids
- Manufacturer-specific tools depending on the concern
- Machine lifting equipment, if applicable

Some Safety Issues to Consider

- Activities require you to measure electrical values. Always ensure that the instructor/supervisor checks test instrument connections prior to connecting power or taking measurements. High current flows can be dangerous; avoid accidental short circuits or grounding a battery's positive connections.

- Activities may require test driving the equipment on the school grounds, which carry severe risks. Attempt this task only with full permission from your supervisor/instructor, and follow all the guidelines exactly.
- Lifting equipment such as equipment jacks and stands, hoists, and engine hoists are important tools that increase productivity and make the job easier. However, they can also cause severe injury or death if used improperly. Make sure you follow the manufacturer's operation procedures. Also make sure you have your supervisor/instructor's permission to use any particular type of lifting equipment.
- Comply with personal and environmental safety practices associated with clothing; eye protection; hand tools; power equipment; proper ventilation; and the handling, storage, and disposal of chemicals/materials in accordance with federal, state, and local regulations.
- Always wear the correct protective eyewear and clothing and use the appropriate safety equipment, as well as fender covers, seat protectors, and floor mat protectors.
- Make sure you understand and observe all legislative and personal safety procedures when carrying out practical assignments. If you are unsure of what these are, ask your supervisor/instructor.

Name _________________________ Date __________ Class _________________

Vehicle used for this activity:

Year _____________ Make _______________________ Model _________________

Odometer __________ Hour meter __________ VIN _____________________

▶ **TASK** Know the basic structure of conductors, insulators, and semiconductors.

AED
2.1a

Time off_________________

Time on_________________

Total time_________________

CDX Tasksheet Number: E0035

Cu

A

Ar

B

Si

C

1. **Using the figure above, which atom would be considered an insulator?**

   ____

      a. **Atom A.**
      b. **Atom B.**
      c. **Atom C.**
      d. **None of the above.**

2. **Using the figure above, which atom would be considered a conductor?**

   ____

      a. **Atom A.**
      b. **Atom B.**
      c. **Atom C.**
      d. **None of the above.**

3. **Using the figure above, which atom would be considered a semiconductor?**

   ____

      a. **Atom A.**
      b. **Atom B.**
      c. **Atom C.**
      d. **None of the above.**

4. **In your own words describe what makes an atom a good conductor.**

   ____

   ____

   ____

   ____

5. In your own words describe what makes an atom a good insulator.

____

____

____

____

6. In your own words describe what makes an atom a good semiconductor.

____

____

____

____

7. In your own words describe the types of materials that are often used as conductors in off-road machines.

____

____

____

____

8. In your own words describe the types of materials that are often used as insulators in off-road machines.

____

____

____

____

9. In your own words describe the types of materials that are often used as semiconductors in off-road machines.

____

____

____

____

10. Instructor Comments:

____

____

____

____

Performance Rating

CDX Tasksheet Number: E0035

0	1	2	3	4

Supervisor/instructor signature ______________________________________ Date ____________

▶ TASK Know the reaction of like and unlike charges.

AED 2.1

Time off_______________

Time on_______________

Total time_______________

CDX Tasksheet Number: E0036

1. Forces of repulsion and attraction between electrons and protons are termed electrostatic force and are the primary type of energy contained in electricity to perform work.

2. Research how the forces of repulsion and attraction between electrons and protons occur and answer the questions below.

    a. In your own words describe how unlike charges are attracted to one another (positive to negative).

    ____
    ____
    ____
    ____

    b. In your own words describe how like charges are repelled by one another (positive repels positive, negative repels negative).

    ____
    ____
    ____
    ____

    c. In your own words describe how the energy of electrostatic force contained in electricity is used to perform work.

    ____
    ____
    ____
    ____

3. **Instructor Comments:**

____

____

____

____

Name _________________________________ Date ___________ Class _____________________

Vehicle used for this activity:

Year _______________ Make _________________________ Model _____________________

Odometer ____________ Hour meter ____________ VIN _________________________________

▶ **TASK** Describe the differences of conventional and electron theory current flow.

AED
2.1b

Time off_________________

Time on_________________

Total time_______________

CDX Tasksheet Number: E0037

1. **Electron theory and conventional theory of electron flow each convey the idea of current flow, and each may in some instances be helpful when performing electrical diagnostic work.**

2. **Research electron theory and conventional theory of electron flow and answer the questions below.**

    a. **In your own words describe how electrons would flow through a circuit using the electron theory.**

    ____

    ____

    ____

    ____

    b. **In your own words describe how electrons would flow through a circuit using the conventional theory.**

    ____

    ____

    ____

    ____

    c. **In your own words describe when you may encounter the need to understand the electron theory when performing work.**

    ____

    ____

    ____

    ____

    d. **In your own words describe when you may encounter the need to understand the conventional theory when performing work.**

    ____

    ____

    ____

    ____

▶ TASK Define resistance and its effect on current flow.

AED 2.1b

Time off_________________

Time on_________________

Total time_________________

CDX Tasksheet Number: E0038

1. Electrical resistance is similar to the concept of friction. Resistance in an electrical circuit can slow down electron speed.

2. Research electrical resistance and its effect on an electrical circuit and answer the questions below.

   a. In your own words describe the ways resistance can occur in an electrical circuit.

   ____

   ____

   ____

   ____

   b. In your own words describe how resistance can affect voltage and amperage in an electrical circuit.

   ____

   ____

   ____

   ____

   c. In your own words describe how resistance can be measured in an electrical circuit.

   ____

   ____

   ____

   ____

   d. In your own words describe when you may encounter the need to measure electrical resistance when trouble shooting an electrical problem.

   ____

   ____

   ____

   ____

3. Instructor Comments:

Student/Intern information:

Name _________________________________ Date ___________ Class _________________________

Vehicle used for this activity:

Year _______________ Make _____________________________ Model _______________________________

Odometer _____________ Hour meter _____________ VIN __________________________________

▶ TASK Demonstrate the principles of operation and the correct usage of the various types of meters to measure volts, amps, and ohms.

AED 2.1b

Time off_____________________

Time on_____________________

CDX Tasksheet Number: E0039

Total time_____________________

> **NOTE:** This tasksheet will require the use of a machine or simulator with an electrical fault. Ask your instructor which machine or simulator you are to use.

1. **Using the appropriate service information for the machine/simulator you are working on, research how to check applied voltages, circuit voltages, and voltage drops in electrical/electronic circuits. List the circuits, including the lead connections.**

   a. **Applied voltages**
   b. **Circuit voltages**
   c. **Voltage drops**

2. **Select a DVOM and prepare it to measure DC volts.**

   a. **What are the steps necessary in preparing a DVOM to measure voltage?**

   ____

   ____

   ____

   ____

   ____

3. **Have your supervisor/instructor verify your research.**

   Supervisor/instructor's initials: ___________

4. **Using the DVOM, check applied voltages, circuit voltages, and voltage drops in electrical/electronic circuits. Ask your instructor/supervisor for a machine/simulator and circuits to check.**

   a. **Applied voltages: List the circuit being checked**

   i. **Are the results within the specifications?**
      Yes: _________ **No:** _________

   If no, what are the recommendations?

    b. Circuit voltages: List the circuit being checked.

        i. Are the results within the specifications? Yes: ______ No: ______

        If no, what are the recommendations?

    c. Voltage drops: List the circuit being checked.

        i. Are the results within the specifications? Yes: ______ No: ______

        If no, what are the recommendations?

5. Using the appropriate service information for the machine/simulator you are working on, research how to check current flow in electrical/electronic circuits and components. List the circuits, including the lead connections.

    a. DVOM current measurements: Draw a diagram showing the lead connection and circuit.

    b. DVOM with inductive clamp current measurements: Draw a diagram showing the lead connection and circuit.

6. Prepare the DVOM to measure DC current.

    a. What are the steps necessary in preparing a DVOM to measure current?

7. Have your supervisor/instructor verify your research.

    Supervisor/instructor's initials: __________

8. Using the appropriate service information, check current flow in electrical/electronic circuits and components. Ask your instructor/supervisor for a machine/simulator and circuits to check. Caution: Ensure that the meter is connected correctly when measuring current draw. Damage may occur to circuits and the test equipment if connections are made incorrectly.

    a. List the circuit being checked

        i. Are the results within the specifications? Yes: ______ No: ______

        If no, what are the recommendations?

    b. List the circuit being checked

        i. Are the results within the specifications? Yes: ______ No: ______

        If no, what are the recommendations?

    c. List the circuit being checked

        i. Are the results within the specifications? Yes: ______ No: ______

        If no, what are the recommendations?

____

____

____

____

9. **Using the appropriate service information for the machine/simulator you are working on, research how to check resistance in electrical/electronic circuits and components. List the circuits, including the lead connections.**

   a. **DVOM resistance measurements: Draw a diagram showing the lead connection and circuit.**

10. **Prepare the DVOM to measure resistance.**

   a. **What are the steps necessary in preparing your DVOM to measure resistance?**

11. **Explain why resistance measurements should only be undertaken without power connected to the circuit under test.**

12. **Have your supervisor/instructor verify your research.**

   Supervisor/instructor's initials: _____________

13. **Using the appropriate service information, check resistance in electrical/ electronic circuits and components. Ask your instructor/supervisor for a machine/simulator and circuits to check.**

   a. **List the circuit being checked**

   i. **Are the results within the specifications? Yes: _________ No: _________**

   If no, what are the recommendations?

   b. **List the circuit being checked**

   i. **Are the results within the specifications? Yes: _________ No: _________**

   If no, what are the recommendations?

   c. **List the circuit being checked**

   i. **Are the results within the specifications? Yes: _________ No: _________**

   If no, what are the recommendations?

14. **Return machine/simulator to its beginning condition, and return any tools you used to their proper locations.**

15. Discuss your findings with the instructor.

____

____

____

____

▶ **TASK** Demonstrate knowledge of the laws governing permanent magnets, electromagnets, and magnetic fields.

AED
2.1b

Time off_______________

Time on_______________

Total time_______________

CDX Tasksheet Number: E0040

1. **The laws governing permanent magnets, electromagnets, and magnetic fields are important for electricity to perform work.**

2. **Research the laws governing permanent magnets, electromagnets, and magnetic fields and answer the questions below.**

   a. **In your own words describe the three basic types of magnets, and what they are typically made from.**

   ____
   ____
   ____
   ____

   b. **In your own words describe what makes up a magnetic field.**

   ____
   ____
   ____
   ____

   c. **In your own words describe how electromagnetic induction occurs.**

   ____
   ____
   ____
   ____

3. **Instructor Comments:**

____
____
____
____

Student/Intern information:

Name _________________________________ Date ____________ Class _______________________

Vehicle used for this activity:

Year ______________ Make _________________________ Model ___________________________

Odometer ____________ Hour meter ____________ VIN ___________________________________

▶ TASK Demonstrate knowledge of the effects of magnetic forces
on current carrying conductors.

AED 2.1b

Time off________________

Time on________________

CDX Tasksheet Number: E0041

Total time________________

1. **Research the effects of magnetic forces on current carrying conductors and answer the questions below.**

   a. **In your own words describe how magnetism can be useful in electrical/electronic circuits.**

   ____
   ____
   ____
   ____

   b. **In your own words describe how magnetism can affect electrical/ electronic circuits in a negative way.**

   ____
   ____
   ____
   ____

   c. **List below some electrical/electronic components that rely on magnetism to function:**

   ____
   ____
   ____
   ____

2. **Using a compass and electrical circuits and components observe the effects of magnetism. Ask your instructor/supervisor for a machine/simulator and circuits to check.**

   a. **Turn the circuit on.**

   Is the circuit working? Yes: ________ No: ________

   b. **Hold the compass near the electrical circuit. Which way is the compass needle pointing? ____
   _____________.**

   c. **Turn the circuit off and place the compass near the electrical circuit. Which way is the compass needle pointing? _______________________________
   _____________.**

3. Why did the compass needle move when the circuit was on then off?

    ____

    ____

    ____

    ____

4. **Instructor Comments:**

    ____

    ____

    ____

    ____

Performance Rating   CDX Tasksheet Number: E0041

| 0 | 1 | 2 | 3 | 4 |

Supervisor/instructor signature ______________________________________ Date __________

▶ TASK Know the basic parts and operation of the basic types of storage batteries.

AED 2.1b

Time off_______________

Time on_______________

Total time_______________

CDX Tasksheet Number: E0042

1. Ask your instructor for a lab cutaway of a typical storage battery used in MORE applications. Identify the components that makeup the battery and answer questions below.

   a. Locate the case of the battery. What is the case of a typical battery made from? _________________________________

   b. Locate the negative plates of the battery. What are the negative plates typically made from and what is their purpose in battery function?

   ____

   ____

   ____

   ____

   c. Locate the positive plates of the battery. What are the positive plates typically made from and what is their purpose in battery function?

   ____

   ____

   ____

   ____

   d. Locate the separator plates of the battery. What are the separator plates typically made from and what is their purpose in battery function?

   ____

   ____

   ____

   ____

   e. Locate the cell straps of the battery. What are the cell straps typically made from and what is their purpose in battery function?

   ____

   ____

   ____

   ____

   f. Locate the positive and negative posts of the battery. What are the posts typically made from and what is their purpose in battery function?

   ____

   ____

   ____

   ____

g. Locate the vent(s) of the battery. What is the function of the battery vent?

____

____

____

____

2. What are the typical types and classifications of batteries used in MORE applications?

____

____

____

____

   a. Are all batteries constructed the same and out of the same materials?

____

____

____

____

   b. Ask your instructor for various types of batteries and using the manufacturers label on the battery answer the information below:

	Battery 1	Battery 2	Battery 3	Battery 4
SLI				
Deep cycle				
CCA				
CA				
RC				
Maintenance free				
Size type				
Post type				

3. Instructor Comments:

____

____

____

____

Performance Rating

CDX Tasksheet Number: E0042

0	1	2	3	4

Supervisor/instructor signature _______________________________ Date _________

Ohm's Law

Student/Intern information:

Name _______________________________ Date ____________ Class _______________________

Vehicle used for this activity:

Year _____________ Make _______________________ Model _______________________

Odometer ____________ Hour meter ____________ VIN _______________________

Learning Objective/Task	CDX Tasksheet Number	2014 Edition Rev2/12/16 AED Standard
• Demonstrate the mathematical relationship of the various terms in Ohm's Law as they pertain to series, parallel, and series-parallel circuits.	E0043	2.2a
• Demonstrate the ability to set up and measure the voltage, amperage, and resistance values in series, parallel, and series/parallel DC circuits.	E0044	2.2b

Time off_______________

Time on_______________

Total time_______________

Materials Required

- Machine or simulator with electrical circuit concerns
- Equipment manufacturer's workshop manual including schematic wiring diagrams
- Digital volt-ohmmeter (DVOM), ammeter, current clamp
- Electrical spare parts, including fuses, circuit breakers, relays, and solenoids
- Manufacturer-specific tools depending on the concern
- Machine lifting equipment, if applicable

Some Safety Issues to Consider

- Activities require you to measure electrical values. Always ensure that the instructor/supervisor checks test instrument connections prior to connecting power or taking measurements. High current flows can be dangerous; avoid accidental short circuits or grounding a battery's positive connections.
- Activities may require test driving the equipment on the school grounds, which carry severe risks. Attempt this task only with full permission from your supervisor/instructor, and follow all the guidelines exactly.
- Lifting equipment such as equipment jacks and stands, hoists, and engine hoists are important tools that increase productivity and make the job easier. However, they can also cause severe injury or death if used improperly. Make sure you follow the manufacturer's operation procedures. Also make sure you have your supervisor/instructor's permission to use any particular type of lifting equipment.
- Comply with personal and environmental safety practices associated with clothing; eye protection; hand tools; power equipment; proper ventilation; and the handling, storage, and disposal of chemicals/materials in accordance with federal, state, and local regulations.

- Always wear the correct protective eyewear and clothing and use the appropriate safety equipment, as well as fender covers, seat protectors, and floor mat protectors.
- Make sure you understand and observe all legislative and personal safety procedures when carrying out practical assignments. If you are unsure of what these are, ask your supervisor/instructor.

Student/Intern information:

Name _________________________________ Date _____________ Class _________________________

Vehicle used for this activity:

Year _______________ Make _____________________________ Model _________________________

Odometer ____________ Hour meter ____________ VIN _________________________________

▶ TASK Demonstrate the mathematical relationship of the various terms in Ohm's Law as they pertain to series, parallel, and series-parallel circuits.

AED
2.2a

Time off_______________

Time on_______________

Total time_______________

CDX Tasksheet Number: E0043

1. **Using Ohm's Law answer the problems below:**

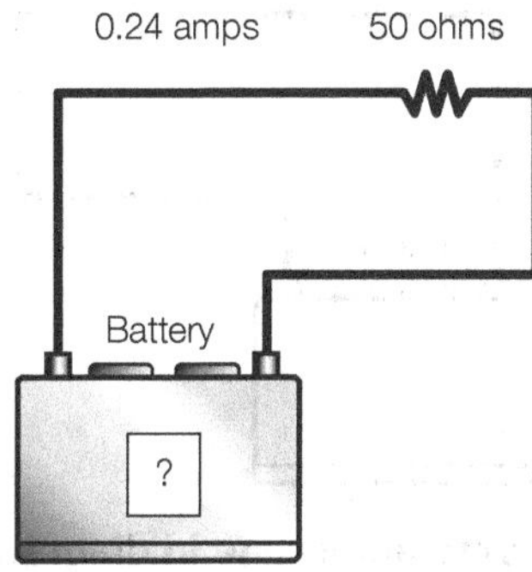

a. **Calculate missing voltage in diagram above (show your calculations).**

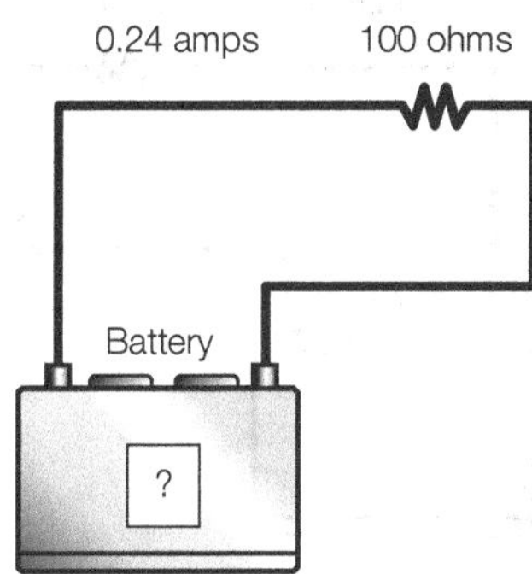

b. **Calculate missing voltage in diagram above (show your calculations).**

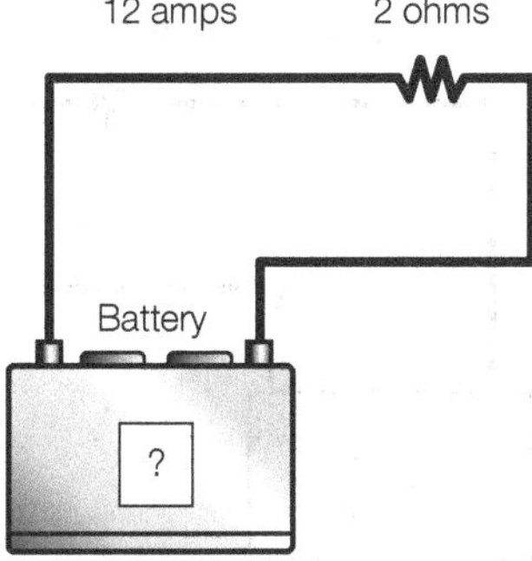

c. Calculate missing voltage in diagram above (show your calculations).

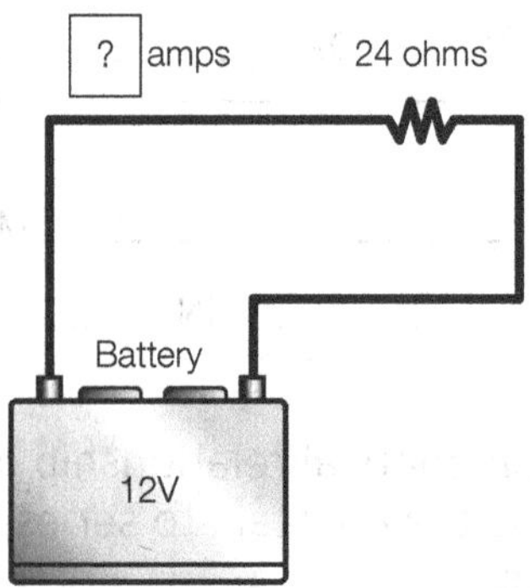

d. Calculate missing amperage in diagram above (show your calculations).

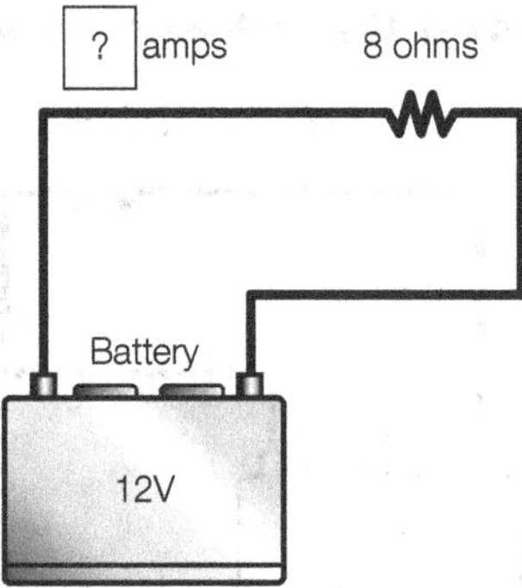

e. Calculate missing amperage in diagram above (show your calculations).

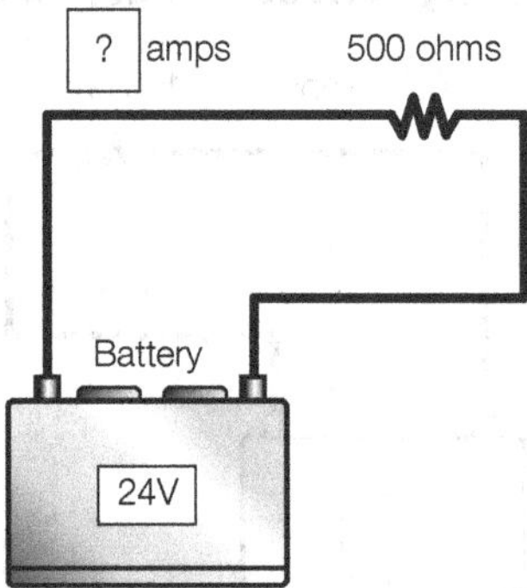

f. Calculate missing amperage in diagram above (show your calculations).

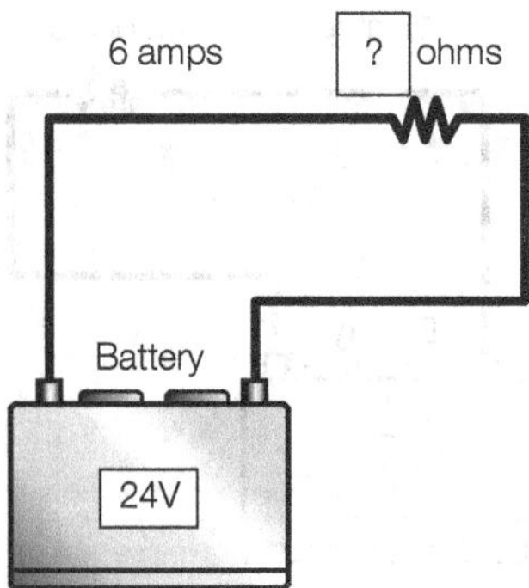

g. Calculate missing resistance in diagram above (show your calculations).

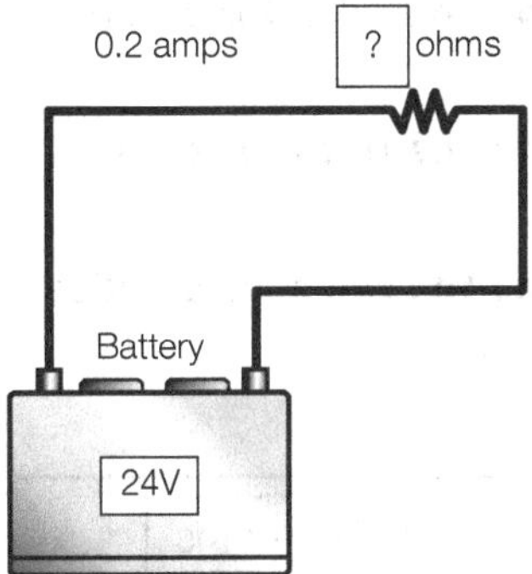

h. Calculate missing resistance in diagram above (show your calculations).

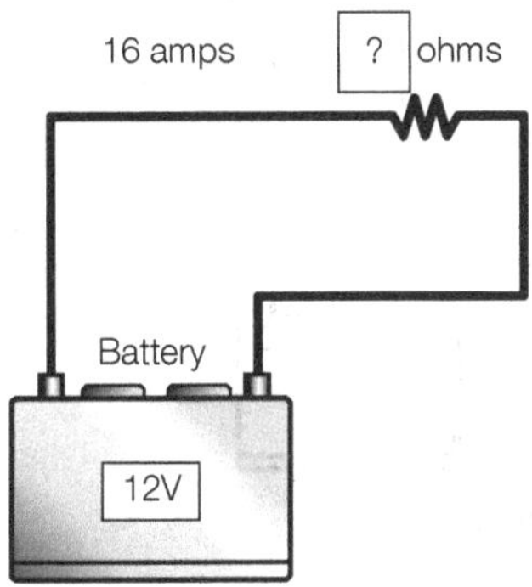

i. Calculate missing resistance in diagram above (show your calculations).

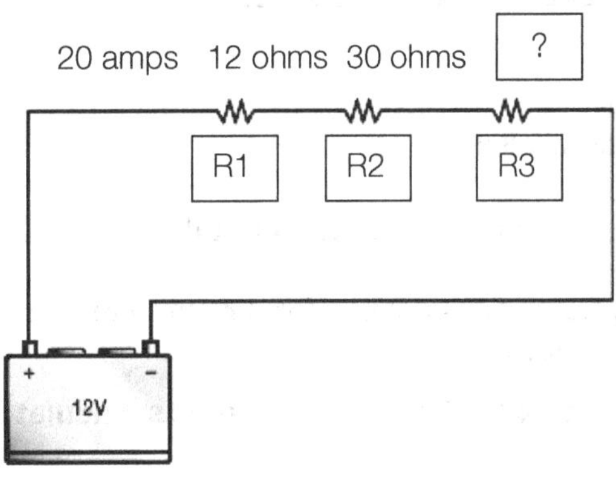

Figure 1

Using the circuit above answer the questions below.

2. What type of circuit is shown above?

    a. Series
    b. Parallel
    c. Series-parallel

3. What is the supply voltage to the circuit? ____________

4. Using Ohm's Law and the rules for circuits calculate the total current flow through the circuit. ____________

5. Using Ohm's Law and the rules for circuits calculate the total circuit resistance. ___________

6. Using Ohm's Law and the rules for circuits calculate the voltage drop across R2. ___________

7. Using Ohm's Law and the rules for circuits calculate the resistance of R3. ___________

8. Using Ohm's Law and the rules for circuits calculate the current flow through R1. ___________

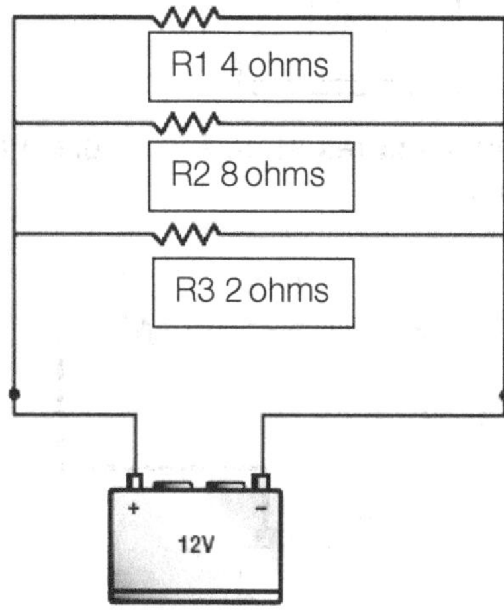

Figure 2

Using the circuit above answer the questions below.

9. What type of circuit is shown above?

    a. Series
    b. Parallel
    c. Series-parallel

10. What is the supply voltage to the circuit? ___________

11. Using Ohm's Law and the rules for circuits calculate the total current flow through the circuit. ___________

12. Using Ohm's Law and the rules for circuits calculate the total circuit resistance. ___________

13. Using Ohm's Law and the rules for circuits calculate the voltage drop across R1. ___________

14. Using Ohm's Law and the rules for circuits calculate the voltage drop across R3. ___________

15. Using Ohm's Law and the rules for circuits calculate the current flow through R1. ___________

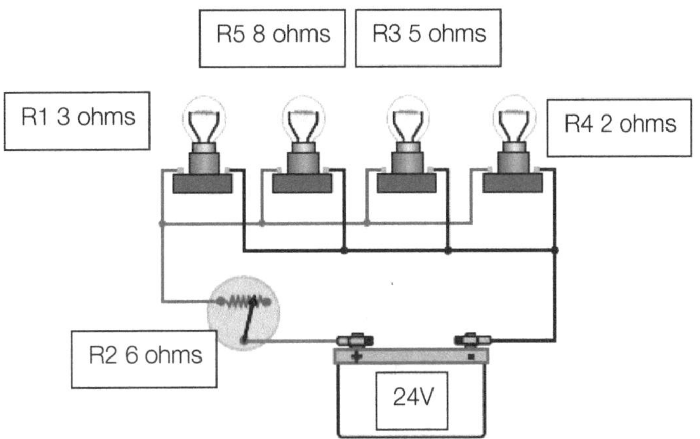

Figure 3

Using the circuit above answer the questions below.

16. **What type of circuit is shown above?**

    a. **Series**
    b. **Parallel**
    c. **Series-parallel**

17. **What is the supply voltage to the circuit?** _____________

18. **Using Ohm's Law and the rules for circuits calculate the total current flow through the circuit.** _____________

19. **Using Ohm's Law and the rules for circuits calculate the total circuit resistance.** _____________

20. **Using Ohm's Law and the rules for circuits calculate the voltage drop across R1.** _____________

21. **Using Ohm's Law and the rules for circuits calculate the voltage drop across R3.** _____________

22. **Using Ohm's Law and the rules for circuits calculate the current flow through R1.** _____________

23. **Instructor Comments:**

    ____
    ____
    ____
    ____

Performance Rating

CDX Tasksheet Number: E0043

0	1	2	3	4

Supervisor/instructor signature _________________________________ Date _____________

▶ TASK Demonstrate the ability to set up and measure the voltage, amperage, and resistance values in series, parallel, and series/parallel DC circuits.

AED 2.2b

Time off________________

Time on________________

CDX Tasksheet Number: E0044

Total time________________

> **NOTE:** This tasksheet will require the use of an electrical training aid simulator.

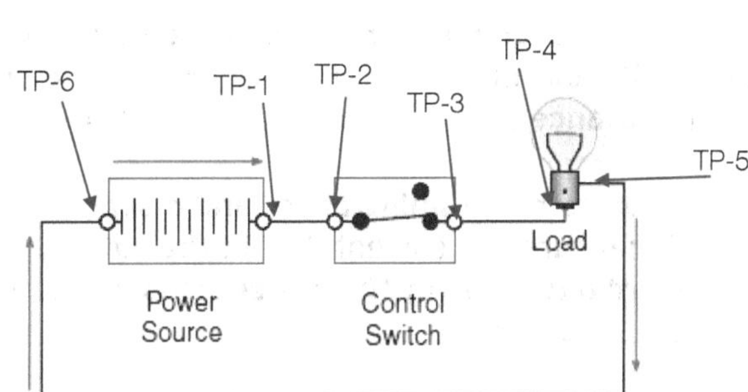

Figure 4

1. **Connect the training aid as shown in the figure above, and then complete the tasks below.**

    a. **With the training aid in the off position verify the lamps are not illuminated. Yes ___________ No ___________**

    b. **Using a DVOM measure the resistance (TP2-TP3) and record. ___________ ohms**

    c. **Using a DVOM measure the resistance across the switch (TP2-TP5) and record. ___________ ohms**

    d. **Using a DVOM measure the resistance across the lamp (TP4-TP5) and record. ___________ ohms**

2. **Connect the training aid as shown in the figure above, and then complete the tasks below.**

    a. **Turn the training aid on and verify the lamp is illuminated.**

       Yes ___________ No ___________

    b. **Using a DVOM measure the supply voltage and record. ___________ volts**

c. Using a DVOM measure the voltage drop (TP2-TP3) and record.
_____________ volts

d. Using a DVOM measure the voltage drop across the switch (TP2-TP5) and record. _____________ volts

e. Using a DVOM measure the voltage drop across the lamp (TP4-TP5) and record. _____________ volts

f. What type of circuit is shown above.

   i. Series

   ii. Parallel

   iii. Series-parallel

3. Turn the power off to the training board and disconnect the wire lead at TP2. Using a DVOM measure the current flow in the circuit. Turn the power on to the training board and connect the meter leads at the wire lead and TP2 and record. _____________ amps

   a. Use the amperage measurement above to calculate the total resistance in the circuit. The only resistance in the circuit should be the total resistance of the lamp. _____________ ohms

4. Turn the power off to the training board and disconnect the wire lead at TP6. Using a DVOM measure the current flow in the circuit. Turn the power on to the training board and connect the meter leads at the wire lead and TP6 and record. _____________ amps

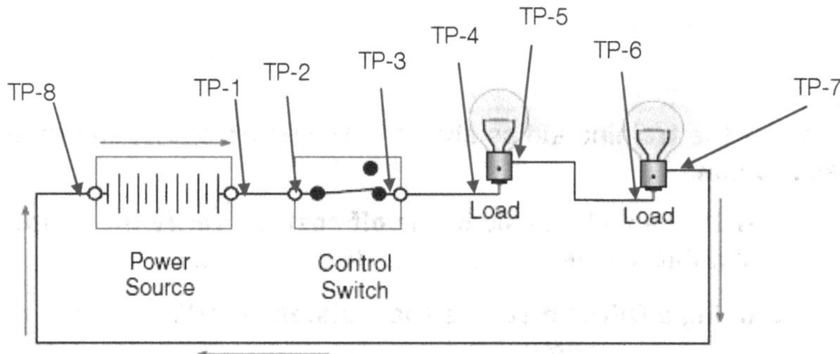

Figure 5

5. Connect the training aid as shown in the figure above, and then complete the tasks below.

   a. With the training aid in the off position verify the lamps are not illuminated. Yes _____________ No _____________

   b. Using a DVOM measure the resistance (TP1-TP4) and record. _____________ ohms

   c. Using a DVOM measure the resistance across lamp (TP4-TP5) and record. _____________ ohms

d. Using a DVOM measure the resistance across lamp (TP6-TP7) and record. ___________ ohms

   Using a DVOM measure the resistance across both lamps (TP3-TP8) and record. ___________ ohms

   Turn the training aid on and verify the lamps are illuminated.

   Yes ___________ No ___________

e. Using a DVOM measure the supply voltage and record. ___________ volts

f. Using a DVOM measure the voltage drop (TP1-TP4) and record. ___________ volts

g. Using a DVOM measure the voltage drop across lamp (TP4-TP5) and record. ___________ volts

h. Using a DVOM measure the voltage drop across lamp (TP6-TP7) and record. ___________ volts

i. Using a DVOM measure the voltage drop across both lamps (TP3-TP8) and record. ___________ volts

j. What type of circuit is shown above?
   i. Series
   ii. Parallel
   iii. Series-parallel

6. Turn the power off to the training board and disconnect the wire lead at TP2. Using a DVOM measure the current flow in the circuit. Turn the power on to the training board and connect the meter leads at the wire lead and TP2 and record. ___________ amps

   a. Use the amperage measurement above to calculate the total resistance in the circuit. The only resistance in the circuit should be the total resistance of the lamps. ___________ ohms

7. Turn the power off to the training board and disconnect the wire lead at TP7. Using a DVOM measure the current flow in the circuit. Turn the power on to the training board and connect the meter leads at the wire lead and TP7 and record. ___________ amps

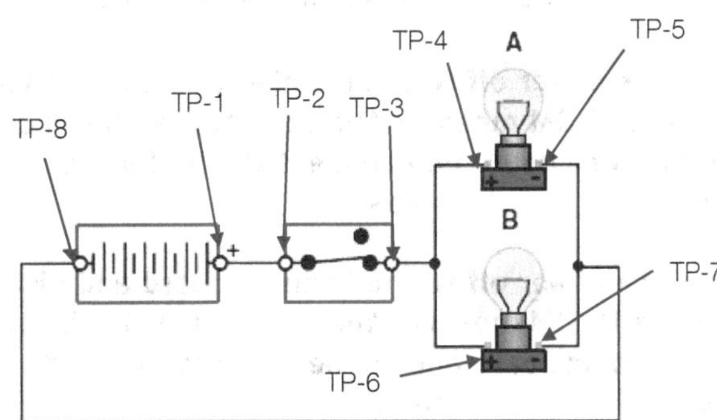

Figure 6

8. Connect the training aid as shown in the figure above and then complete the tasks below.

   a. Turn the training aid to the off position. Yes ___________
No ___________

   b. Using a DVOM measure the resistance (TP1-TP5) and record.
___________ ohms

   c. Using a DVOM measure the resistance across resistor (TP5-TP6) and record. ___________ ohms

   d. Using a DVOM measure the resistance across resistor (TP3-TP4) and record. ___________ ohms

   e. Using a DVOM measure the resistance across resistor (TP7-TP8) and record. ___________ ohms

   f. Using a DVOM measure the resistance across (TP1-TP2) and record.
___________ ohms

   g. Turn the training aid to the on position. Yes ___________
No ___________

   h. Using a DVOM measure the supply voltage and record.
___________ volts

   i. Using a DVOM measure the voltage drop (TP1-TP5) and record.
___________ volts

   j. Using a DVOM measure the voltage drop across (TP5-TP6) and record. ___________ volts

   k. Using a DVOM measure the voltage drop across (TP3-TP4) and record. ___________ volts

   l. Using a DVOM measure the voltage drop across (TP7-TP8) and record. ___________ volts

   m. What type of circuit is shown above?

      i. Series

      ii. Parallel

      iii. Series-parallel

9. Turn the power off to the training board and disconnect the wire lead at TP1. Using a DVOM measure the current flow in the circuit. Turn the power on to the training board and connect the meter leads at the wire lead and TP1 and record. ___________ amps

   a. Use the amperage measurement above to calculate the total resistance in the circuit. The only resistance in the circuit should be the total resistance of the resistors. ___________ ohms

10. Turn the power off to the training board and disconnect the wire lead at TP2. Using a DVOM measure the current flow in the circuit. Turn the power on to the training board and connect the meter leads at the wire lead and TP2 and record. ___________ amps

11. Turn the power off to the training board and disconnect the wire lead at TP3. Using a DVOM measure the current flow in the circuit. Turn the power on to the training board and connect the meter leads at the wire lead and TP3 and record. ___________ amps

12. Turn the power off to the training board and disconnect the wire lead at TP5. Using a DVOM measure the current flow in the circuit. Turn the power on to the training board and connect the meter leads at the wire lead and TP5 and record. _____________ amps

13. Turn the power off to the training board and disconnect the wire lead at TP7. Using a DVOM measure the current flow in the circuit. Turn the power on to the training board and connect the meter leads at the wire lead and TP7 and record. _____________ amps

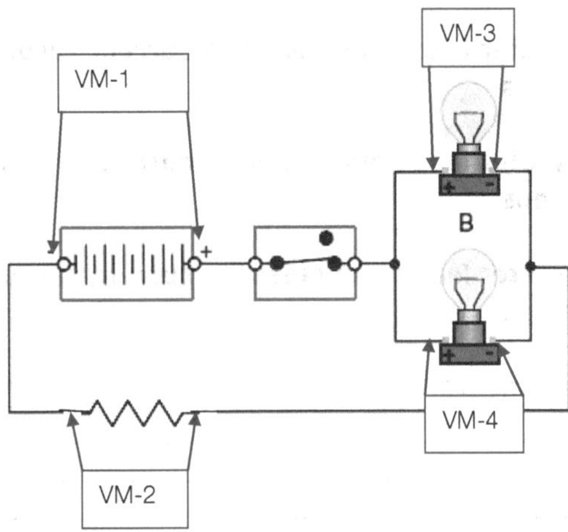

Figure 7

14. Connect the training aid as shown in the figure above and then complete the tasks below.
    a. Turn the training aid to the off position. Yes _________ No _________
    b. Using a DVOM measure the resistance at V1 and record.

      _____________ ohms
    c. Using a DVOM measure the resistance across V2 and record.

      _____________ ohms
    d. Using a DVOM measure the resistance across V3 and record.

      _____________ ohms
    e. Using a DVOM measure the resistance across V4 and record.

      _____________ ohms
    f. What is the total resistance in the circuit? _____________ ohms
    g. Turn the training aid to the on position. Yes _________ No _________
    h. Using a DVOM measure the supply voltage and record. _____________ volts
    i. Using a DVOM measure the voltage drop V1 and record.

      _____________ volts
    j. Using a DVOM measure the voltage drop across V2 and record.

      _____________ volts
    k. Using a DVOM measure the voltage drop across V3 and record.

      _____________ volts

l. Using a DVOM measure the voltage drop across V4 and record.
___________ volts

m. What type of circuit is shown above?

   i. Series

   ii. Parallel

   iii. Series-parallel

15. Connect the meter leads and measure current through VM-1 and record.
___________ amps

16. Connect the meter leads and measure current through VM-2 and record.
___________ amps

17. Connect the meter leads and measure current through VM-3 and record.
___________ amps

18. Connect the meter leads and measure current through VM-4 and record.
___________ amps

19. Instructor Comments:

____
____
____
____

Performance Rating

CDX Tasksheet Number: E0044

| 0 | 1 | 2 | 3 | 4 |

Supervisor/instructor signature ____ Date __________

Battery

Student/Intern information:

Name _________________________________ Date _____________ Class _________________________

Vehicle used for this activity:

Year _______________ Make _____________________________ Model _______________________

Odometer ______________ Hour meter ____________ VIN _________________________________

Learning Objective/Task	CDX Tasksheet Number	2014 Edition Rev2/12/16 AED Standard
• Identify battery type; perform appropriate battery load test; determine needed action.	E0045	N/A
• Determine battery state of charge using an open circuit voltage test.	E0046	N/A
• Inspect, clean, and service battery; replace as needed.	E0047	N/A
• Inspect and clean battery boxes, mounts, and hold downs; repair or replace as needed.	E0048	N/A
• Charge battery using appropriate method for battery type.	E0049	N/A
• Inspect, test, and clean battery cables and connectors; repair or replace as needed.	E0050	N/A
• Jump start a machine using jumper cables and a booster battery or appropriate auxiliary power supply using proper safety procedures.	E0051	N/A
• Perform battery capacitance test; determine needed action.	E0052	N/A

Time off_________________

Time on_________________

Total time_________________

Materials Required

- Machine or simulator with battery concern
- Equipment manufacturer's workshop manual
- Battery hydrometer, digital volt-ohmmeter (DVOM), conductance/capacitance tester, slow and fast chargers, jumper cables, booster battery
- Electrical spare parts, including fuses, circuit breakers, relays, and solenoids
- Manufacturer-specific tools depending on the concern
- Personal protective equipment (PPE)

Some Safety Issues to Consider

- Diagnosis of this fault may require running the engine and managing an environment of dangerous gases and chemicals that carry severe risks. Attempt this task only with full permission from your supervisor/instructor, and follow all the guidelines exactly.
- Use extreme caution when working around batteries. Immediately remove any electrolyte that you may come in contact with. Electrolyte is a mixture of sulfuric acid and water. Batteries may produce explosive mixtures of gas containing hydrogen; avoid creating any sparks around batteries. Please consult with the shop safety and emergency procedures when working with or around batteries.
- Activities may require test driving the equipment on the school grounds, which carry severe risks. Attempt this task only with full permission from your supervisor/instructor, and follow all the guidelines exactly.
- Lifting equipment such as equipment jacks and stands, hoists, and engine hoists are important tools that increase productivity and make the job easier. However, they can also cause severe injury or death if used improperly. Make sure you follow the manufacturer's operation procedures. Also make sure you have your supervisor/instructor's permission to use any lifting equipment.
- Comply with personal and environmental safety practices associated with clothing; eye protection; hand tools; power equipment; proper ventilation; and the handling, storage, and disposal of chemicals/materials in accordance with federal, state, and local regulations.
- Always wear the correct protective eyewear and clothing and use the appropriate safety equipment, as well as fender covers, seat protectors, and floor mat protectors.
- Make sure you understand and observe all legislative and personal safety procedures when carrying out practical assignments. If you are unsure of what these are, ask your supervisor/instructor.

Performance Standard

0—No exposure: No information or practice provided during the program; complete training required

1—Exposure only: General information provided with no practice time; close supervision needed; additional training required

2—Limited practice: Has practiced job during training program; additional training required to develop skill

3—Moderately skilled: Has performed job independently during training program; limited additional training may be required

4—Skilled: Can perform job independently with no additional training

▶ TASK Identify battery type; perform appropriate battery load test; determine needed action.

AED

Time off__________

Time on__________

Total time__________

CDX Tasksheet Number: E0045

1. **Using the appropriate service manual, research and confirm proper battery capacity for the machine application. List the specifications:**

____
____
____
____

2. **Perform the battery load test following the testing equipment's test instructions. List your observations. Note: Battery load testing is used to determine if the battery has the capacity to handle all of the electrical requirements while the machine is being operated.**

____
____
____
____

3. **Does the system meet the manufacturer's specifications? Yes: ________ No: ________**

4. **If no, what repairs are necessary to rectify the system?**

____
____
____
____

5. **Return machine to its beginning condition, and return any tools you used to their proper locations.**

6. **Discuss your findings with the instructor.**

Student/Intern information:

Name _________________________________ Date _____________ Class _________________________

Vehicle used for this activity:

Year _______________ Make _____________________________ Model _______________________

Odometer ______________ Hour meter _____________ VIN _______________________________

▶ **TASK** Determine battery state of charge using an open circuit voltage test.

AED

Time off_______________

Time on_______________

CDX Tasksheet Number: EO046

Total time_______________

> **NOTE:** This tasksheet will require the use of a machine or simulator with an electrical fault. Ask your instructor which machine or simulator you are to use.

1. **Locate "Open Circuit Voltage Test" in the assigned machine service manual. List the steps for this procedure. Note: Open circuit voltage testing should not be used to make a final determination for battery replacement.**

   ____

   ____

   ____

2. **Make sure the engine is off and the battery is stabilized. If the battery has just been recharged, you must load test the battery to remove the surface charge. Turning the headlights on for 30 seconds can be used to remove the surface charge of a battery. Wait at least 10 minutes after the load test before measuring the open circuit voltage. Follow the manufacturer's recommendations closely.**

3. **Prepare the DVOM to measure voltage Measure the battery voltage with the meter. Place the red lead on the positive post/terminal and the black lead on the negative post/terminal.**

   What is the measured voltage (open circuit voltage) of the battery?
   _________________ **volts**

4. **The chart below represents the open circuit voltage of the battery. Please mark the state of charge of the battery as it relates to the voltage measured.**

12.6V or greater	100% charge	
12.4-12.6V	75-100% charge	
12.2-12.4V	50-75% charge	
12.0-12.2V	25-50% charge	
11.7-12.0V	0-25% charge	
11.7-0.0V	0% or no charge	

5. **Determine any necessary action(s).**

6. **Return machine to its beginning condition, and return any tools you used to their proper locations.**

7. **Discuss your findings with the instructor.**

Performance Rating

CDX Tasksheet Number: E0046

| 0 | 1 | 2 | 3 | 4 |

Supervisor/instructor signature ____ Date _______________

▶ TASK Inspect, clean, and service battery; replace as needed. _______ **_AED_**

CDX Tasksheet Number: E0047

Time off____________

Time on____________

1. **Prepare the machine for the task.**

2. **Research the recommended process in the appropriate service manual or other available data. List the steps:**

   ____
   ____
   ____

Total time____________

3. **Inspect, clean, fill, and/or replace battery, as required. Note your observations:**

   ____
   ____
   ____

4. **Determine any necessary further action(s):**

   ____
   ____
   ____

5. **What are the steps necessary in preparing your DVOM to measure voltage?**

   ____
   ____
   ____

6. **Return machine to its beginning condition, and return any tools you used to their proper locations.**

7. **Discuss your findings with the instructor.**

Performance Rating   **CDX Tasksheet Number: E0047**

0	1	2	3	4

Supervisor/instructor signature _______________________________ Date ______________

Student/Intern information:

Name _______________________________ Date _____________ Class _______________________________

Vehicle used for this activity:

Year _______________ Make _______________________________ Model _______________________________

Odometer _____________ Hour meter _____________ VIN _______________________________

▶ TASK Inspect and clean battery boxes, mounts, and hold downs; repair as needed.

AED

Time off_______________

Time on_______________

Total time_______________

CDX Tasksheet Number: E0048

1. **Research inspecting and cleaning battery boxes, mounts, and hold-downs.**

2. **List the steps for inspecting and cleaning battery boxes, mounts, and hold-downs:**

   ____

   ____

   ____

3. **Determine whether the battery should be removed to allow this task to be performed. Follow the manufacturer's recommendations.**

4. **Inspect the battery hold-down hardware and the battery tray. List your findings and any recommendations you may have:**

   ____

   ____

   ____

5. **Clean the battery, battery tray, and hold-down hardware with a suitable cleaner or by mixing baking soda and water. Be careful when using baking soda as it is an acid neutralizer if introduced into the cells.**

> **NOTE:** The consistency of the baking soda and water should be like a paste. The use of a small brush will help the cleaning process.

   Have your supervisor/instructor check your work.

   Supervisor/instructor's initials: _________

6. **Rinse the components with clean water. Wipe the components dry. Do not use compressed air.**

7. **Install the battery (if removed) and hold-down hardware. Reconnect the cables as outlined in the service information.**

8. Return machine to its beginning condition, and return any tools you used to their proper locations.

9. Discuss your findings with the instructor.

Vehicle used for this activity:

Year _______________ Make _________________________ Model _____________________________

Odometer _____________ Hour meter ____________ VIN _______________________________

▶ TASK Charge battery using appropriate method for battery type. `AED`

CDX Tasksheet Number: E0049

Time off_______________

Time on_______________

Total time_______________

> **NOTE:** You may use either a machine with a discharged battery or an assigned battery that is out of a machine. The method of recharging a battery differs from manufacturer to manufacturer. It is important that you follow the recharging steps recommended by the manufacturer of the battery that is assigned to you. Be careful when attempting to charge a battery. Hooking the charger up incorrectly could cause injury to eyes and skin. Using the appropriate service information for the machine you are working on, research how to check resistance in electrical/electronic circuits and components. List the circuits, including the lead connections.

1. **Using the appropriate service manual, research slow and/or fast battery charging for this machine. Follow all directions.**

    i. **List the steps for recharging the battery:**

    ____
    ____
    ____
    ____.

2. **What method is required for recharging the battery: slow charge or fast charge?**

    ____
    ____
    ____
    ____

3. **List the steps for charging at the rate determined in Step 2:**

    ____
    ____
    ____
    ____

4. **Have your supervisor/instructor check your steps for recharging the battery. Supervisor/instructor's initials:** ___________

5. Charge the battery according to the manufacturer's recommendations.

    i. How long did the battery take to charge?

    ii. What was the lowest amperage setting during charging?
_______________ amps

    iii. How did you determine the battery was fully charged?

6. Determine any necessary action(s):

7. Return machine to its beginning condition, and return any tools you used to their proper locations.

8. Discuss your findings with the instructor.

Performance Rating

| 0 | 1 | 2 | 3 | 4 |

Supervisor/instructor signature _______________________________________ Date ___________

Student/Intern information:

Name ________________________________ Date ____________ Class ________________________

Vehicle used for this activity:

Year ______________ Make __________________________ Model ____________________________

Odometer ____________ Hour meter ____________ VIN ________________________________

▶ TASK Inspect, test, and clean battery cables and connectors; repair or replace as needed.

AED

Time off________________

Time on________________

Total time________________

CDX Tasksheet Number: E0050

1. **Research inspecting, testing, cleaning, and repair or replacement of battery cables and connectors. List the steps:**

   ____

   ____

   ____

   ____

2. **Determine whether the battery should be removed to allow this task to be performed. Follow the manufacturer's recommendations.**

3. **Inspect the battery cables and connectors. List your findings and any recommendations you may have:**

   ____

   ____

   ____

   ____

4. **Clean the battery cables and connectors with a suitable cleaner or by mixing baking soda and water. Note: The consistency of the baking soda and water should be like a paste. The use of a small brush will help the cleaning process.**

5. **Repair or replace the battery cables and connectors with suitable replacements as per the manufacturer's specifications.**

6. Return machine to its beginning condition, and return any tools you used to their proper locations.

7. Discuss your findings with the instructor.

Performance Rating

CDX Tasksheet Number: E0050

0	1	2	3	4

Supervisor/instructor signature ____ Date _____________

Student/Intern information:

Name _________________________ Date __________ Class _________________________

Vehicle used for this activity:

Year ____________ Make ___________________ Model _________________________

Odometer __________ Hour meter __________ VIN _________________________

▶ TASK Jump start a machine using jumper cables and a booster battery or appropriate auxiliary power supply using proper safety procedures.   **AED**

CDX Tasksheet Number: EOO51

Time off__________

Time on__________

Total time__________

1. **Locate "starting a machine with a dead battery" or "jump-starting procedures" in the appropriate service information for the machine you are working on.**

2. **List the steps as outlined in the service information:**

   ____
   ____
   ____
   ____

3. **Have your supervisor/instructor verify the steps for the machine assigned to you. Supervisor/instructor's initials:** _______________________

4. **Connect the jumper cables as outlined in the service information or connect the auxiliary power supply (jump box) as was outlined in the service information.**

5. **Explain why the last connection to the dead battery is away from the battery, preferably on a nonpainted metal component on the engine or frame.**

   ____
   ____
   ____

6. **Start the engine.**

7. **Remove the cables in the reverse order as they were installed.**

8. **Return machine to its beginning condition, and return any tools you used to their proper locations.**

9. **Discuss your findings with the instructor.**

▶ **TASK** Perform battery capacitance test; determine needed action. **AED**

Time off_________________

Time on_________________

CDX Tasksheet Number: E0052

1. **Using the appropriate service manual, research and confirm proper battery capacity for the machine application. List the specifications:**

   ____

   ____

   ____

Total time_________________

2. **Perform the battery capacity test (or conductance test) following the testing equipment's test instructions. List your observations. Always be mindful of safety when working in or around the battery area. Wear appropriate clothing and eye protection.**

   ____

   ____

   ____

3. **Determine any necessary actions:**

   ____

   ____

   ____

4. **Return machine to its beginning condition, and return any tools you used to their proper locations.**

5. **Discuss your findings with the instructor.**

Performance Rating

CDX Tasksheet Number: E0052

0	1	2	3	4

Supervisor/instructor signature _________________________________ Date _____________

12/24 Volt Cranking Circuits

Student/Intern information:

Name _______________________________ Date _____________ Class _____________________

Vehicle used for this activity:

Year _______________ Make _________________________ Model _____________________

Odometer _____________ Hour meter ___________ VIN _________________________________

<table>
<tr><td rowspan="2">

Learning Objective/Task
</td><td>

CDX Tasksheet Number
</td><td>

2014 Edition Rev2/12/16 AED Standard
</td></tr>
<tr><td></td><td></td></tr>
<tr><td>

• Know the basic components that make up the various types of 12/24 volt cranking systems.
</td><td>E0053</td><td>2.3a</td></tr>
<tr><td>

• Demonstrate the sequence of operation of the components contained within a cranking system. The emphasis is on how each component effects the system's overall operation.
</td><td>E0054</td><td>2.3b</td></tr>
<tr><td>

• Demonstrate the ability to isolate problems using voltage drops and other diagnostic methods. The proper use of testing equipment is paramount.
</td><td>E0055</td><td>2.3c</td></tr>
<tr><td>

• Demonstrate the ability to properly test, evaluate, and replace the following components using manufacturers' service publications and specifications.
1. Conductors
2. Relays/Solenoids
3. Starters
</td><td>E0056</td><td>2.3d</td></tr>
</table>

Time off_______________

Time on_______________

Total time_______________

Materials Required

- Machine or simulator with starter concern
- Equipment manufacturer's workshop manual
- Digital volt-ohmmeter (DVOM), ammeter for starter circuits, test lamp, starter circuit spare parts
- Manufacturer-specific tools depending on the concern
- Personal protective equipment (PPE)

Some Safety Issues to Consider

- Activities may require running the engine and managing an environment of rotating equipment and large current draw, which carry severe risks. Attempt this task only with full permission from your supervisor/instructor, and follow all the guidelines exactly.

- Use extreme caution when working around batteries. Immediately remove any electrolyte that you may come in contact with. Electrolyte is a mixture of sulfuric acid and water. Batteries may produce explosive mixtures of gas containing hydrogen; avoid creating any sparks around batteries. Please consult with the shop safety and emergency procedures when working with or around batteries.
- Lifting equipment such as equipment jacks and stands, hoists, and engine hoists are important tools that increase productivity and make the job easier. However, they can also cause severe injury or death if used improperly. Make sure you follow the manufacturer's operation procedures. Also make sure you have your supervisor/instructor's permission to use any particular type of lifting equipment.
- Comply with personal and environmental safety practices associated with clothing; eye protection; hand tools; power equipment; proper ventilation; and the handling, storage, and disposal of chemicals/materials in accordance with federal, state, and local regulations.
- Always wear the correct protective eyewear and clothing and use the appropriate safety equipment, as well as fender covers, seat protectors, and floor mat protectors.
- Make sure you understand and observe all legislative and personal safety procedures when carrying out practical assignments. If you are unsure of what these are, ask your supervisor/instructor.

Performance Standard

0–No exposure: No information or practice provided during the program; complete training required

1–Exposure only: General information provided with no practice time; close supervision needed; additional training required

2–Limited practice: Has practiced job during training program; additional training required to develop skill

3–Moderately skilled: Has performed job independently during training program; limited additional training may be required

4–Skilled: Can perform job independently with no additional training

Name _______________________ Date ___________ Class _______________________

Vehicle used for this activity:

Year _____________ Make _______________________ Model _______________________

Odometer ___________ Hour meter ___________ VIN _______________________

▶ TASK Know the basic components that make up the various types of 12/24 volt cranking systems.

AED 2.3a

Time off_______________

Time on_______________

Total time_______________

CDX Tasksheet Number: E0053

> **NOTE:** This tasksheet will require the use of a machine or simulator. Ask your instructor which machine or simulator you are to use.

1. **Using the machine or simulator service manual, wiring diagrams, and schematics locate and tag the components that make up the various types of 12/24 volt cranking systems below. If the machine does not utilize a component listed in the table below mark the box N/A. Enter tag number in each empty box.**

Component	Machine #1	Machine #2	Machine #3
Battery			
Load cables			
Starter			
Starter solenoid			
Mega-fuse			
Ignition switch			
Starter relay			
Neutral safety switch			
Overcrank protection			
ADLO			

2. **Discuss your findings with the instructor.**

Performance Rating

CDX Tasksheet Number: E0053

0	1	2	3	4

Supervisor/instructor signature _______________________ Date ___________

Student/Intern information:

Name _________________________________ Date ____________ Class _________________________

Vehicle used for this activity:

Year _______________ Make _____________________________ Model _________________________

Odometer ____________ Hour meter ____________ VIN _________________________________

▶ **TASK** Demonstrate the sequence of operation of the components contained within a cranking system. The emphasis is on how each component effects the system's overall operation.

AED 2.36

Time off_______________

Time on_______________

Total time_______________

CDX Tasksheet Number: E0054

> **NOTE:** This tasksheet will require the use of a machine or simulator. Ask your instructor which machine or simulator you are to use.

1. **Using the machine or simulator service manual, wiring diagrams, and schematics determine the purpose and operation of the components below. Write a short description of the purpose and operation of the components. Note: Understanding how a component is designed to operate within a circuit will make problems easier to diagnose.**

Component	Purpose	Operation
Battery		
Load cables		
Starter		
Starter solenoid		
Mega-fuse		
Ignition switch		
Starter relay		
Neutral safety switch		

Overcrank protection		
ADLO		

2. Discuss your findings with the instructor.

Name ___________________________ Date ___________ Class ___________________________

Vehicle used for this activity:

Year _____________ Make ___________________ Model ___________________________

Odometer ___________ Hour meter ___________ VIN ___________________________

▶ **TASK** Demonstrate the ability to isolate problems using voltage drops and other diagnostic methods. The proper use of testing equipment is paramount.

AED 2.6

Time off___________

Time on___________

CDX Tasksheet Number: E0055

Perform starter circuit cranking voltage and voltage drop tests; determine needed action.

Total time___________

1. **Locate "cranking voltage test" in the assigned machine service manual and list the steps for this procedure.**

   ____

   ____

   ____

2. **Prepare the DVOM to measure voltage.**

3. **Measure the battery voltage at the starter motor with the meter while the engine is cranking. Place the red lead on the main battery terminal of the solenoid and the black lead on the ground terminal of the starter motor. Note: If the starter is not easily accessible, cranking voltage may be measured at the battery. Remember that any voltage drop between the battery and the starter may affect the test results.**

   a. **What is the measured voltage (cranking circuit voltage) of the battery?**

      _________ volts

4. **Does the system meet the manufacturer's specifications?**
   Yes: _________ No: _________

5. **If no, what repairs are necessary to rectify the system?**

   ____

   ____

   ____

6. **Have your supervisor/instructor verify these steps.**

   Supervisor/instructor's initials: _________

7. Using the appropriate service information for the machine you are working on, locate starter circuit voltage drop tests. List the steps outlined:

____

____

____

8. What is the maximum voltage drop specification(s) for this test?

    a. Positive side: _____________ volts

    b. Negative side, if specified: _____________ volts

9. To conduct this test, it will be required that the engine cranks but does not start. Follow the manufacturer's recommendations for disabling the engine so it does not start. Identify the steps you will take to prevent the engine from starting.

____

____

____

10. Have your supervisor/instructor verify these steps.

    Supervisor/instructor's initials: _____________

11. Starter circuit voltage drop test: Positive/feed side. List the voltmeter connection points in the circuit.

    a. Black lead: _____________________________________

    b. Red lead: _____________________________________

12. Conduct the starter circuit voltage drop test.

    a. What is the voltage drop on the negative side?

       _____________ volts

    b. Is this reading within specifications? Yes: _______ No: _______

    c. If no, what repairs are necessary to rectify the system?

____

____

____

____

13. Return the vehicle engine to normal operating condition.

14. Start the vehicle and verify proper operation of the starting system and engine.

15. Discuss your findings with the instructor.

16. Have your supervisor/instructor verify these steps.

    Supervisor/instructor's initials: _____________

Inspect and test components (key switch, push button, and/or magnetic switch) and wires and harnesses in the starter control circuit; replace as needed.

1. Locate the wiring diagram for the component of the starting system that you are testing.

2. Determine the purpose and operation of the components.

3. Write a short description of the purpose and operation of the components. Note: Understanding how a component is designed to operate within a circuit will make problems easier to diagnose.

   ____

   ____

   ____

4. Using the appropriate service information, determine the steps necessary for testing the components. List the suggested steps for diagnosing the components in question:

   ____

   ____

   ____

5. Determine any necessary action(s).

6. Return machine to its beginning condition, and return any tools you used to their proper locations.

7. Discuss your findings with the instructor.

Inspect and test starter relays and solenoids/switches; replace as needed.

1. Using the appropriate service information for the machine you are working on, locate the starter relay and/or solenoid testing procedure Research the recommended process in the appropriate service manual or other available data. List the steps:

   ____

   ____

   ____

2. What is the maximum voltage drop specification for this test?

   _______________________ volts

3. To conduct this test, it will be required that the engine cranks but does not start. Follow the manufacturer's recommendations on disabling the engine so it does not start. Identify the steps you will take to prevent the engine from starting.

4. **Determine any necessary further action(s):**

____

____

____

   a. **Have your supervisor/instructor verify these steps.**

   Supervisor/instructor's initials: _______________

5. **List the voltmeter connection points in the circuit:**

   a. **Black lead:** _______________________________

   b. **Red lead:** _______________________________

6. **Conduct the test.**

   a. **What is the voltage drop?** _______________________

   b. **Is this reading within specifications?** _______________________

7. **Determine any necessary further action(s):**

____

____

____

8. **Return the vehicle engine to normal operating condition.**

9. **Start the vehicle and verify proper operation of the starting system and engine.**

10. **Discuss your findings with the instructor.**

11. **Have your supervisor/instructor verify these steps.**

   Supervisor/instructor's initials: _______________

Performance Rating   **CDX Tasksheet Number: E0055**

0	1	2	3	4

Supervisor/instructor signature _______________________________ Date _____________

▶ TASK Demonstrate the ability to properly test, evaluate, and replace the
following components using manufacturers' service publications
and specifications.
1. Conductors
2. Relays/Solenoids
3. Starters

**AED
2.3d**

Time off_________________

Time on_________________

Total time_________________

CDX Tasksheet Number: E0056

> **NOTE:** This tasksheet will require the use of a machine or simulator. Ask your
> instructor which machine or simulator you are to use.

1. **Using the machine or simulator service manual, wiring diagrams, and
   schematics troubleshoot the operator complaint below.**

   a. **Complaint: Engine cranks over very slow.**

2. **Verify: Does this condition exist?**

   Yes: ________ No: ________

3. **Using the appropriate service information for the machine you are working
   on, locate the testing procedure related to this complaint and research the
   recommended process in the appropriate service manual or other available
   data. List the steps for the tests you will perform:**

   ____

   ____

   ____

   ____

4. **Conduct these tests and report below.**

   a. **Did the tests fall out of manufacturer's specifications?**
      Yes: ________ No: ________
   b. **If yes, what part of the system including component did not pass the
      test? ____________________________**
   c. **Repair or replace the component according to manufacturer's
      specification.**
   d. **Try starting the machine.**
   e. **Does the machine still crank slow? Yes: ________ No: ________**

5. Determine any necessary further action(s):

____

____

____

____

   a. Have your supervisor/instructor verify these steps.

      Supervisor/instructor's initials: _______________

6. Using the machine or simulator service manual, wiring diagrams, and schematics troubleshoot the operator complaint below.

   a. Complaint: Engine will not crank.

7. Verify: Does his condition exist? Yes: _________ No: _________

8. Using the appropriate service information for the machine you are working on, locate the testing procedure related to this complaint and research the recommended process in the appropriate service manual or other available data. List the steps for the tests you will perform:

____

____

____

____

9. Conduct these tests and report below.

   a. Did the tests fall out of manufacturer's specifications?
      Yes: _________ No: _________
   b. If yes, what part of the system including component did not pass the test? _______________________________
   c. Repair or replace the component according to manufacturer's specification.
   d. Try starting the machine.
   e. Does the machine still not crank? Yes: _________ No: _________

10. Determine any necessary further action(s):

____

____

____

____

   a. Have your supervisor/instructor verify these steps.

      Supervisor/instructor's initials: _______________

Remove and replace starter; inspect flywheel ring gear or flex plate.

1. Using the appropriate service manual, research the starter removal and installation procedure for this machine.

   a. List the procedures and tools required to perform this task:

____

____

____

____

2. **List any torque specifications for this procedure.**

   ___________________________________ **(lb-ft/N·m)**

3. **Have your supervisor/instructor verify these steps.**

   Supervisor/instructor's initials: ___________________________________

4. **Remove the starter following the manufacturer's procedures.**

   a. **Have your supervisor/instructor verify that the starter is removed.**

      Supervisor/instructor's initials: ___________________________________

5. **Using the appropriate service manual, research the manufacturer's procedure for inspection of the flywheel ring gear or flex plate for this machine.**

   a. **List the checks that should be made:**

      ____

      ____

      ____

      ____

   b. **Perform the inspection of the flywheel ring gear or flex plate.**

   c. **List the result of the inspection:**

      ____

      ____

      ____

      ____

6. **Have your supervisor/instructor check your work.**

   Supervisor/instructor's initials: ___________________________________

7. **Install the starter following the manufacturer's procedures.**

   Have your supervisor/instructor verify the starter installation.

   Supervisor/instructor's initials: _______________

8. **Return machine to its beginning condition, and return any tools you used to their proper locations.**

9. **Discuss your findings with the instructor.**

Performance Rating

CDX Tasksheet Number: E0056

0	1	2	3	4

Supervisor/instructor signature ___________________________________ Date ___________

12/24 Volt Charging Circuits

Learning Objective/Task	CDX Tasksheet Number	2014 Edition Rev2/12/16 AED Standard
• Know the basic components that make up the various types of 12/24 volt charging systems.	E0057	2.4a
• Demonstrate the sequence of operation of the components contained within a charging system. The emphasis is on how each component effects the system's overall operation.	E0058	2.4b
• Demonstrate the ability to isolate problems using voltage drops and other diagnostic methods. The proper use of testing equipment is paramount.	E0059	2.4c
• Demonstrate the ability to properly test, evaluate, and replace the following components using manufacturers' service publications and specifications. 1. Conductors 2. Alternators 3. Regulators	E0060	2.4d

Time off_________________

Time on_________________

Total time_______________

Materials Required

- Machine or simulator with possible alternator concern, including cable, wiring, or connector faults
- Equipment manufacturer's workshop manual
- Digital volt-ohmmeter (DVOM), ammeter for starter circuits, test lamp alternator testing equipment such as load banks, oscilloscope, wire, cable connectors, electrical hand tools
- Manufacturer-specific tools depending on the concern
- Personal protective equipment (PPE)

Some Safety Issues to Consider

- Activities may require running the engine and managing an environment of rotating equipment and large current draw, which carry severe risks. Attempt this task only with full permission from your supervisor/ instructor, and follow all the guidelines exactly.
- Do not run the alternator without a load connected or allow the output voltage to exceed the manufacturer's specified maximum.

- Use extreme caution when working around batteries. Immediately remove any electrolyte that may come in contact with you. Electrolyte is a mixture of sulfuric acid and water. Batteries may produce explosive mixtures of gas containing hydrogen; avoid creating any sparks around batteries. Please consult with the shop safety and emergency procedures when working with or around batteries.
- Lifting equipment such as equipment jacks and stands, hoists, and engine hoists are important tools that increase productivity and make the job easier. However, they can also cause severe injury or death if used improperly. Make sure you follow the manufacturer's operation procedures. Also make sure you have your supervisor/instructor's permission to use any particular type of lifting equipment.
- Comply with personal and environmental safety practices associated with clothing; eye protection; hand tools; power equipment; proper ventilation; and the handling, storage, and disposal of chemicals/materials in accordance with federal, state, and local regulations.
- Always wear the correct protective eyewear and clothing and use the appropriate safety equipment, as well as fender covers, seat protectors, and floor mat protectors.
- Make sure you understand and observe all legislative and personal safety procedures when carrying out practical assignments. If you are unsure of what these are, ask your supervisor/instructor.

Performance Standard

0—No exposure: No information or practice provided during the program; complete training required

1—Exposure only: General information provided with no practice time; close supervision needed; additional training required

2—Limited practice: Has practiced job during training program; additional training required to develop skill

3—Moderately skilled: Has performed job independently during training program; limited additional training may be required

4—Skilled: Can perform job independently with no additional training

Student/Intern information:

Name _______________________________ Date _____________ Class _________________________

Vehicle used for this activity:

Year _______________ Make _________________________ Model _________________________

Odometer _____________ Hour meter _____________ VIN _________________________

▶ **TASK** Know the basic components that make up the various types of 12/24 volt charging systems.

AED
2.4a

Time off_____________

Time on_____________

Total time_____________

CDX Tasksheet Number: E0057

> **Note:** This tasksheet will require the use of a machine or simulator. Ask your instructor which machine or simulator you are to use.

1. **Using the machine or simulator service manual, wiring diagrams, and schematics locate and tag the components that make up the various types of 12/24 volt charging systems below. If the machine does not utilize a component listed in the table below, mark the box N/A. Enter tag number in each empty box.**

Component	Machine #1	Machine #2	Machine #3
Battery			
Load cables			
Alternator			
Regulator			
Drive belts			
Ignition switch			
Warning lamp			
Volt gauge			
ECM control			

2. **Discuss your findings with the instructor.**

Performance Rating

CDX Tasksheet Number: E0057

| 0 | 1 | 2 | 3 | 4 |

Supervisor/instructor signature _________________________________ Date _____________

Name _________________________________ Date ____________ Class _________________________

Vehicle used for this activity:

Year _______________ Make _________________________ Model _________________________

Odometer ____________ Hour meter ____________ VIN _________________________________

▶ TASK Demonstrate the sequence of operation of the components contained within a charging system. The emphasis is on how each component effects the system's overall operation.

AED 2.4b

Time off_________________

Time on_________________

Total time_________________

CDX Tasksheet Number: EOO58

> **Note:** This tasksheet will require the use of a machine or simulator. Ask your instructor which machine or simulator you are to use.

1. **Using the machine or simulator service manual, wiring diagrams, and schematics determine the purpose and operation of the components below. Write a short description of the purpose and operation of the components. Note: Understanding how a component is designed to operate within a circuit will make problems easier to diagnose.**

Component	Purpose	Operation
Battery		
Load cables		
Alternator		
Regulator		
Drive belts		
Ignition switch		
Warning lamp		

Volt gauge		
ECM control		

2. Discuss your findings with the instructor.

▶ TASK Demonstrate the ability to isolate problems using voltage drops and other diagnostic methods. The proper use of testing equipment is paramount.

AED 2.4c

Time off_________________

Time on_________________

CDX Tasksheet Number: E0059

Total time_________________

Test instrument panel-mounted voltmeters and/or indicator lamps; determine needed action.

1. **Using the appropriate service information for the machine you are working on, locate the test for panel-mounted voltmeters and/or indicator lamps.**

2. **List the inspection and test procedures for panel-mounted voltmeters and/ or indicator lamps:**

   ____

   ____

   ____

3. **Have your supervisor/instructor verify these steps.**

   Supervisor/instructor's initials: _______________

4. **Using the appropriate service information, inspect and test panel-mounted voltmeters and/or indicator lamps.**

   a. **List the results of conducting your tests:**

      ____

      ____

      ____

      ____

   b. **Determine and list any necessary corrective action(s):**

      ____

      ____

      ____

      ____

5. **Check corrective actions with your supervisor/instructor.**

   Supervisor/instructor's initials: _______________

Identify causes of no charge, low charge, or overcharge problems; determine needed action.

6. Using the appropriate service information for the instructor-designated machine you are working on, research the identification of causes and repair action of no-charge, low-charge, or overcharge problems.

   a. Potential causes, test procedures, and repair action of no charge are:

   ____

   ____

   ____

   ____

   b. Potential causes, test procedures, and repair action of low charge are:

   ____

   ____

   ____

   ____

   c. Potential cause, test procedures, and repair action of overcharge are:

   ____

   ____

   ____

   ____

7. Have your supervisor/instructor verify your research.

   Supervisor/instructor's initials: _______________

8. Ask your supervisor for a machine exhibiting one of the conditions above, and perform the necessary test procedures.

9. Compare your results to the manufacturer's specifications. List your observations:

   ____

   ____

   ____

10. Determine any necessary action(s):

   ____

   ____

   ____

11. Return machine to its beginning condition, and return any tools you used to their proper locations.

12. Supervisor/instructor's initials: _______________

Inspect and replace alternator drive belts, pulleys, fans, tensioners, and mounting brackets; adjust drive belts and check alignment.

13. Using the appropriate service information for the machine you are working on, locate "inspecting, adjusting, and/or replacing alternator drive belts, pulleys, fans, tensioners, and mounting brackets; check pulley and belt alignment."

    a. List the steps outlined in the service information:

    b. Write down the specified tension of the alternator drive belt.

       _______________ lb/Kg

    c. List faults to look for when inspecting drive belts, pulleys, tensioners, and mounting brackets:

    d. Describe how to check correct pulley and belt alignment:

    e. Inspect drive belts, pulleys, tensioners, and mounting brackets on machine, and list faults below:

    f. Remove belts from machine and inspect drive belts, pulleys, tensioners, and mounting brackets, and list faults below:

    g. Following manufacturers procedures along with prescribed tooling inspect pulleys for proper alignment. List faults below:

    h. Have your supervisor/instructor verify these steps.
       Supervisor/instructor's initials: _______________

14. Refit the machine drive belts using appropriate service information and set to proper tension.

    a. Have your supervisor/instructor verify accuracy of these steps.
       Supervisor/instructor's initials: _______________

Perform charging system voltage and amperage output tests; perform AC ripple test; determine needed action.

15. Using the appropriate service manual, research and list the steps of the test for system voltage, amperage, and AC ripple as outlined in the service information for the machine you have been assigned.

   ____

   ____

   ____

   ____

16. What is the specified charging system output?

   _____________ amps at _____________ volts at _____________ RPM, AC ripple: _____________
   Have your supervisor/instructor verify these steps.

   Supervisor/instructor's initials: _____________

17. Connect the test instruments as outlined in the appropriate service information or as listed in Step 15.

   a. Have your supervisor/instructor verify your test procedure and connections. Supervisor/instructor's initials: _____________

18. Conduct the charging system output test. List the measured results at the maximum output:

   _____________ amps at _____________ volts at _____________ RPM, AC ripple: _____________

19. Compare your results to the manufacturer's specifications. List your observations.

   ____

   ____

   ____

   ____

20. Have your supervisor/instructor check your work.

   Supervisor/instructor's initials: _____________

Perform charging circuit voltage drop tests; determine needed action.

21. Locate "perform charging circuit voltage drop" tests in the service information for the machine you are working on; determine necessary action.

   a. List the procedure as outlined in the service information to perform charging circuit voltage drop tests:

      ____

      ____

      ____

      ____

   b. List the maximum specified charging circuit voltage drop allowable.

      _____________ volts

22. **Connect the tester as outlined in the appropriate service information or as listed in Step 21a.**

    a. **Have your supervisor/instructor verify these steps.**

       Supervisor/instructor's initials: _______________

23. **Connect the test instruments as outlined in the appropriate service information or as listed in Step 21a.**

    a. **Have your supervisor/instructor verify your test procedure and connections. Supervisor/instructor's initials:** _______________

24. **Conduct the charging system voltage drop test. Repeat the tests as many times as required to test all parts of the charging circuit as described in Step 21a. List the measured results:**

    a. **Voltage drop between __________ and __________ is __________ volts at __________ amps.**

    b. **Voltage drop between __________ and __________ is __________ volts at __________ amps.**

    c. **List the total voltage drop for the charging circuit: __________ volts.**

25. **Compare your results to the manufacturer's specifications. List your observations:**

____

____

____

____

26. **Have your supervisor/instructor check your work.**

    Supervisor/instructor's initials: __________

Performance Rating

CDX Tasksheet Number: E0059

0	1	2	3	4

Supervisor/instructor signature ____ Date _______________

Student/Intern information:

Name ___________________________________ Date ___________ Class ___________________________

Vehicle used for this activity:

Year _______________ Make _______________________________ Model _________________________

Odometer ____________ Hour meter ____________ VIN ____

▶ **TASK** Demonstrate the ability to properly test, evaluate, and replace the
following components using manufacturers' service publications
and specifications.
1. Conductors
2. Alternators
3. Regulators

**AED
2.4d**

Time off___________________

Time on___________________

Total time___________________

CDX Tasksheet Number: E0060

> **NOTE:** This tasksheet will require the use of a machine or simulator. Ask your
> instructor which machine or simulator you are to use.

1. **Using the machine or simulator service manual, wiring diagrams, and
   schematics troubleshoot the operator complaint below.**

   a. Complaint: Alternator not charging (warning lamp stays on).

2. **Verify: Does this condition exist? Yes: _____ No: _____**

3. **Using the appropriate service information for the machine you are working
   on, locate the testing procedure related to this complaint and research the
   recommended process in the appropriate service manual or other available
   data. List the steps for the tests you will perform.**

   ____

   ____

   ____

   ____

4. **Conduct the tests and report below.**

   a. **Did the tests fall out of manufacturer's specifications?**
      Yes: _____ No: _____
   b. **If yes, what part of the system including component did not pass the
      tests? _____________________________________.**
   c. **Repair or replace the component according to manufacturer's
      specification.**
   d. **Try starting the machine.**
   e. **Does the machine complaint still exist? Yes: _____ No: _____**

5. **Determine any necessary further action(s):**

____

____

____

____

   a. **Have your supervisor/instructor verify these steps.**

   Supervisor/instructor's initials: _______________

6. **Using the machine or simulator service manual, wiring diagrams, and schematics troubleshoot the operator complaint below.**

   a. Complaint: Alternator low charging (voltmeter reads low).

7. **Verify: Does this condition exist? Yes: ____ No: ____**

8. **Using the appropriate service information for the machine you are working on, locate the testing procedure related to this complaint and research the recommended process in the appropriate service manual or other available data. List the steps for the tests you will perform:**

____

____

____

____

9. **Conduct the tests and report below.**

   a. **Did the tests fall out of manufacturer's specifications?**
   Yes: ____ No: ____

   b. **If yes, what part of the system including component did not pass the test? _______________________________**

   c. **Repair or replace the component according to manufacturer's specification.**

   d. **Try starting the machine.**

   e. **Does the machine complaint still exist? Yes: ____ No: ____**

____

10. **Determine any necessary further action(s):**

____

____

____

____

   a. **Have your supervisor/instructor verify these steps.**

   Supervisor/instructor's initials: _______________

Remove and replace alternator.

11. **Locate "remove and replace alternator" in the appropriate service information for the machine you are working on.**

    a. **List the steps as outlined in the service information to remove alternator:**

    ____
    ____
    ____
    ____

12. **Remove the alternator as per the service information.**

13. **List the steps outlined in the service information to install the alternator:**

    ____
    ____
    ____

    a. **Install the alternator as per the service information.**

    b. **Start the vehicle and check the alternator output and voltage.**

       i. **Voltage output: _______________ volts**

       ii. **Amperage output: _______________ amps**

    c. **Have your supervisor/instructor inspect your work.**

    d. **Supervisor/instructor's initials: _______________**

14. **Discuss your findings with the instructor.**

Performance Rating

CDX Tasksheet Number: E0060

0	1	2	3	4

Supervisor/instructor signature ____ Date ____________

SAE Computer CANbus Standards

Student/Intern information:

Name _______________________________ Date ____________ Class _________________________

Vehicle used for this activity:

Year ______________ Make _________________________ Model __________________________

Odometer ____________ Hour meter ____________ VIN ___________________________________

Learning Objective/Task	CDX Tasksheet Number	2014 Edition Rev2/12/16 AED Standard
• Demonstrate the knowledge of the different systems used to communicate on computer controlled machinery. SAE J1587 and J1939.	E0061	2.7a
• Understanding the importance of twisted and shielded wire systems.	E0062	2.7a
• Demonstrate the knowledge of the codes to identify errors within the different systems.	E0063	2.7b

Time off______________

Time on______________

Total time______________

Materials Required

- Machines or simulators with onboard computer systems and related electrical faults such as data bus communications problems
- Machine manufacturer's workshop manual including schematic wiring diagrams
- Digital volt-ohmmeter (DVOM), ammeter, current clamp, PC-based software, and/or data scan tools
- Electrical spare parts, including fuses, circuit breakers, relays, wires, connectors, terminating resistors
- Manufacturer-specific tools depending on the concern
- Vehicle-lifting equipment if applicable

Some Safety Issues to Consider

- Activities may require running the engine and managing an environment of rotating equipment and large current draw, which carry severe risks. Attempt this task only with full permission from your supervisor/instructor, and follow all the guidelines exactly.
- Activities require you to measure electrical values. Always ensure that the instructor/supervisor checks test instrument connections prior to connecting power or taking measurements. High current flows can be dangerous; avoid accidental short circuits or grounding the battery's positive connections.

- Use extreme caution when working around batteries. Immediately remove any electrolyte that you may come in contact with. Electrolyte is a mixture of sulfuric acid and water. Batteries may produce explosive mixtures of gas containing hydrogen; avoid creating any sparks around batteries. Please consult with the shop safety and emergency procedures when working with or around batteries.
- Lifting equipment such as equipment jacks and stands, hoists, and engine hoists are important tools that increase productivity and make the job easier. However, they can also cause severe injury or death if used improperly. Make sure you follow the manufacturer's operation procedures. Also make sure you have your supervisor/instructor's permission to use any particular type of lifting equipment.
- Comply with personal and environmental safety practices associated with clothing; eye protection; hand tools; power equipment; proper ventilation; and the handling, storage, and disposal of chemicals/materials in accordance with federal, state, and local regulations.
- Always wear the correct protective eyewear and clothing and use the appropriate safety equipment, as well as fender covers, seat protectors, and floor mat protectors.
- Make sure you understand and observe all legislative and personal safety procedures when carrying out practical assignments. If you are unsure of what these are, ask your supervisor/instructor.

Performance Standard

0—No exposure: No information or practice provided during the program; complete training required

1—Exposure only: General information provided with no practice time; close supervision needed; additional training required

2—Limited practice: Has practiced job during training program; additional training required to develop skill

3—Moderately skilled: Has performed job independently during training program; limited additional training may be required

4—Skilled: Can perform job independently with no additional training

Name _________________________________ Date _____________ Class _________________________

Vehicle used for this activity:

Year _______________ Make _______________________________ Model _________________________

Odometer _____________ Hour meter _____________ VIN _______________________________

▶ **TASK** Demonstrate the knowledge of the different systems used to communicate on computer controlled machinery. SAE J1587 and J1939.

AED
2.7a

Time off________________

Time on________________

CDX Tasksheet Number: E0061

Total time________________

> **NOTE:** This tasksheet will require the use of a machine or simulator. Ask your instructor which machine or simulator you are to use.

1. **Using Chapter 19 in your textbook research the construction of onboard communication networks. In your own words answer the questions below that describe the construction of onboard communication networks. Topics should include the following:**

   a. **What is typology?**

   ____
   ____
   ____
   ____

   b. **What is the physical layer?**

   ____
   ____
   ____
   ____

   c. **What is network protocol?**

   ____
   ____
   ____
   ____

   d. **What is centralized or distributed control?**

   ____
   ____
   ____
   ____

2. **Using Chapter 19 in your textbook research the construction of onboard communication networks. In your own words answer the questions below.**

   a. **Explain how time division multiplexing (TDM) is like a phone call where modules share a phone extension.**

   ____
   ____
   ____
   ____

    b. What is the rule of TDM multiplexing?

    c. Explain how a gateway module enables communication between different onboard networks.

3. Using Chapter 19 in your textbook research the purpose, function, and application of distributed network control systems. In your own words answer the questions below.

    a. Describe the term controlled area networks (CAN) as it relates to mobile off-road equipment (MORE).

    b. Describe the construction and function of J-1708/1587 controlled area networks.

    c. Describe the construction and function of J-1939 controlled area network.

    d. Describe the term serial communication as it relates to MORE.

    e. Describe why the CAN protocol is important when sending network messages.

f. Describe how messages over the CAN network are formatted.

____

____

____

____

g. Describe how data bus arbitration works.

____

____

____

____

4. Discuss your findings with the instructor.

Performance Rating

CDX Tasksheet Number: E0061

0	1	2	3	4

Supervisor/instructor signature _______________________________________ Date ___________

▶ TASK Understanding the importance of twisted and shielded wire systems.

AED
2.7a

CDX Tasksheet Number: E0062

> **NOTE:** This tasksheet will require the use of a machine or simulator. Ask your instructor which machine or simulator you are to use.

1. **Using Chapter 19 in your textbook research the construction, function, and operation of twisted wire pair data buses. In your own words answer the questions below.**

    a. **Describe the construction of twisted wire pair data buses used in CAN systems.**

    ____

    ____

    ____

    ____

    b. **Describe the operation of twisted wire pair data buses used in CAN systems.**

    ____

    ____

    ____

    ____

    c. **Describe the function of twisted wire pair data buses used in CAN systems.**

    ____

    ____

    ____

    ____

    d. **Explain why the J-1939 CANbus uses a terminating resistor.**

    ____

    ____

    ____

    ____

2. Using manufacturer's service information, research the troubleshooting steps to diagnose problems with twisted wire pair data buses and list them.

____

____

____

____

3. Following manufacturer's procedures conduct the tests and list the result below:

a. Test ______________ Specs. ______________ Results ______________

b. Test ______________ Specs. ______________ Results ______________

c. Test ______________ Specs. ______________ Results ______________

d. Test ______________ Specs. ______________ Results ______________

4. Discuss your findings with the instructor.

____

____

____

____

Performance Rating

CDX Tasksheet Number: E0062

| 0 | 1 | 2 | 3 | 4 |

Supervisor/instructor signature ____ Date ______________

Student/Intern information:

Name _________________________________ Date _____________ Class _____________________

Vehicle used for this activity:

Year _______________ Make _________________________ Model _________________________

Odometer ____________ Hour meter ____________ VIN _________________________________

▶ TASK Demonstrate the knowledge of the codes to identify errors within the different systems.

AED 2.7b

Time off_____________

Time on_____________

CDX Tasksheet Number: E0063

Total time_____________

> **NOTE:** This tasksheet will require the use of a machine or simulator. Ask your instructor which machine or simulator you are to use.

1. **Using Chapters 19 and 20 in your textbook and the equipment manufacturer's service information, research the fundamentals of onboard diagnostics (OBD). In your own words answer the questions below.**

   a. **Describe the development of OBD and the two meanings OBD has for technicians.**

   ____
   ____
   ____
   ____

   b. **Describe the self-diagnostic capabilities and approaches of OBD systems.**

   ____
   ____
   ____
   ____

   c. **Describe the terms associated with faults when they occur in the system and how they are classified.**

      i. **When the OBD system reports an active fault, what does that mean?**

   ____
   ____
   ____

      ii. **When the OBD system reports a historical fault, what does that mean?**

   ____
   ____
   ____

      iii. **When the OBD system reports an intermittent fault, what does that mean?**

   ____
   ____
   ____

iv. When the OBD system reports an incipient fault, what does that mean?

____

____

____

d. J-1587 fault codes are constructed into four parts to help technicians isolate problems. Name the four parts of construction:

i. _______________________________

ii. _______________________________

iii. _______________________________

iv. _______________________________

e. J-1939 fault codes are constructed similar to J-1587 fault codes but more comprehensive. List the similarities and differences between the two below:

____

____

____

____

2. Using Chapters 19 and 20 in your textbook and the equipment manufacturer's service information, wiring diagrams, and schematics, research the construction, function, and operation of data link connector. In your own words answer the questions below.

a. Describe the construction of the data link connector used in CAN systems.

____

____

____

____

b. Describe the operation of the data link connector used in CAN systems.

____

____

____

____

c. Describe the function of data link connector used in CAN systems.

____

____

____

3. Following manufacturer's procedures connect to the OBD system using a scan tool or Laptop and locate the following. Document below.

a. List active fault codes:

____

____

____

b. List historic fault codes:

____

____

____

c. List which module in the system, component, and fault description of each fault code you found in the table below:

Fault Code	System Module	Component	Fault Description	How Often Did It Fail?
128-110-000	Engine	Coolant temperature	Above normal operating range	2

4. Discuss your findings with the instructor.

Performance Rating

CDX Tasksheet Number: E0063

□	□	□	□	□
0	1	2	3	4

Supervisor/instructor signature ____ Date _____________

Diagnostics System Troubleshooting

Learning Objective/Task	CDX Tasksheet Number	2014 Edition Rev2/12/16 AED Standard
• Understand the complaint prior to beginning diagnostic tests.	E0064	2.8
• Demonstrate the ability to perform a diagnostic procedure.	E0065	2.8
• Demonstrate the ability to reason with regard to a specific malfunction in the system.	E0066	2.8

Time off_______________

Time on_______________

Total time_______________

Materials Required

- Machine or simulator with possible alternator concern, including cable, wiring, or connector faults
- Equipment manufacturer's workshop manual
- Digital volt-ohmmeter (DVOM), ammeter for starter circuits, test lamp alternator testing equipment such as load banks, oscilloscope, wire, cable connectors, electrical hand tools
- Manufacturer-specific tools depending on the concern
- Personal protective equipment (PPE)

Some Safety Issues to Consider

- Activities may require running the engine and managing an environment of rotating equipment and large current draw, which carry severe risks. Attempt this task only with full permission from your supervisor/ instructor, and follow all the guidelines exactly.
- Do not run the alternator without a load connected or allow the output voltage to exceed the manufacturer's specified maximum.
- Use extreme caution when working around batteries. Immediately remove any electrolyte that may come in contact with you. Electrolyte is a mixture of sulfuric acid and water. Batteries may produce explosive mixtures of gas containing hydrogen; avoid creating any sparks around batteries. Please consult with the shop safety and emergency procedures when working with or around batteries.
- Lifting equipment such as equipment jacks and stands, hoists, and engine hoists are important tools that increase productivity and make the job easier. However, they can also cause severe injury or death if used improperly. Make sure you follow the manufacturer's operation procedures. Also make sure you have your supervisor/instructor's permission to use any particular type of lifting Tequipment.

- Comply with personal and environmental safety practices associated with clothing; eye protection; hand tools; power equipment; proper ventilation; and the handling, storage, and disposal of chemicals/materials in accordance with federal, state, and local regulations.
- Always wear the correct protective eyewear and clothing and use the appropriate safety equipment, as well as fender covers, seat protectors, and floor mat protectors.
- Make sure you understand and observe all legislative and personal safety procedures when carrying out practical assignments. If you are unsure of what these are, ask your supervisor/ instructor.

Performance Standard

0—No exposure: No information or practice provided during the program; complete training required

1—Exposure only: General information provided with no practice time; close supervision needed; additional training required

2—Limited practice: Has practiced job during training program; additional training required to develop skill

3—Moderately skilled: Has performed job independently during training program; limited additional training may be required

4—Skilled: Can perform job independently with no additional training

Student/Intern information:

Name ________________________________ Date ____________ Class ________________________

Vehicle used for this activity:

Year ______________ Make __________________________ Model ____________________________

Odometer ______________ Hour meter ____________ VIN ________________________________

▶ TASK Understand the complaint prior to beginning diagnostic tests.　**AED 2.8**

CDX Tasksheet Number: E0064

Time off________________

Time on________________

1. **Research the methods required to ask the customer proper questions when troubleshooting a complaint.**

   a. **List general questions you would ask the customer for a complaint:**

      i. ________________________________

      ii. ________________________________

      iii. ________________________________

   Total time________________

   b. **List some specific questions you would ask for the given complaints below.**

      i. **Engine cranks over slow:**

      ____
      ____
      ____
      ____

      ii. **Machine needs a jump start each morning:**

      ____
      ____
      ____
      ____

      iii. **Check engine light comes on and goes off:**

      ____
      ____
      ____
      ____

   c. **In your own words describe how asking the customer proper questions can aid you in the troubleshooting process.**

      ____
      ____
      ____
      ____
      ____
      ____

2. Discuss your findings with the instructor.

Student/Intern information:

Name _______________________________ Date _____________ Class _______________________________

Vehicle used for this activity:

Year _______________ Make _______________________ Model _______________________________

Odometer _____________ Hour meter _____________ VIN _______________________________

▶ TASK Demonstrate the ability to perform a diagnostic procedure.

AED 2.8

Time off_______________

Time on_______________

CDX Tasksheet Number: E0065

> **NOTE:** This tasksheet will require the use of a machine or simulator. Ask your instructor which machine or simulator you are to use.

Total time_______________

1. **Using the machine or simulator service manual, wiring diagrams, and schematics perform the diagnostic procedures necessary to determine the root cause of system malfunction.**

    a. **Ask your instructor for the customer's complaint and interview the customer.**

        i. **Complaint:**

        ____

        ____

        ii. **Interview questions:**

        ____

        ____

    b. **Verify that the complaint exists.**

    c. **Using the manufacturer's service information troubleshoot and diagnose the system malfunction. List the steps you took to determine the root cause:**

        ____

        ____

        ____

        ____

        ____

        ____

        ____

        ____

        ____

        ____

    d. **List the recommended repair needed to fix the problem:**

        ____

        ____

        ____

        ____

2. **Discuss your findings with the instructor.**

Performance Rating

CDX Tasksheet Number: E0065

| 0 | 1 | 2 | 3 | 4 |

Supervisor/instructor signature ____ Date _____________

Name _________________________________ Date ___________ Class _______________________

Vehicle used for this activity:

Year _______________ Make _____________________________ Model _____________________________

Odometer _____________ Hour meter ____________ VIN _______________________________

▶ **TASK** Demonstrate the ability to reason with regard to a specific malfunction in the system.

AED
2.8

Time off___________________

Time on___________________

Total time___________________

CDX Tasksheet Number: E0066

> **NOTE:** This tasksheet will require the use of a machine or simulator. Ask your instructor which machine or simulator you are to use.

1. **Using the machine or simulator service manual, wiring diagrams, and schematics perform the diagnostic procedures necessary to determine the root cause of system malfunction.**

2. **For this task you are simulating a real-life scenario technician encounter frequently in the shop and field. Read the technicians note that she left for you at the end of her shift and begin the diagnostic process.**

3. **Technicians note: Customer complains that the engine cranks. Verified complaint and engine does crank slow. Tested battery open circuit voltage and was at 12.60 volts. Tested voltage at battery while cranking was at 8.60 volts. Recommend replacing batteries.**

4. **Do you think the technician above has diagnosed the problem accurately?**

   ____

   ____

   ____

   ____

   a. **Using the manufacturers, service information troubleshoot and diagnose the system malfunction. List the steps you took to determine the root cause:**

      ____

      ____

      ____

      ____

      ____

      ____

      ____

      ____

      ____

      ____

   b. List the recommended repair needed to fix the problem:

     ____

     ____

     ____

     ____

5. Have your supervisor/instructor check your work.

    Supervisor/instructor's initials: ___________

Performance Rating

CDX Tasksheet Number: E0066

0	1	2	3	4
☐	☐	☐	☐	☐

Supervisor/instructor signature ____ Date ___________

Section A3: Hydraulics/Hydrostatics

CONTENTS

Hydraulic and Hydrostatic: Theory and Operation

Student/Intern information:

Name _________________________________ Date ____________ Class _____________________

Vehicle used for this activity:

Year _______________ Make _____________________________ Model _____________________

Odometer _____________ Hour meter ____________ VIN _________________________________

Learning Objective/Task	CDX Tasksheet Number	2014 Edition Rev2/12/16 AED Standard
• Demonstrate knowledge that fluids have no shape of their own, are practically incompressible, apply equal pressure in all directions, and provide great increases in work force.	E0067	3.1

Materials Required

- Basic hydraulic training unit
- Equipment manufacturer's workshop manual including schematic hydraulic diagrams
- Hydraulic pressure gauges
- Hydraulic spare parts, including fittings, hoses, and fluid
- Manufacturer-specific tools depending on the concern
- Machine lifting equipment, if applicable
- Machine lockout/tagout equipment, if applicable

Some Safety Issues to Consider

- Activities require you to measure hydraulic pressures. Always ensure that the instructor/supervisor checks test instrument connections prior to connecting power or taking measurements. High fluid pressures can be dangerous; avoid opening hydraulic circuits until system is deenergized.
- Activities may require test driving the equipment on the school grounds, which carry severe risks. Attempt this task only with full permission from your supervisor/instructor, and follow all the guidelines exactly.
- Lifting equipment such as equipment jacks and stands, hoists, and engine hoists are important tools that increase productivity and make the job easier. However, they can also cause severe injury or death if used improperly. Make sure you follow the manufacturer's operation procedures. Also make sure you have your supervisor/instructor's permission to use any particular type of lifting equipment.
- Comply with personal and environmental safety practices associated with clothing; eye protection; hand tools; power equipment; proper ventilation; and the handling, storage, and disposal of chemicals/materials in accordance with federal, state, and local regulations.

- Always wear the correct protective eyewear and clothing and use the appropriate safety equipment, as well as fender covers, seat protectors, and floor mat protectors.
- Make sure you understand and observe all legislative and personal safety procedures when carrying out practical assignments. If you are unsure of what these are, ask your supervisor/instructor.

Student/Intern information:

Name ________________________________ Date ____________ Class ________________________________

Vehicle used for this activity:

Year ______________ Make ________________________________ Model ________________________________

Odometer ____________ Hour meter ____________ VIN ________________________________

▶ TASK Demonstrate knowledge that fluids have no shape of their own, are practically incompressible, apply equal pressure in all directions, and provide great increases in work force.

AED 3.1

Time off____________

Time on____________

Total time____________

CDX Tasksheet Number: E0067

1. **Using Pascal's Law and the principles of hydraulics answer the problems below.**

   a. **Using various equipment in the shop, identify the various systems that utilize hydraulic principles in their design. List the systems for each piece of equipment below:**

Machine Type	System	System	System
Back hoe	implement	steering	power train

2. **Why do the systems use a liquid to transfer energy? Could they use a gas instead? What are the advantages and disadvantages of using a liquid instead of a gas?**

____

____

____

____

____

____

____

____

3. In your own words describe Pascal's Law.

____

____

____

____

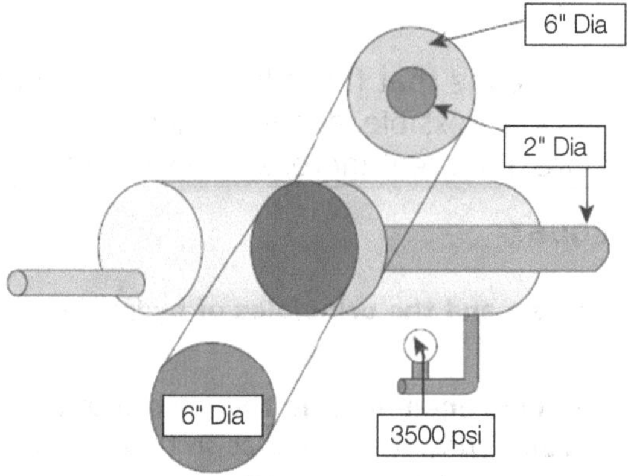

Figure 1

4. Using the figure above answer the questions below.

   a. What is the area of the piston head end? _______________

   b. What is the area of the piston rod end? _______________

   c. What is the output force when extended in figure 1? _______________

   d. In your own words describe how the output force of an actuator can be increased.

   ____

   ____

   ____

   ____

5. Have your instructor review your answers before continuing. Instructor Comments:

____

____

____

> **NOTE:** The exercise below requires the use of a hydraulic training unit. Make sure you have been properly trained on the safe and proper use of the unit before moving on.

6. **Using the hydraulic training unit manufacturer's instructions make the following connections:**

   a. **Connect the unit's hydraulic pump to the main relief valve.**

   b. **Connect a suitable pressure gauge (1) to main relief valve manifold.**

   c. **Connect relief valve reservoir port to unit's tank return manifold.**

   d. **Connect a suitable pressure gauge (2) to pressure manifold.**

   e. **Connect 2 pressure gauges (3), (4) in series to pressure manifold.**

   f. **Turn unit on and adjust relief valve so gauge 1 reads 200 psi.**

   g. **Next read the psi on the remaining gauges 2-4 and list below:**

      i. __________ psi

      ii. __________ psi

      iii. __________ psi

      iv. __________ psi

   h. **Did all gauges read the same? Yes __________ No __________**

   i. **If yes, answer why.**

   ____
   ____
   ____

   j. **If no, see your instructor.**

7. **Instructor Comments:**

   ____
   ____
   ____

Performance Rating

CDX Tasksheet Number: E0067

0	1	2	3	4

Supervisor/instructor signature ____________________________________ Date ____________

Hydraulic Theory

Student/Intern information:

Name _____________________________ Date ___________ Class _____________________

Vehicle used for this activity:

Year _______________ Make _________________________ Model _____________________

Odometer ______________ Hour meter ____________ VIN _____________________________

Learning Objective/Task	CDX Tasksheet Number	2014 Edition Rev2/12/16 AED Standard
• Demonstrate the understanding of the function of a reservoir, pump, filters, relief valve, control valve, and cylinder in relation to each other.	E0068	3.1
• Know that open and closed center systems are determined by one or all of the following: (a) the type of control valve, (b) the type of pump, (c) use of unloading valve, (d) path of oil return to reservoir from pump.	E0069	3.1
• Describe a basic, but complete, open center hydraulic system, explaining the operation of the system, the route of fluid during the use of a function, and the route of the fluid while the machine is running when no hydraulic function is being used.	E0070	3.1
• Describe a basic, but complete, closed center load sensing hydraulic system, explaining the operation of the system, the route of fluid during the use of a function, and the route of the fluid while the machine is running when no hydraulic function is being used.	E0071	3.1
• Identify applications and the benefits of those applications on construction equipment.	E0072	3.1

Time off_______________

Time on_______________

Total time_______________

Materials Required
- Basic hydraulic training unit or machine
- Equipment manufacturer's workshop manual including schematic hydraulic diagrams
- Hydraulic pressure gauges
- Hydraulic spare parts, including fittings, hoses, and fluid
- Manufacturer-specific tools depending on the concern
- Machine lifting equipment, if applicable
- Machine lockout/tagout equipment, if applicable

Some Safety Issues to Consider

- Activities require you to measure hydraulic pressures. Always ensure that the instructor/supervisor checks test instrument connections prior to connecting power or taking measurements. High fluid pressures can be dangerous; avoid opening hydraulic circuits until system is deenergized.
- Activities may require test driving the equipment on the school grounds, which carry severe risks. Attempt this task only with full permission from your supervisor/instructor, and follow all the guidelines exactly.
- Lifting equipment such as equipment jacks and stands, hoists, and engine hoists are important tools that increase productivity and make the job easier. However, they can also cause severe injury or death if used improperly. Make sure you follow the manufacturer's operation procedures. Also make sure you have your supervisor/instructor's permission to use any particular type of lifting equipment.
- Comply with personal and environmental safety practices associated with clothing; eye protection; hand tools; power equipment; proper ventilation; and the handling, storage, and disposal of chemicals/materials in accordance with federal, state, and local regulations.
- Always wear the correct protective eyewear and clothing and use the appropriate safety equipment, as well as fender covers, seat protectors, and floor mat protectors.
- Make sure you understand and observe all legislative and personal safety procedures when carrying out practical assignments. If you are unsure of what these are, ask your supervisor/instructor.

Student/Intern information:

Name _________________________________ Date _____________ Class _____________________

Vehicle used for this activity:

Year _______________ Make _________________________ Model _____________________________

Odometer ______________ Hour meter ______________ VIN _______________________________

▶ TASK Demonstrate the understanding of the function of a reservoir, pump, filters, relief valve, control valve, and cylinder in relation to each other.

AED
3.1

Time off________________

Time on________________

CDX Tasksheet Number: E0068

Total time________________

1. **Using an instructor-identified machine locate the reservoir, pump, filters, relief valve, control valve, and cylinder that make up the hydraulic system. List the function of each and how each contributes to the hydraulic system's operation below:**

Component	Function	System Operation
Reservoir		
Pump		
Filters		
Relief valve		
Control valve		
Cylinder		

2. Instructor Comments:

____

____

____

Performance Rating

CDX Tasksheet Number: E0068

0	1	2	3	4

Supervisor/instructor signature _________________________________ Date _____________

Student/Intern information:

Name _________________________________ Date ____________ Class _________________________________

Vehicle used for this activity:

Year _______________ Make _______________________ Model _______________________

Odometer ____________ Hour meter ____________ VIN _______________________________

▶ **TASK** Know that open and closed center systems are determined
by one or all of the following:

    a) the type of control valve
    b) the type of pump
    c) use of unloading valve
    d) path of oil return to reservoir from pump

AED 3.1

Time off____________

Time on____________

Total time____________

CDX Tasksheet Number: E0069

1. **Research the primary components that make up a closed center hydraulic system. List the components, component type, and their function below:**

Component	Component Type	Function

2. **Research the primary components that make up an open center hydraulic system. List the components, component type, and their function below:**

Component	Component Type	Function

3. **What are the differences between a closed center and open center hydraulic system?**

____
____
____
____
____
____

4. What machine applications might we see open center and closed center systems?

____

____

____

____

5. Instructor Comments:

____

____

____

Performance Rating

CDX Tasksheet Number: E0069

0	1	2	3	4

Supervisor/instructor signature ____ Date _____________

Name _________________________________ Date ____________ Class _________________________

Vehicle used for this activity:

Year _______________ Make _______________________ Model _______________________

Odometer ____________ Hour meter ____________ VIN _________________________________

▶ TASK Describe a basic, but complete, open center hydraulic system, explaining the operation of the system, the route of fluid during the use of a function, and the route of the fluid while the machine is running when no hydraulic function is being used.

AED
3.1

Time off________________

Time on________________

Total time________________

CDX Tasksheet Number: E0070

1. **Using an instructor-designated machine perform the steps below.**

    a. **Using machine model and serial number locate the service information including diagrams and schematics for the hydraulic system.**

    b. **Using the service information determine if the hydraulic system is an open center or closed center system.**

        i. **Open center __________ Closed center __________**

    c. **Using the service information explain the basic operation of the system.**

        ____
        ____
        ____
        ____
        ____

    d. **Using the service information explain the route of fluid during the use of a function.**

        i. **Function operated _________________________**

        Fluid route

        ____
        ____
        ____
        ____
        ____

    e. **Using the service information explain the route of fluid while the machine is running when no hydraulic function is being used.**

        ____
        ____
        ____
        ____
        ____

2. **Instructor Comments:**

Name _________________________________ Date _____________ Class _________________________

Vehicle used for this activity:

Year _______________ Make _________________________ Model _________________________

Odometer _____________ Hour meter _____________ VIN _________________________________

▶ TASK Describe a basic, but complete, closed center load sensing hydraulic system, explaining the operation of the system, the route of fluid during the use of a function, and the route of the fluid while the machine is running when no hydraulic function is being used.

AED
3.1

Time off_________________

Time on_________________

Total time_________________

CDX Tasksheet Number: E0071

1. **Using an instructor-designated machine perform the steps below.**

    a. **Using machine model and serial number locate the service information including diagrams and schematics for the hydraulic system.**

    b. **Using the service information determine if the hydraulic system is an open center or closed center system.**

        i. **Open center _____ Closed center _____ Closed center load sensing _____**

    c. **Using the service information explain the basic operation of the system.**

        ____
        ____
        ____
        ____
        ____

    d. **Using the service information explain the route of fluid during the use of a function.**

        i. **Function operated ___________________**

        Fluid route

        ____
        ____
        ____
        ____
        ____

    e. **Using the service information explain the route of fluid while the machine is running when no hydraulic function is being used.**

        ____
        ____
        ____
        ____
        ____

2. **Instructor Comments:**

Student/Intern information:

Name ___________________________ Date ___________ Class ___________________________

Vehicle used for this activity:

Year _____________ Make _____________________ Model _________________________

Odometer ____________ Hour meter ____________ VIN _______________________________

▶ **TASK** Identify applications and the benefits of those applications on construction equipment.

AED
3.1

Time off________________

Time on________________

Total time________________

CDX Tasksheet Number: E0072

1. **Research the applications and benefits of open center and closed center hydraulic system used on various types of construction equipment.**

2. **List below the applications and benefits you have discovered on the various construction equipment:**

Equipment Type	Application	Benefit
Skid steer implement	Open center	Simple design/fixed displacement pump

3. **Instructor Comments:**

____

____

____

Performance Rating

CDX Tasksheet Number: E0072

0	1	2	3	4

Supervisor/instructor signature ___________________________________ Date ___________

Hydrostatic Theory

Student/Intern information:

Name _________________________________ Date ____________ Class ___________________________

Vehicle used for this activity:

Year _______________ Make ________________________________ Model ____________________________

Odometer _____________ Hour meter ___________ VIN _______________________________________

Learning Objective/Task	CDX Tasksheet Number	2014 Edition Rev2/12/16 AED Standard
• Demonstrate knowledge of hydrostatic systems, including closed-loop and open-loop systems.	E0073	3.1
• Understand the various types of cooling circuits.	E0074	3.1
• Explain the different characteristics between various types of pumps; exhibit the ability to follow the oil flow through each pump both while using a hydraulic function and with no hydraulic function being used.	E0075	3.1

Time off________________

Time on________________

Total time________________

Materials Required

- Basic hydraulic training unit or machine
- Equipment manufacturer's workshop manual including schematic hydraulic diagrams
- Hydraulic pressure gauges
- Hydraulic spare parts, including fittings, hoses, and fluid
- Manufacturer-specific tools depending on the concern
- Machine lifting equipment, if applicable
- Machine lockout/tagout equipment, if applicable

Some Safety Issues to Consider

- Activities require you to measure hydraulic pressures. Always ensure that the instructor/ supervisor checks test instrument connections prior to connecting power or taking measurements. High fluid pressures can be dangerous; avoid opening hydraulic circuits until system is deenergized.
- Activities may require test driving the equipment on the school grounds which carry severe risks. Attempt this task only with full permission from your supervisor/instructor, and follow all the guidelines exactly.
- Lifting equipment such as equipment jacks and stands, hoists, and engine hoists are important tools that increase productivity and make the job easier. However, they can also cause severe injury or death if used improperly. Make sure you follow the manufacturer's operation procedures. Also make sure you have your supervisor/instructor's permission to use any particular type of lifting equipment.

- Comply with personal and environmental safety practices associated with clothing; eye protection; hand tools; power equipment; proper ventilation; and the handling, storage, and disposal of chemicals/materials in accordance with federal, state, and local regulations.
- Always wear the correct protective eyewear and clothing and use the appropriate safety equipment, as well as fender covers, seat protectors, and floor mat protectors.
- Make sure you understand and observe all legislative and personal safety procedures when carrying out practical assignments. If you are unsure of what these are, ask your supervisor/instructor.

Student/Intern information:

Name _________________________________ Date _____________ Class _________________________________

Vehicle used for this activity:

Year _______________ Make _________________________________ Model _________________________________

Odometer _____________ Hour meter _____________ VIN _________________________________

▶ TASK Demonstrate knowledge of hydrostatic systems, including closed-loop and open-loop systems.

AED 3.1

Time off_________________________

Time on_________________________

Total time_________________________

CDX Tasksheet Number: E0073

1. **Research closed-loop hydraulic systems and identify and understand the components, operation, and fluid flow.**

   a. **List the components that make up a closed-loop hydraulic system in the table below.**

   b. **List the function of the components that make up a closed-loop hydraulic system in the table below.**

Component	Function

2. Using an instructor-assigned machine, its service information, diagrams, and schematics, locate the major components that makeup a closed-loop hydraulic system.

3. Using an instructor-assigned machine, its service information, diagrams, and schematics, describe the flow of fluid through a closed-loop hydraulic system.

4. Instructor Comments:

5. Research an open-loop hydraulic system and identify and understand the components, operation, and fluid flow.

    a. List the components that make up an open-loop hydraulic system in the table below.

    b. List the function of the components that make up an open-loop hydraulic system in the table below.

Component	Function

6. Using an instructor-assigned machine, its service information, diagrams, and schematics, locate the major components that make up an open-loop hydraulic system.

7. Using an instructor-assigned machine, its service information, diagrams, and schematics, describe the flow of fluid through an open-loop hydraulic system.

8. Instructor Comments:

____

____

____

9. What are the differences between a closed-loop and open-loop hydraulic system?

____

____

____

____

____

____

10. Instructor Comments:

____

____

____

Performance Rating

CDX Tasksheet Number: E0073

| 0 | 1 | 2 | 3 | 4 |

Supervisor/instructor signature ___________________________ Date ____________

▶ TASK Understand the various types of cooling circuits.

AED
3.1

Time off_____________

Time on_____________

Total time_____________

CDX Tasksheet Number: E0074

1. **Research the various types of cooling circuits used in hydrostatic systems. List the components and their functions below:**

Component	Function

2. **Using an instructor-assigned machine, its service information, diagrams, and schematics, locate the major components that makeup cooling circuits used in hydrostatic systems.**

3. **Using an instructor-assigned machine, its service information, diagrams, and schematics, describe the flow of fluid through a cooling circuit used in hydrostatic systems.**

4. Instructor Comments:

Student/Intern information:

Name _______________________________ Date _____________ Class _______________________

Vehicle used for this activity:

Year _______________ Make _______________________ Model _______________________

Odometer _____________ Hour meter _____________ VIN _______________________

▶ **TASK** Explain the different characteristics between various types of pumps; exhibit the ability to follow the oil flow through each pump both while using a hydraulic function and with no hydraulic function being used.

AED
3.1

Time off___________

Time on___________

Total time___________

CDX Tasksheet Number: E0075

1. **Research the various types of hydraulic pumps used in mobile heavy equipment applications.**

   a. **In your own words describe the purpose of hydraulic pumps in a hydraulic system.**

   ____
   ____
   ____
   ____

      i. **List below the two general categories of hydraulic pumps:**

         1. _______________________________________
         2. _______________________________________

   b. **In your own words describe the design characteristics of nonpositive displacement pumps.**

   ____
   ____
   ____
   ____
   ____

   c. **In your own words describe the design characteristics of positive displacement pumps.**

   ____
   ____
   ____
   ____
   ____

   d. **In your own words describe what type of pump (nonpositive displacement or positive displacement) is best suited for mobile heavy equipment hydraulic systems.**

   ____
   ____
   ____
   ____
   ____

e. In your own words describe the general characteristics of fixed displacement type pumps.

____

____

____

____

____

f. In your own words describe the general characteristics of variable displacement type pumps.

____

____

____

____

____

2. In the table below list the types of fixed displacement pumps UNDER EACH CATEGORY and how each creates pump flow.

Pump Type	Function (How Flow Is Created)
Gear	
Vane	
Piston	

3. In the table below list the types of variable displacement pumps UNDER EACH CATEGORY and how each creates variable pump flow.

Pump Type	Function (How Variable Flow Is Created)
Vane	
Piston	

4. Describe how fluid flows through a typical fixed displacement pump.

    a. When using a hydraulic function.

    b. With no hydraulic function being used.

5. Describe how fluid flows through a typical variable displacement pump.

    a. When using a hydraulic function.

b. With no hydraulic function being used

____

____

____

____

____

6. Instructor Comments:

____

____

____

<table>
<tr><td>Performance Rating</td><td colspan="4">CDX Tasksheet Number: E0075</td></tr>
<tr><td>☐</td><td>☐</td><td>☐</td><td>☐</td><td>☐</td></tr>
<tr><td>0</td><td>1</td><td>2</td><td>3</td><td>4</td></tr>
</table>

Supervisor/instructor signature _________________________________ Date _____________

Pump Identification and Operation

Student/Intern information:

Name _________________________________ Date _____________ Class _____________________

Vehicle used for this activity:

Year _______________ Make _________________________ Model _____________________

Odometer ____________ Hour meter ___________ VIN _________________________________

Learning Objective/Task	CDX Tasksheet Number	2014 Edition Rev2/12/16 AED Standard
• Identify a gear pump, name all the parts, follow the oil flow through the pump, identify inlet and outlet ports, and identify the direction of rotation of the pump.	E0076	3.1
• Identify a vane pump, name all the parts, follow the oil flow through the pump, identify inlet and outlet ports, and identify the direction of rotation of the pump. Explain how a vane pump can be changed to operate in the opposite direction, when applicable.	E0077	3.1
• Identify various piston pumps, name all the parts, follow the oil flow through the pump, identify inlet and outlet ports (both variable and fixed), and identify the direction of rotation of the pump.	E0078	3.1
• Identify types of swash plate control (manual, servo piston, electronic, etc.).	E0079	3.1

Time off____________

Time on____________

Total time____________

Materials Required

- Various gear pumps, vane pumps, piston pumps
- Equipment manufacturer's, pump manufacturer's workshop manual including schematic hydraulic diagrams
- Basic mechanics tool set
- Manufacturer-specific tools depending on the concern
- Lifting equipment, if applicable

Some Safety Issues to Consider

- Activities require you to measure hydraulic pressures. Always ensure that the instructor/supervisor checks test instrument connections prior to connecting power or taking measurements. High fluid pressures can be dangerous; avoid opening hydraulic circuits until system is deenergized.
- Activities may require test driving the equipment on the school grounds, which carries severe risks. Attempt this task only with full permission from your supervisor/instructor, and follow all the guidelines exactly.

- Lifting equipment such as equipment jacks and stands, hoists, and engine hoists are important tools that increase productivity and make the job easier. However, they can also cause severe injury or death if used improperly. Make sure you follow the manufacturer's operation procedures. Also make sure you have your supervisor/instructor's permission to use any particular type of lifting equipment.
- Comply with personal and environmental safety practices associated with clothing; eye protection; hand tools; power equipment; proper ventilation; and the handling, storage, and disposal of chemicals/materials in accordance with federal, state, and local regulations.
- Always wear the correct protective eyewear and clothing and use the appropriate safety equipment, as well as fender covers, seat protectors, and floor mat protectors.
- Make sure you understand and observe all legislative and personal safety procedures when carrying out practical assignments. If you are unsure of what these are, ask your supervisor/instructor.

Student/Intern information:

Name _________________________________ Date ____________ Class _________________________

Vehicle used for this activity:

Year _______________ Make _______________________ Model _______________________

Odometer ___________ Hour meter ___________ VIN _______________________________

▶ **TASK** Identify a gear pump, name all the parts, follow the oil flow through the pump, identify inlet and outlet ports, and identify the direction of rotation of the pump.

AED 3.1

Time off_______________

Time on_______________

Total time_______________

CDX Tasksheet Number: E0076

1. **Disassemble an instructor-designated** external gear pump **using manufacturer's service information. Complete the questions below.**

    a. **Identify and list the components that make up the gear pump:**

        1.   6.

        2.   7.

        3.   8.

        4.   9.

        5.

    b. **Identify the fluid inlet and outlet ports and be prepared to show your instructor.**

    c. **Describe the fluid flow through the pump starting at the inlet.**

        ____
        ____
        ____
        ____
        ____
        ____

    d. **Is this pump designed for right hand (clockwise) or left hand (counterclockwise) rotation?**

        ____
        ____
        ____
        ____
        ____

    e. **Reassemble gear pump following manufacturer's procedures.**

2. **Disassemble an instructor-designated** internal gear pump **using manufacturer's service information. Complete the questions below.**

    a. **Identify and list the components that makeup the gear pump:**

        1.   6.

        2.   7.

        3.   8.

        4.   9.

        5.

    b. **Identify the fluid inlet and outlet ports and be prepared to show your instructor.**

    c. **Describe the fluid flow through the pump starting at the inlet.**

____
____
____
____
____

    d. **Is this pump designed for right hand (clockwise) or left hand (counterclockwise) rotation?**

____
____
____
____

    e. **Reassemble gear pump following manufacturer's procedures.**

3. **Instructor Comments:**

____
____
____
____

Performance Rating

CDX Tasksheet Number: E0076

| 0 | 1 | 2 | 3 | 4 |

Supervisor/instructor signature _______________________________ Date __________

Name _______________________________ Date ____________ Class _______________________

Vehicle used for this activity:

Year ______________ Make _______________________________ Model _______________________

Odometer ____________ Hour meter ____________ VIN _________________________________

▶ TASK Identify a vane pump, name all the parts, follow the oil flow through the pump, identify inlet and outlet ports, and identify the direction of rotation of the pump. Explain how a vane pump can be changed to operate in the opposite direction, when applicable.

AED 3.1

Time off_________________

Time on_________________

Total time_________________

CDX Tasksheet Number: E0077

1. **Disassemble an instructor-designated** vane pump **using manufacturer's service information. Complete the questions below.**

   a. **Identify and list the components that make up the pump:**

1.	6.
2.	7.
3.	8.
4.	9.
5.	

   b. **Identify the fluid inlet and outlet ports and be prepared to show your instructor.**

   c. **Describe the fluid flow through the pump starting at the inlet.**

   ____

   ____

   ____

   ____

   ____

   d. **Is the pump a balanced or unbalanced vane pump?**

   ____

   e. **Describe the difference between a balanced and unbalanced vane pump.**

   ____

   ____

   ____

   ____

   ____

   f. **Is this pump designed for right hand (clockwise) or left hand (counterclockwise) rotation?**

   ____

   ____

   ____

   ____

   ____

g. Explain how a vane pump can be changed to operate in the opposite direction, when applicable.

____

____

____

____

____

2. **Disassemble an instructor-designated** variable displacement vane pump **using manufacturer's service information. Complete the questions below.**

   a. **Identify and list the components that make up the pump:**

   1.   6.

   2.   7.

   3.   8.

   4.   9.

   5.

   b. **Identify the fluid inlet and outlet ports and be prepared to show your instructor.**

   c. **Describe the fluid flow through the pump starting at the inlet.**

   ____

   ____

   ____

   ____

   ____

   ____

   d. **Is the pump a balanced or unbalanced vane pump?**

   ____

   e. **Is this pump designed for right hand (clockwise) or left hand (counterclockwise) rotation?**

   ____

   ____

   ____

   ____

Performance Rating

CDX Tasksheet Number: E0077

| 0 | 1 | 2 | 3 | 4 |

Supervisor/instructor signature ____ Date _____________

Student/Intern information:

Name _________________________________ Date ____________ Class _________________________

Vehicle used for this activity:

Year _______________ Make ___________________________ Model ________________________

Odometer ______________ Hour meter ____________ VIN _________________________________

> **▶ TASK** Identify various piston pumps, name all the parts, follow the oil flow through the pump, identify inlet and outlet ports (both variable and fixed), and identify the direction of rotation of the pump.

AED 3.1

Time off________________

Time on________________

Total time________________

CDX Tasksheet Number: E0078

1. **Disassemble an instructor-designated** axial piston pump **using manufacturer's service information. Complete the questions below.**

    a. **Identify and list the components that makeup the pump:**

1.	6.
2.	7.
3.	8.
4.	9.
5.	

    b. **Identify the fluid inlet and outlet ports and be prepared to show your instructor.**

    c. **Describe the fluid flow through the pump starting at the inlet.**

    ____
    ____
    ____
    ____
    ____

    d. **Is the pump a fixed or variable displacement pump?**

    ____

    e. **Describe the difference between a fixed or variable displacement piston pump.**

    ____
    ____
    ____
    ____

    f. **Is this pump designed for right hand (clockwise) or left hand (counterclockwise) rotation or is it bidirectional?**

    ____
    ____
    ____
    ____

2. **Disassemble an instructor-designated** bent axis piston pump **using manufacturer's service information. Complete the questions below.**

   a. **Identify and list the components that make up the pump:**

   1.   6.

   2.   7.

   3.   8.

   4.   9.

   5.

   b. **Identify the fluid inlet and outlet ports and be prepared to show your instructor.**

   c. **Describe the fluid flow through the pump starting at the inlet.**

   ____
   ____
   ____
   ____
   ____

   d. **Is the pump a fixed or variable displacement pump?**

   ____

   e. **Describe the difference between a fixed or variable displacement piston pump.**

   ____
   ____
   ____
   ____

   f. **Is this pump designed for right hand (clockwise) or left hand (counterclockwise) rotation or is it bidirectional?**

   ____
   ____
   ____
   ____

Performance Rating

CDX Tasksheet Number: E0078

| 0 | 1 | 2 | 3 | 4 |

Supervisor/instructor signature ____ Date ___________

Name _______________________________ Date ____________ Class _____________________

Vehicle used for this activity:

Year _______________ Make _________________________ Model _____________________

Odometer ____________ Hour meter ____________ VIN _______________________________

▶ **TASK** Identify types of swash plate control (manual, servo piston, electronic, etc.).

AED
3.1

Time off____________

Time on____________

Total time____________

CDX Tasksheet Number: E0079

1. **Using instructor-designated pumps and the appropriate manufacturer's service information. Complete the questions below.**

   a. **Identify and list the type of swash plate control for pump #1.**

      i. _______________ **type**

   b. **Describe how the swash plate control works for the pump above.**

      ____
      ____
      ____
      ____
      ____

   c. **Identify and list the type of swash plate control for pump #2.**

      i. _______________ **type**

   d. **Describe how the swash plate control works for the pump above.**

      ____
      ____
      ____
      ____
      ____

   e. **Identify and list the type of swash plate control for pump #3.**

      i. _______________ **type**

f. Describe how the swash plate control works for the pump above.

____

____

____

____

____

Motor Identification and Operation

Student/Intern information:

Name _________________________________ Date _____________ Class _____________________

Vehicle used for this activity:

Year _______________ Make _________________________ Model _____________________

Odometer _____________ Hour meter ___________ VIN _________________________________

Learning Objective/Task	CDX Tasksheet Number	2014 Edition Rev2/12/16 AED Standard
• Identify a gear motor, name all the parts, follow the oil flow through the motor, identify inlet and outlet ports, and identify the direction of rotation of the motor.	E0080	3.1
• Identify a vane motor, name all the parts, follow the oil flow through the motor, identify inlet and outlet ports, and identify the direction of rotation of the motor.	E0081	3.1
• Identify radial and axial piston motors, name all parts of these motors, follow the oil flow through these motors, identify inlet and outlet ports (both variable and fixed), and identify the direction of rotation of the motors.	E0082	3.1
• Identify a gerotor motor, name all parts, and understand its operation.	E0083	3.1

Time off_______________

Time on_______________

Total time_______________

Materials Required

- Various gear motors, vane motors, piston motors
- Equipment manufacturer's, motor manufacturer's workshop manual including schematic hydraulic diagrams
- Basic mechanics tool set
- Manufacturer-specific tools depending on the concern
- Lifting equipment, if applicable

Some Safety Issues to Consider

- Activities require you to measure hydraulic pressures. Always ensure that the instructor/supervisor checks test instrument connections prior to connecting power or taking measurements. High fluid pressures can be dangerous; avoid opening hydraulic circuits until system is deenergized.
- Activities may require test driving the equipment on the school grounds, which carry severe risks. Attempt this task only with full permission from your supervisor/instructor, and follow all the guidelines exactly.

- Lifting equipment such as equipment jacks and stands, hoists, and engine hoists are important tools that increase productivity and make the job easier. However, they can also cause severe injury or death if used improperly. Make sure you follow the manufacturer's operation procedures. Also make sure you have your supervisor/instructor's permission to use any particular type of lifting equipment.
- Comply with personal and environmental safety practices associated with clothing; eye protection; hand tools; power equipment; proper ventilation; and the handling, storage, and disposal of chemicals/materials in accordance with federal, state, and local regulations.
- Always wear the correct protective eyewear and clothing and use the appropriate safety equipment, as well as fender covers, seat protectors, and floor mat protectors.
- Make sure you understand and observe all legislative and personal safety procedures when carrying out practical assignments. If you are unsure of what these are, ask your supervisor/instructor.

Student/Intern information:

Name ________________________________ Date ___________ Class ___________________________

Vehicle used for this activity:

Year _______________ Make _________________________ Model ___________________________

Odometer _____________ Hour meter ____________ VIN _________________________________

▶ **TASK** Identify a gear motor, name all parts, follow the oil flow through the motor, identify inlet and outlet ports, and identify the direction of rotation of the motor.

AED
3.1

Time off_______________

Time on_______________

CDX Tasksheet Number: EO080

Total time_______________

1. **Disassemble an instructor-designated** gear motor **using manufacturer's service information. Complete the questions below.**

    a. **Identify and list the components that make up the motor:**

    1.   6.

    2.   7.

    3.   8.

    4.   9.

    5.

    b. **Identify the fluid inlet and outlet ports and be prepared to show your instructor.**

    c. **Describe the fluid flow through the motor starting at the inlet.**

    ____

    ____

    ____

    ____

    ____

    d. **Is this motor designed for right hand (clockwise) or left hand (counterclockwise) rotation?**

    ____

    ____

    ____

    ____

    e. **Re-assemble gear motor following manufacturer's procedures.**

2. **Instructor Comments:**

____

____

____

Student/Intern information:

Name ________________________________ Date ____________ Class ________________________

Vehicle used for this activity:

Year ______________ Make ____________________________ Model ________________________

Odometer ____________ Hour meter ____________ VIN ________________________________

▶ **TASK** Identify a vane motor, name all parts of a vane motor, follow the oil flow through a vane motor, identify inlet and outlet ports of a vane motor, and identify the direction of rotation of the motor.

AED 3.1

Time off____________

Time on____________

Total time____________

CDX Tasksheet Number: E0081

1. **Disassemble an instructor-designated** vane motor **using manufacturer's service information. Complete the questions below.**

    a. **Identify and list the components that make up the motor:**

1.	6.
2.	7.
3.	8.
4.	9.
5.	

    b. **Identify the fluid inlet and outlet ports and be prepared to show your instructor.**

    c. **Describe the fluid flow through the motor starting at the inlet.**

    ____
    ____
    ____
    ____
    ____

    d. **Is this motor designed for right hand (clockwise) or left hand (counterclockwise) rotation?**

    ____
    ____
    ____
    ____

e. Explain how a vane motor can be changed to operate in the opposite direction, when applicable.

____

____

____

____

Student/Intern information:

Name _______________________________ Date _____________ Class _____________________________

Vehicle used for this activity:

Year _______________ Make _____________________________ Model _____________________________

Odometer ____________ Hour meter ____________ VIN _______________________________

▶ **TASK** Identify radial and axial piston motors, name all parts of these motors, follow the oil flow through these motors, identify inlet and outlet ports (both variable and fixed), and identify the direction of rotation of the motors.

AED
3.1

Time off______________

Time on______________

Total time______________

CDX Tasksheet Number: E0082

1. **Disassemble an instructor-designated** radial piston motor **using manufacturer's service information. Complete the questions below.**

   a. **Identify and list the components that make up the motor:**

   1.   6.

   2.   7.

   3.   8.

   4.   9.

   5.

   b. **Identify the fluid inlet and outlet ports and be prepared to show your instructor.**

   c. **Describe the fluid flow through the motor starting at the inlet.**

   ____
   ____
   ____
   ____
   ____

   d. **Is this motor designed for right hand (clockwise) or left hand (counterclockwise) rotation?**

   ____
   ____
   ____
   ____

2. **Disassemble an instructor-designated** axial piston motor **using manufacturer's service information. Complete the questions below.**

   a. **Identify and list the components that make up the motor:**

      1.   6.

      2.   7.

      3.   8.

      4.   9.

      5.

   b. **Identify the fluid inlet and outlet ports and be prepared to show your instructor.**

   c. **Describe the fluid flow through the motor starting at the inlet.**

   d. **Is this motor designed for right hand (clockwise) or left hand (counterclockwise) rotation?**

Performance Rating

CDX Tasksheet Number: E0082

| 0 | 1 | 2 | 3 | 4 |

Supervisor/instructor signature _________________________________ Date _____________

▶ **TASK** Identify a gerotor motor, name all parts, and understand its operation.

AED 3.1

Time off_______________

Time on_______________

Total time_______________

CDX Tasksheet Number: E0083

1. **Disassemble an instructor-designated** gerotor motor **using manufacturer's service information. Complete the questions below.**

    a. **Identify and list the components that make up the motor:**

    1.   6.

    2.   7.

    3.   8.

    4.   9.

    5.

    b. **Identify the fluid inlet and outlet ports and be prepared to show them to your instructor.**

    c. **Describe the fluid flow through the motor starting at the inlet.**

    ____

    ____

    ____

    ____

    ____

    d. **Is this motor designed for right hand (clockwise) or left hand (counterclockwise) rotation?**

    ____

    ____

    ____

    ____

Performance Rating

CDX Tasksheet Number: E0083

0	1	2	3	4

Supervisor/instructor signature _________________________________ Date ___________

Function and Operation of Hydraulic Valves

____

Student/Intern information:

Name ___________________________ Date __________ Class _______________

Vehicle used for this activity:

Year ____________ Make _________________________ Model _______________

Odometer __________ Hour meter __________ VIN _________________________

Learning Objective/Task	CDX Tasksheet Number	2014 Edition Rev2/12/16 AED Standard
• Exhibit the differences between these three major types: a) Pressure control valves b) Directional control valves c) Volume control valves	E0084	3.1
• Exhibit knowledge of the uses and functions of the following valves: a) Direct-acting relief valves b) Pilot operated relief valves c) Cartridge relief valves d) Pilot operated valves e) Sequence valves f) Unloading valves g) Multi-function valves h) Counterbalance valves i) Pressure reducing valves j) Pressure limiting valves	E0085	3.1

Materials Required
- Various pressure control valves, directional control valves, volume control valves
- Equipment manufacturer's, motor manufacturer's workshop manual including schematic hydraulic diagrams
- Basic mechanics tool set
- Manufacturer-specific tools depending on the concern
- Lifting equipment, if applicable

Some Safety Issues to Consider
- Activities require you to measure hydraulic pressures. Always ensure that the instructor/supervisor checks test instrument connections prior to connecting power or taking measurements. High fluid pressures can be dangerous; avoid opening hydraulic circuits until system is deenergized.

- Activities may require test driving the equipment on the school grounds, which carries severe risks. Attempt this task only with full permission from your supervisor/instructor, and follow all the guidelines exactly.
- Lifting equipment such as equipment jacks and stands, hoists, and engine hoists are important tools that increase productivity and make the job easier. However, they can also cause severe injury or death if used improperly. Make sure you follow the manufacturer's operation procedures. Also make sure you have your supervisor/instructor's permission to use any particular type of lifting equipment.
- Comply with personal and environmental safety practices associated with clothing; eye protection; hand tools; power equipment; proper ventilation; and the handling, storage, and disposal of chemicals/materials in accordance with federal, state, and local regulations.
- Always wear the correct protective eyewear and clothing and use the appropriate safety equipment, as well as fender covers, seat protectors, and floor mat protectors.
- Make sure you understand and observe all legislative and personal safety procedures when carrying out practical assignments. If you are unsure of what these are, ask your supervisor/instructor.

▶ **TASK** Exhibit the differences between these three major types:
a) Pressure control valves
b) Directional control valves
c) Volume control valves

AED
3.1

Time off_______________

Time on_______________

Total time_______________

CDX Tasksheet Number: E0084

1. **Research the three major types of valves used in hydraulic systems and explain the function and operation in the table below.**

VALVE TYPE	FUNCTION	OPERATION
Pressure control valves		
Directional control valves		
Volume control valves		

2. **Describe the locations where pressure control valves are likely to be found in a hydraulic system.**

____

____

____

____

3. Describe the locations where directional control valves are likely to be found in a hydraulic system.

____
____
____
____

4. Describe the locations where volume control valves are likely to be found in a hydraulic system.

____
____
____
____

5. Instructor Comments:

____
____
____

Performance Rating

CDX Tasksheet Number: E0084

| 0 | 1 | 2 | 3 | 4 |

Supervisor/instructor signature ____ Date __________

Name _________________________________ Date _____________ Class _________________________

Vehicle used for this activity:

Year _______________ Make _________________________ Model _________________________

Odometer _____________ Hour meter _____________ VIN _________________________

▶ **TASK** Exhibit knowledge of the uses and functions of the following valves:
    a) Direct-acting relief valves
    b) Pilot operated relief valves
    c) Cartridge relief valves
    d) Pilot operated valves
    e) Sequence valves
    f) Unloading valves
    g) Multi-function valves
    h) Counterbalance valves
    i) Pressure reducing valves
    j) Pressure limiting valves

Time off_________________

Time on_________________

Total time_________________

AED
3.1

CDX Tasksheet Number: E0085

1. **Research the various types of pressure control valves used in hydraulic systems and explain the function and uses in the table below.**

VALVE TYPE	FUNCTION	USES
Directacting relief valves		
Pilot operated relief valves		
Cartridge relief valves		
Pilot operated valves		
Sequence valves		

Unloading valves		
Multi-function valves		
Counterbalance valves		
Pressure reducing valves		
Pressure limiting valves		

Function and Operation of Hydraulic Valves: Electro-Hydraulics

Student/Intern information:

Name _________________________________ Date ___________ Class _________________________

Vehicle used for this activity:

Year _______________ Make _________________________ Model _____________________

Odometer ____________ Hour meter ___________ VIN _____________________________

Learning Objective/Task	CDX Tasksheet Number	2014 Edition Rev2/12/16 AED Standard
• Exhibit knowledge of the uses and functions of the following valves: a. Check valves b. Rotary valves c. Spool valves d. Pilot controlled poppet valves e. Electro-hydraulic valves f. Electro-hydraulic control systems g. Pulse-width modulated valves	E0086	3.1

• Exhibit knowledge of the uses and functions of the following valves:
a. Flow control valves
  1. Compensated
  2. Non-compensated
b. Flow divider valves
  1. Priority
  2. Non-priority
  3. Proportional
	E0087	3.1

Time off_______________

Time on_______________

Total time_______________

Materials Required

- Various pressure control valves, directional control valves, volume control valves
- Equipment manufacturer's, motor manufacturer's workshop manual including schematic hydraulic diagrams
- Basic mechanics tool set
- Manufacturer-specific tools depending on the concern
- Lifting equipment, if applicable

Some Safety Issues to Consider

- Activities require you to measure hydraulic pressures. Always ensure that the instructor/supervisor checks test instrument connections prior to connecting power or taking measurements. High fluid pressures can be dangerous; avoid opening hydraulic circuits until system is deenergized.
- Activities may require test driving the equipment on the school grounds, which carry severe risks. Attempt this task only with full permission from your supervisor/instructor, and follow all the guidelines exactly.
- Lifting equipment such as equipment jacks and stands, hoists, and engine hoists are important tools that increase productivity and make the job easier. However, they can also cause severe injury or death if used improperly. Make sure you follow the manufacturer's operation procedures. Also make sure you have your supervisor/instructor's permission to use any particular type of lifting equipment.
- Comply with personal and environmental safety practices associated with clothing; eye protection; hand tools; power equipment; proper ventilation; and the handling, storage, and disposal of chemicals/materials in accordance with federal, state, and local regulations.
- Always wear the correct protective eyewear and clothing and use the appropriate safety equipment, as well as fender covers, seat protectors, and floor mat protectors.
- Make sure you understand and observe all legislative and personal safety procedures when carrying out practical assignments. If you are unsure of what these are, ask your supervisor/instructor.

▶ **TASK** Exhibit knowledge of the uses and functions of the following valves:

    a. Check valves
    b. Rotary valves
    c. Spool valves
    d. Pilot controlled poppet valves
    e. Electro-hydraulic valves
    f. Electro-hydraulic control systems
    g. Pulse-width modulated valves

Time off_____________

Time on_____________

Total time_____________

AED
3.1

CDX Tasksheet Number: E0086

1. **Research the various types of directional control valves used in hydraulic systems and explain the function and uses in the table below.**

VALVE TYPE	FUNCTION	USES
Check valves		
Rotary valves		
Pilot controlled poppet valves		
Electro-hydraulic valves		
Electro-hydraulic control systems		

Pulse-width modulated valves		
Multi-function valves		

▶ **TASK** Exhibit knowledge of the uses and functions of the following valves:

    a. Flow control valves
       1. Compensated
       2. Non-compensated
    b. Flow divider valves
       1. Priority
       2. Non-priority
       3. Proportional

Time off __________

Time on __________

Total time __________

AED 3.1

CDX Tasksheet Number: E0087

1. **Research the various types of volume control valves used in hydraulic systems and explain the function and uses in the table below.**

VALVE TYPE	FUNCTION	USES
Compensated flow control valves		
Non-compensated flow control valves		
Priority flow divider valves		

Non-priority flow divider valves		
Proportional flow divider valves		

Cylinder Identification and Operation

Student/Intern information:

Name _________________________________ Date _____________ Class _________________________

Vehicle used for this activity:

Year _______________ Make _______________________________ Model _____________________________

Odometer _____________ Hour meter ___________ VIN ____

Learning Objective/Task	CDX Tasksheet Number	2014 Edition Rev2/12/16 AED Standard
• Explain the uses and movements of the two types of cylinders.	E0088	3.1
• Identify a single acting cylinder, name all of its parts, and follow the oil flow through the cylinder.	E0089	3.1
• Understand operation of a cushioned cylinder.	E0090	3.1
• Identify a double acting cylinder, name all of its parts, and follow the oil flow through the cylinder.	E0091	3.1

Time off_____________

Time on_____________

Total time_____________

Materials Required

- Various single acting cylinders, double acting cylinders
- Equipment manufacturer's, motor manufacturer's workshop manual including schematic hydraulic diagrams
- Basic mechanics tool set
- Manufacturer-specific tools depending on the concern
- Lifting equipment, if applicable

Some Safety Issues to Consider

- Activities require you to measure hydraulic pressures. Always ensure that the instructor/supervisor checks test instrument connections prior to connecting power or taking measurements. High fluid pressures can be dangerous; avoid opening hydraulic circuits until system is deenergized.
- Activities may require test driving the equipment on the school grounds, which carries severe risks. Attempt this task only with full permission from your supervisor/instructor, and follow all the guidelines exactly.
- Lifting equipment such as equipment jacks and stands, hoists, and engine hoists are important tools that increase productivity and make the job easier. However, they can also cause severe injury or death if used improperly. Make sure you follow the manufacturer's operation procedures. Also make sure you have your supervisor/instructor's permission to use any particular type of lifting equipment.

- Comply with personal and environmental safety practices associated with clothing; eye protection; hand tools; power equipment; proper ventilation; and the handling, storage, and disposal of chemicals/materials in accordance with federal, state, and local regulations.
- Always wear the correct protective eyewear and clothing and use the appropriate safety equipment, as well as fender covers, seat protectors, and floor mat protectors.
- Make sure you understand and observe all legislative and personal safety procedures when carrying out practical assignments. If you are unsure of what these are, ask your supervisor/instructor.

▶ TASK Explain the uses and movements of the two types of cylinders.

AED 3.1

Time off_________________

Time on_________________

Total time_________________

CDX Tasksheet Number: E0088

1. **Research the two major types of cylinders used in hydraulic systems and explain the function and operation in the table below.**

CYLINDER TYPE	FUNCTION	USES
Single acting		
Double acting		

2. **Describe the locations where cylinders are likely to be found in a hydraulic system.**

____

____

____

____

3. **Instructor Comments:**

____

____

____

Performance Rating

CDX Tasksheet Number: E0088

0	1	2	3	4

Supervisor/instructor signature ____ Date ___________

Name _________________________________ Date ____________ Class _________________________

Vehicle used for this activity:

Year _______________ Make _______________________ Model _________________________

Odometer ____________ Hour meter ____________ VIN ___________________________________

▶ **TASK** Identify a single acting cylinder, name all of its parts, and follow the oil flow through the cylinder.

AED
3.1

Time off__________________

Time on__________________

CDX Tasksheet Number: E0089

Total time__________________

1. **Disassemble an instructor-designated** single acting cylinder **using manufacturer's service information. Complete the questions below.**

   a. **Identify and list the components that make up the cylinder:**

   1.   6.

   2.   7.

   3.   8.

   4.   9.

   5.

   b. **Identify the fluid inlet and outlet ports and be prepared to show your instructor.**

   c. **Describe the fluid flow through the cylinder.**

   ____
   ____
   ____
   ____
   ____

   d. **Reassemble cylinder following manufactures procedures.**

   ____

2. **Instructor Comments:**

   ____
   ____
   ____
   ____

Performance Rating

CDX Tasksheet Number: E0089

0	1	2	3	4

Supervisor/instructor signature _________________________________ Date ____________

▶ TASK Understand operation of a cushioned cylinder.

AED
3.1

Time off_________________

Time on_________________

CDX Tasksheet Number: E0090

1. **Research cushioned cylinders used in hydraulic systems and explain the function and operation in the table below.**

Total time_________________

Cylinder Type	Function	Uses

2. **Describe the locations where cushioned cylinders are likely to be found in a hydraulic system.**

____
____
____
____

3. **Instructor Comments:**

____
____
____

Performance Rating

CDX Tasksheet Number: E0090

0	1	2	3	4

Supervisor/instructor signature ____ Date ____________

Student/Intern information:

Name _________________________ Date ___________ Class _________________________

Vehicle used for this activity:

Year _____________ Make _________________________ Model _________________________

Odometer ___________ Hour meter ___________ VIN _________________________

▶ **TASK** Identify a double acting cylinder, name all of its parts, and follow the oil flow through the cylinder.

AED
3.1

Time off___________

Time on___________

CDX Tasksheet Number: E0091

Total time___________

1. **Disassemble an instructor-designated** double acting cylinder **using manufacturer's service information. Complete the questions below.**

   a. **Identify and list the components that make up the cylinder:**

   1.   6.

   2.   7.

   3.   8.

   4.   9.

   5.

   b. **Identify the fluid inlet and outlet ports and be prepared to show your instructor.**

   c. **Describe the fluid flow through the cylinder.**

   ____
   ____
   ____
   ____
   ____

   d. **Reassemble cylinder following manufacturer's procedures.**

2. **Instructor Comments:**

   ____
   ____
   ____

Performance Rating

CDX Tasksheet Number: E0091

| 0 | 1 | 2 | 3 | 4 |

Supervisor/instructor signature ____ Date ___________

Accumulator Identification and Operation

Student/Intern information:

Name _________________________ Date ___________ Class _________________________

Vehicle used for this activity:

Year _______________ Make _______________________ Model _________________________

Odometer ___________ Hour meter ___________ VIN _________________________________

Learning Objective/Task	CDX Tasksheet Number	2014 Edition Rev2/12/16 AED Standard
• Explain how accumulators store energy, absorb shocks, build pressure, and maintain a constant pressure within a system.	E0092	3.1
• Explain where and why gas, pneumatic, spring-loaded, and weighted accumulators are used.	E0093	3.1
• Explain and practice all accumulator safety.	E0094	3.1

Time off_______________

Time on_______________

Total time_______________

Materials Required

- Various accumulators
- Equipment manufacturer's, motor manufacturer's workshop manual including schematic hydraulic diagrams
- Basic mechanics tool set
- Manufacturer-specific tools depending on the concern
- Lifting equipment, if applicable

Some Safety Issues to Consider

- Activities require you to measure hydraulic pressures. Always ensure that the instructor/ supervisor checks test instrument connections prior to connecting power or taking measurements. High fluid pressures can be dangerous; avoid opening hydraulic circuits until system is deenergized.
- Activities may require test driving the equipment on the school grounds, which carries severe risks. Attempt this task only with full permission from your supervisor/instructor, and follow all the guidelines exactly.
- Lifting equipment such as equipment jacks and stands, hoists, and engine hoists are important tools that increase productivity and make the job easier. However, they can also cause severe injury or death if used improperly. Make sure you follow the manufacturer's operation procedures. Also make sure you have your supervisor/instructor's permission to use any particular type of lifting equipment.

- Comply with personal and environmental safety practices associated with clothing; eye protection; hand tools; power equipment; proper ventilation; and the handling, storage, and disposal of chemicals/materials in accordance with federal, state, and local regulations.
- Always wear the correct protective eyewear and clothing and use the appropriate safety equipment, as well as fender covers, seat protectors, and floor mat protectors.
- Make sure you understand and observe all legislative and personal safety procedures when carrying out practical assignments. If you are unsure of what these are, ask your supervisor/instructor.

Student/Intern information:

Name ________________________________ Date ____________ Class ________________________

Vehicle used for this activity:

Year ______________ Make ________________________ Model ________________________

Odometer ____________ Hour meter ____________ VIN ________________________

▶ TASK Explain how accumulators store energy, absorb shocks, build pressure, and maintain a constant pressure within a system.

AED 3.1

Time off_______________

Time on_______________

CDX Tasksheet Number: E0092

Total time_______________

1. **Research the function of accumulators used in hydraulic systems and explain how accumulators store energy, absorb shocks, build pressure, and maintain a constant pressure within a system in the table below.**

Accumulator Function	How Function Works
Store energy	
Absorb shocks	
Build pressure	
Maintain a constant pressure	

2. **Describe the locations where accumulators are likely to be found in a hydraulic system.**

____

____

____

____

3. **Instructor Comments:**

____

____

____

Student/Intern information:

Name _________________________________ Date ____________ Class _________________________

Vehicle used for this activity:

Year _______________ Make _____________________________ Model _________________________

Odometer ____________ Hour meter ____________ VIN _____________________________________

▶ TASK Explain where and why gas, pneumatic, spring-loaded, and weighted accumulators are used.

AED
3.1

Time off____________________

Time on____________________

Total time____________________

CDX Tasksheet Number: E0093

1. **Research the types of accumulators used in hydraulic systems and where and why gas, pneumatic, spring-loaded, and weighted accumulators are used. List them in the table below.**

Accumulator Type	Application(s) Used in
Gas	
Pneumatic	
Spring-loaded	
Weighted	

2. **Locate accumulators used in machine hydraulic systems on instructor-assigned machines using manufacturer's service information.**

   a. **Machine #1**

   i. **Locate accumulator in the system using manufacturer's service information and answer below.**

   A. **Describe type.**

   ____

   ____

   ____

   ____

B. Describe location on machine.

____
____
____
____

b. Machine #2

i. Locate accumulator in the system using manufacturer's service information and answer below.

A. Describe type.

____
____
____
____

B. Describe location on machine.

____
____
____
____

c. Machine #3

i. Locate accumulator in the system using manufacturer's service information and answer below.

A. Describe type.

____
____
____
____

B. Describe location on machine.

____
____
____
____

3. Instructor Comments:

____
____
____
____

Performance Rating

CDX Tasksheet Number: E0093

0	1	2	3	4
☐	☐	☐	☐	☐

Supervisor/instructor signature _______________________________ Date __________

Student/Intern information:

Name _______________________________ Date ____________ Class _____________________

Vehicle used for this activity:

Year _______________ Make _____________________ Model _______________________

Odometer ____________ Hour meter ____________ VIN _______________________________

▶ **TASK** Explain and practice all accumulator safety.

**AED
3.1**

Time off_______________

Time on_______________

CDX Tasksheet Number: E0094

1. **Research the hazards and safety precautions that must be understood and followed when working on hydraulic systems using accumulators. In the table below list the hazards and precautions.**

Total time_______________

Hazards	Precautions

<table>
<tr><td></td><td></td></tr>
<tr><td></td><td></td></tr>
<tr><td></td><td></td></tr>
<tr><td></td><td></td></tr>
<tr><td></td><td></td></tr>
</table>

2. **Demonstrate the procedures to safely deenergize an accumulator used in machine hydraulic systems on instructor-assigned machines using manufacturer's service information.**

    a. **List the safety steps.**

    b. **Demonstrate procedures under instructor supervision.**

3. **Instructor Comments:**

____
____
____

CDX Tasksheet Number: E0094

| 0 | 1 | 2 | 3 | 4 |

Supervisor/instructor signature _________________________________ Date _____________

Fluids, Transfer Components, and Filtering

Student/Intern information:

Name _________________________________ Date ____________ Class _________________________

Vehicle used for this activity:

Year ______________ Make _________________________ Model ___________________

Odometer ____________ Hour meter ___________ VIN _________________________

Learning Objective/Task	CDX Tasksheet Number	2014 Edition Rev2/12/16 AED Standard
• Exhibit the ability to select the proper hose for a given function, taking into consideration the flow needed, pressures to be used, routing, clamping, fittings required, and pulsating of lines.	E0095	3.2
• Exhibit knowledge of the understanding of hydraulic fittings, the importance of selecting the proper fitting, and their relationship to noise and vibration.	E0096	3.2
• Demonstrate the ability to identify various fittings and thread styles; examples: O-ring boss, NPT, NPTF, British Metric, O-ring flange, ORFS, etc. Proper procedure to torque fittings and flanges.	E0097	3.2
• Demonstrate the ability to crimp hydraulic fittings onto hose.	E0098	3.2
• Know the construction and function of filters used in hydraulic/hydrostatic systems.	E0099	3.2
• Describe the use of various filters in hydraulic and hydrostatic systems.	E0100	3.2

Materials Required

- Various hydraulic hose types, fitting types
- Equipment manufacturer's, motor manufacturer's workshop manual including schematic hydraulic diagrams
- Basic mechanics tool set
- Manufacturer-specific tools depending on the concern
- Lifting equipment, if applicable

Some Safety Issues to Consider

- Activities require you to measure hydraulic pressures. Always ensure that the instructor/supervisor checks test instrument connections prior to connecting power or taking measurements. High fluid pressures can be dangerous; avoid opening hydraulic circuits until system is deenergized.
- Activities may require test driving the equipment on the school grounds, which carries severe risks. Attempt this task only with full permission from your supervisor/instructor, and follow all the guidelines exactly.

- Lifting equipment such as equipment jacks and stands, hoists, and engine hoists are important tools that increase productivity and make the job easier. However, they can also cause severe injury or death if used improperly. Make sure you follow the manufacturer's operation procedures. Also make sure you have your supervisor/instructor's permission to use any particular type of lifting equipment.
- Comply with personal and environmental safety practices associated with clothing; eye protection; hand tools; power equipment; proper ventilation; and the handling, storage, and disposal of chemicals/materials in accordance with federal, state, and local regulations.
- Always wear the correct protective eyewear and clothing and use the appropriate safety equipment, as well as fender covers, seat protectors, and floor mat protectors.
- Make sure you understand and observe all legislative and personal safety procedures when carrying out practical assignments. If you are unsure of what these are, ask your supervisor/instructor.

Student/Intern information:

Name _____________________________ Date ___________ Class _____________________

Vehicle used for this activity:

Year _______________ Make _________________________ Model _____________________

Odometer ____________ Hour meter ___________ VIN _______________________________

▶ **TASK** Exhibit the ability to select the proper hose for a given function, taking into consideration the flow needed, pressures to be used, routing, clamping, fittings required, and pulsating of lines.

AED 3.2

Time off______________

Time on______________

Total time______________

CDX Tasksheet Number: E0095

1. **Research hydraulic hose standards and list the three organizations that set these standards.**

   1. ________________________________

   2. ________________________________

   3. ________________________________

2. **In your own words describe the similarities and difference between the standards above.**

   ____

   ____

   ____

   ____

   ____

   ____

   ____

3. **Describe the layers of construction necessary for the various types of hydraulic hoses.**

   a. **Low-pressure hoses up to 2,500 psi:**

   ____

   ____

   ____

   b. **High-pressure hoses up to 6,000 psi:**

   ____

   ____

   ____

   c. **Describe the construction of an SAE 100R9 hose:**

   ____

   ____

   ____

   d. **Describe the construction of a Type R6 hose:**

   ____

   ____

   ____

4. **Using a conductor sizing nomograph determine the inside hose diameter for the applications below.**

   a. **Pump inlet line size needed for a system with peak flow of 30 gpm.**

      i. Size _______________

   b. **Pressure line size needed for a system with peak flow of 50 gpm.**

      i. Size _______________

   c. **Return line size needed for a system with peak flow of 25 gpm.**

      i. Size _______________

5. **Using an instructor assigned machine determine the proper hose for a given function.**

   a. Function _______________   Proper hose _______________

   b. Function _______________   Proper hose _______________

   c. Function _______________   Proper hose _______________

6. **Instructor Comments:**

   ____
   ____
   ____

Student/Intern information:

Name _______________________________ Date _____________ Class _______________________________

Vehicle used for this activity:

Year _______________ Make _____________________________ Model _______________________________

Odometer _____________ Hour meter _____________ VIN _______________________________

▶ **TASK** Exhibit knowledge of the understanding of hydraulic fittings, the importance of selecting the proper fitting, and their relationship to noise and vibration.

AED 3.2

Time off_______________

Time on_______________

Total time_______________

CDX Tasksheet Number: E0096

1. **Research the types of hydraulic fittings used in hydraulic systems and where and why the various types are used. List in the table below.**

Fitting Type	Application(s) Used In
Pipe-thread	
Flare-type	
Metric	
O-ring boss	
O-ring face	
Split flange	

2. **Instructor Comments:**

____
____
____
____

Student/Intern information:

Name _______________________________ Date ____________ Class ___________________________

Vehicle used for this activity:

Year _______________ Make _____________________________ Model _____________________________

Odometer ____________ Hour meter ____________ VIN ____

▶ TASK Demonstrate the ability to identify various fittings and thread styles; examples: O-ring boss, NPT, NPTF, British Metric, O-ring flange, ORFS, etc. Proper procedure to torque fittings and flanges.

AED 3.2

Time off_______________

Time on_______________

CDX Tasksheet Number: E0097

1. **Using various fittings assigned by your instructor determine the fitting type, fitting size, and fitting torque specification.**

Total time_______________

Fitting Type	Fitting Size	Fitting Torque Spec.
Fitting #1-		
Fitting #2-		
Fitting #3-		
Fitting #4-		
Fitting #5-		
Fitting #6-		
Fitting #7-		
Fitting #8-		
Fitting #9-		
Fitting #10-		

2. **Instructor Comments:**

____
____
____

Performance Rating

CDX Tasksheet Number: E0097

| 0 | 1 | 2 | 3 | 4 |

Supervisor/instructor signature ____ Date ____________

Student/Intern information:

Name _______________________________ Date ____________ Class _____________________

Vehicle used for this activity:

Year ______________ Make _____________________________ Model ____________________

Odometer ____________ Hour meter ____________ VIN _______________________________

▶ **TASK** Demonstrate the ability to crimp hydraulic fittings onto hose.   **AED 3.2**

Time off______

Time on______

CDX Tasksheet Number: E0098

1. **Using a machine assigned by your instructor remove designated hydraulic hose and make a new hose to manufacturer's specification.**

Total time______

2. **All safety precautions must be executed at all times. Assemble the hoses, tubes, connectors, and fittings in accordance with the manufacturers' specifications.**

   a. **Ensure you maintain your working area in a clean state to prevent contamination.**
   b. **Remove hose from machine following all safety procedures.**
   c. **Using the STAMED method determine proper hose for application.**

      i. **Meets the manufacturer's specifications? Yes: ________ No: ________**
      ii. **Instructor approval _______________**

3. **Determine proper fitting type and size.**

      i. **Meets the manufacturer's specifications? Yes: ________ No: ________**
      ii. **If no, do not use. Source the correct hosing.**

4. **Make new hose following manufacturer's recommended procedures.**

5. **Install new hose on machine following manufacturer's procedures.**

6. **Instructor Comments:**

   ____
   ____
   ____

Performance Rating

CDX Tasksheet Number: E0098

0	1	2	3	4

Supervisor/instructor signature _____________________________________ Date ____________

▶ **TASK** Know the construction and function of hydraulic filters used in hydraulic/hydrostatic systems.

AED 3.2

Time off_______________

Time on_______________

CDX Tasksheet Number: E0099

1. **Research the construction of the types of hydraulic filters used in hydraulic systems by location. In the table below describe the construction and function.**

Total time_______________

Filter Type	Construction	Function
Tank breather filter		
Hydraulic screen filter		
Spin-on filter		
Cartridge filter		

2. **Describe the types of media used in hydraulic filters.**

____
____
____
____
____

3. **What is the purpose of a bypass valve in a filter?**

____
____
____
____
____

4. **Cut a filter apart as directed by your instructor and locate the components that make it up.**

5. **Instructor Comments:**

____

____

____

Performance Rating

CDX Tasksheet Number: E0099

0	1	2	3	4

Supervisor/instructor signature ____ Date _____________

Name _________________________________ Date _____________ Class _____________________

Vehicle used for this activity:

Year _______________ Make _________________________ Model _____________________

Odometer _____________ Hour meter ____________ VIN _________________________

▶ TASK Describe the use of various filters in hydraulic and
hydrostatic systems.

AED
3.2

Time off_______________

Time on_______________

CDX Tasksheet Number: E0100

1. **Research the types of hydraulic filters used in hydraulic systems and where and why the various types are used.**

Total time_______________

Filter Type	Location Used in System	Purpose/Function

2. **In your own words describe how the beta ratio is used to determine filter efficiency.**

____

____

____

____

3. **Using a machine assigned by your instructor locate the filters used by the hydraulic system, look up filter part number, and determine the type of filter.**

Filter Location in System	Filter Part #	Filter Type

4. **Instructor Comments:**

____

____

____

Performance Rating

| 0 | 1 | 2 | 3 | 4 |

Supervisor/instructor signature _______________________________ Date ___________

Maintenance Procedures

Learning Objective/Task	CDX Tasksheet Number	2014 Edition Rev2/12/16 AED Standard
• Demonstrate familiarity with, and practice, good hydraulic maintenance/safety practices.	E0101	3.3
• Perform all hydraulic functions and repairs in a clean atmosphere.	E0102	3.3
• Exhibit the proper maintenance techniques to prevent internal and external leaks.	E0103	3.3
• Demonstrate knowledge of overheating conditions. Prevent overheating by keeping the oil at the proper levels, cleaning dirt and mud from around lines and cylinder rods, keep relief valves adjusted properly, do not overload or overspeed systems, and do not hold control valves in a position longer than necessary.	E0104	3.3

Time off____________

Time on____________

Total time____________

Materials Required

- Equipment manufacturer's, motor manufacturer's workshop manual including schematic hydraulic diagrams
- Basic mechanics tool set
- Manufacturer-specific tools depending on the concern
- Lifting equipment, if applicable

Some Safety Issues to Consider

- Activities require you to measure hydraulic pressures. Always ensure that the instructor/supervisor checks test instrument connections prior to connecting power or taking measurements. High fluid pressures can be dangerous; avoid opening hydraulic circuits until system is deenergized.
- Activities may require test driving the equipment on the school grounds, which carries severe risks. Attempt this task only with full permission from your supervisor/instructor, and follow all the guidelines exactly.
- Lifting equipment such as equipment jacks and stands, hoists, and engine hoists are important tools that increase productivity and make the job easier. However, they can also cause severe injury or death if used improperly. Make sure you follow the manufacturer's operation procedures. Also make sure you have your supervisor/instructor's permission to use any particular type of lifting equipment.

- Comply with personal and environmental safety practices associated with clothing; eye protection; hand tools; power equipment; proper ventilation; and the handling, storage, and disposal of chemicals/materials in accordance with federal, state, and local regulations.
- Always wear the correct protective eyewear and clothing and use the appropriate safety equipment, as well as fender covers, seat protectors, and floor mat protectors.
- Make sure you understand and observe all legislative and personal safety procedures when carrying out practical assignments. If you are unsure of what these are, ask your supervisor/ instructor.

Student/Intern information:

Name _________________________________ Date ____________ Class _________________________

Vehicle used for this activity:

Year _______________ Make _________________________________ Model _________________________

Odometer _____________ Hour meter _____________ VIN _________________________________

▶ **TASK** Demonstrate familiarity with, and practice, good hydraulic maintenance/safety practices.

AED 3.3

Time off_____________

Time on_____________

Total time_____________

CDX Tasksheet Number: E0101

1. **Safely prepare a machine for an instructor-designated maintenance procedure or repair.**

2. **Use manufacturer's service information to determine the steps to safely prepare the machine and list below:**

    a. _________________________________

    b. _________________________________

    c. _________________________________

    d. _________________________________

    e. _________________________________

    f. _________________________________

3. **Use manufacturer's service information to determine the necessary personal protection equipment (PPE) required for you to perform the procedure and list below:**

    a. _________________________________

    b. _________________________________

    c. _________________________________

    d. _________________________________

    e. _________________________________

4. **Perform safety steps above on machine under instructor's supervision.**

    ____

    ____

    ____

    ____

    ____

    ____

    ____

5. **Instructor Comments:**

____

____

____

____

▶ TASK Perform all hydraulic functions and repairs in a clean atmosphere.

AED 3.3

Time off_______________

Time on_______________

CDX Tasksheet Number: EO102

Total time_______________

1. **Research equipment manufacturer's standards for shop/worksite cleanliness and describe below the steps that must be taken to maintain cleanliness.**

   ____
   ____
   ____
   ____
   ____
   ____
   ____

2. **Demonstrate the steps above along with school standards on an ongoing basis.**

3. **Instructor Comments:**

   ____
   ____
   ____
   ____

Performance Rating

CDX Tasksheet Number: EO102

0	1	2	3	4

Supervisor/instructor signature _______________________________________ Date ___________

▶ **TASK** Exhibit the proper maintenance techniques to prevent internal and external leaks.

AED 3.3

Time off__________

Time on__________

Total time__________

CDX Tasksheet Number: EO103

1. **Using an instructor-assigned machine and its operation and maintenance manual locate the maintenance procedures that help to prevent internal and external leaks in the hydraulic system. List the procedure(s) below:**

   ____
   ____
   ____
   ____
   ____

2. **Perform the steps above as directed by your instructor.**

3. **Instructor Comments:**

   ____
   ____
   ____

Performance Rating

CDX Tasksheet Number: EO103

0	1	2	3	4

Supervisor/instructor signature _________________________________ Date ___________

Student/Intern information:

Name _____________________________ Date ___________ Class _____________________

Vehicle used for this activity:

Year _______________ Make _______________________ Model _______________________

Odometer ____________ Hour meter ___________ VIN _______________________________

▶ **TASK** Demonstrate knowledge of overheating conditions. Prevent overheating by keeping the oil at the proper levels, cleaning dirt and mud from around lines and cylinder rods, keep relief valves adjusted properly, do not overload or overspeed systems, and do not hold control valves in a position longer than necessary.

AED
3.3

Time off_____________

Time on_____________

Total time_____________

CDX Tasksheet Number: EO104

1. **Research the conditions that can cause overheating in a hydraulic system and list below.**

   ____
   ____
   ____
   ____
   ____
   ____
   ____

2. **Research the steps that can be taken to prevent overheating in a hydraulic system.**

   ____
   ____
   ____
   ____
   ____
   ____
   ____

3. **Instructor Comments:**

   ____
   ____
   ____

Performance Rating

CDX Tasksheet Number: EO104

0	1	2	3	4

Supervisor/instructor signature _________________________________ Date ___________

Component Repair and Replacement

Student/Intern information:

Name _______________________________ Date ____________ Class _______________________

Vehicle used for this activity:

Year ______________ Make _______________________ Model _______________________

Odometer ____________ Hour meter ____________ VIN _______________________

Learning Objective/Task	CDX Tasksheet Number	2014 Edition Rev2/12/16 AED Standard
• Following the proper technical manual/service information, exhibit the ability to remove, disassemble, diagnose failure, evaluate, repair or replace/reinstall, and test operate any given component including but not limited to: a) Gear, vane, and piston pumps b) Gear, vane, and piston motors c) Pressure control valves d) Directional control valves e) Volume control valves f) Single-acting, double-acting cylinders (If OEM recommends or allows: gas, pneumatic, spring- and weight-loaded accumulators.)	E0105	3.4
• Following the proper technical manual/service information, exhibit the ability to remove and replace any given component including but not limited to: a) Gear, vane, and piston pumps b) Gear, vane, and piston motors c) Pressure control valves d) Directional control valves e) Volume control valves f) Single-acting, double-acting cylinders g) Gas, pneumatic, spring- and weight-loaded accumulators h) Hoses, steel lines, and fittings i) Oil coolers j) Reservoirs.	E0106	3.4

Time off_______________

Time on_______________

Total time_______________

Materials Required

- Equipment manufacturer's, motor manufacturer's workshop manual including schematic hydraulic diagrams
- Basic mechanics tool set
- Manufacturer-specific tools depending on the concern
- Lifting equipment, if applicable

Some Safety Issues to Consider

- Activities require you to measure hydraulic pressures. Always ensure that the instructor/supervisor checks test instrument connections prior to connecting power or taking measurements. High fluid pressures can be dangerous; avoid opening hydraulic circuits until system is deenergized.
- Activities may require test driving the equipment on the school grounds, which carries severe risks. Attempt this task only with full permission from your supervisor/instructor, and follow all the guidelines exactly.
- Lifting equipment such as equipment jacks and stands, hoists, and engine hoists are important tools that increase productivity and make the job easier. However, they can also cause severe injury or death if used improperly. Make sure you follow the manufacturer's operation procedures. Also make sure you have your supervisor/instructor's permission to use any particular type of lifting equipment.
- Comply with personal and environmental safety practices associated with clothing; eye protection; hand tools; power equipment; proper ventilation; and the handling, storage, and disposal of chemicals/materials in accordance with federal, state, and local regulations.
- Always wear the correct protective eyewear and clothing and use the appropriate safety equipment, as well as fender covers, seat protectors, and floor mat protectors.
- Make sure you understand and observe all legislative and personal safety procedures when carrying out practical assignments. If you are unsure of what these are, ask your supervisor/instructor.

Student/Intern information:

Name _________________________________ Date ____________ Class _____________________

Vehicle used for this activity:

Year ______________ Make ________________________ Model __________________________

Odometer ___________ Hour meter ____________ VIN __________________________________

▶ TASK Following the proper technical manual/service information, exhibit the ability to remove, disassemble, diagnose failure, evaluate, repair or replace/reinstall, and test operate any given component including but not limited to:
- Gear, vane, and piston pumps
- Gear, vane, and piston motors
- Pressure control valves
- Directional control valves
- Volume control valves
- Single-acting, double-acting cylinders

(If OEM recommends or allows: gas, pneumatic, spring- and weight-loaded accumulators.)

Time off_____________

Time on_____________

Total time_____________

**AED
3.4**

CDX Tasksheet Number: E0105

1. **Safely prepare a machine for an instructor-designated component removal and installation procedure.**

2. **Ask your instructor to assign the component to be removed from the list below:**

   - **Gear, vane, and piston pumps**
   - **Gear, vane, and piston motors**
   - **Pressure control valves**
   - **Directional control valves**
   - **Volume control valves**
   - **Single-acting, double-acting cylinders**
     (If OEM recommends or allows: gas, pneumatic, spring- and weight-loaded accumulators.)

3. **Use manufacturer's service information to determine the necessary personal protection equipment (PPE) required for you to perform the procedure and list below:**

   a. _______________________________

   b. _______________________________

   c. _______________________________

   d. _______________________________

   e. _______________________________

4. **Perform the steps to remove the assigned component above under instructor's supervision.**

5. Instructor Comments:

____

____

____

6. **Perform the steps to disassemble component using manufacturer's disassembly service information.**

7. **Perform the steps to diagnose failure of component using manufacturer's failure analysis service information.**

    a. **List below what you determine cause of failure and make repair recommendations:**

    ____

    ____

    ____

    ____

    ____

8. **Reassemble component following manufacturer's reassembly guidelines.**

9. **Install component on machine following manufacturer's recommended procedures.**

10. **Test operation of installed component to verify machine repair.**

Performance Rating   CDX Tasksheet Number: EO105

0	1	2	3	4

Supervisor/instructor signature ____ Date __________

▶ TASK Following the proper technical manual/service information, exhibit the ability to remove and replace any given component including but not limited to:
a) Gear, vane, and piston pumps
b) Gear, vane, and piston motors
c) Pressure control valves
d) Directional control valves
e) Volume control valves
f) Single-acting, double-acting cylinders
g) Gas, pneumatic, spring- and weight-loaded accumulators
h) Hoses, steel lines, and fittings
i) Oil coolers
j) Reservoirs.

Time off_______________

Time on_______________

Total time_______________

**AED
3.4**

CDX Tasksheet Number: EO106

1. **Safely prepare a machine for an instructor-designated component removal and installation procedure.**

   a. **Ask your instructor to assign the component to be removed from the list below:**

   - **Gear, vane, and piston pumps**
   - **Gear, vane, and piston motors**
   - **Pressure control valves**
   - **Directional control valves**
   - **Volume control valves**
   - **Single-acting, double-acting cylinders**
   - **Gas, pneumatic, spring- and weight-loaded accumulators**
   - **Hoses, steel lines, and fittings**
   - **Oil coolers**

   b. **Use manufacturer's service information to determine the necessary PPE required for you to perform the procedure and list below:**

   a. _________________________________

   b. _________________________________

   c. _________________________________

   d. _________________________________

   e. _________________________________

2. **Perform the steps to remove the assigned component above under instructor's supervision.**

3. **Perform steps necessary to prepare components for installation.**

4. **Install component on machine following manufacturer's recommended procedures.**

5. **Test operation of installed component to verify machine repair.**

6. **Instructor Comments:**

____

____

____

Performance Rating　　　　**CDX Tasksheet Number: EO106**

0	1	2	3	4

Supervisor/instructor signature ___________________________________ Date ___________

Hydraulic Schematics

Student/Intern information:

Name _________________________ Date ___________ Class _______________________

Vehicle used for this activity:

Year _______________ Make _________________________ Model _______________________

Odometer ____________ Hour meter ___________ VIN _______________________

Learning Objective/Task	CDX Tasksheet Number	2014 Edition Rev2/12/16 AED Standard
• Exhibit knowledge of symbol identification through demonstration.	EO107	3.5
• Given a selected schematic, exhibit your knowledge of schematics by using JIC, ISO, and various symbols to identify locations of various components.	EO108	3.5

Time off_______________

Time on_______________

Total time_______________

Materials Required

- Equipment manufacturer's, motor manufacturer's workshop manual including schematic hydraulic diagrams
- Basic mechanics tool set
- Manufacturer-specific tools depending on the concern
- Lifting equipment, if applicable

Some Safety Issues to Consider

- Activities require you to measure hydraulic pressures. Always ensure that the instructor/supervisor checks test instrument connections prior to connecting power or taking measurements. High fluid pressures can be dangerous; avoid opening hydraulic circuits until system is deenergized.
- Activities may require test driving the equipment on the school grounds, which carries severe risks. Attempt this task only with full permission from your supervisor/instructor, and follow all the guidelines exactly.
- Lifting equipment such as equipment jacks and stands, hoists, and engine hoists are important tools that increase productivity and make the job easier. However, they can also cause severe injury or death if used improperly. Make sure you follow the manufacturer's operation procedures. Also make sure you have your supervisor/instructor's permission to use any particular type of lifting equipment.
- Comply with personal and environmental safety practices associated with clothing; eye protection; hand tools; power equipment; proper ventilation; and the handling, storage, and disposal of chemicals/materials in accordance with federal, state, and local regulations.

- Always wear the correct protective eyewear and clothing and use the appropriate safety equipment, as well as fender covers, seat protectors, and floor mat protectors.
- Make sure you understand and observe all legislative and personal safety procedures when carrying out practical assignments. If you are unsure of what these are, ask your supervisor/instructor.

Name _________________________________ Date _____________ Class _________________________

Vehicle used for this activity:

Year ________________ Make _____________________________ Model _________________________

Odometer ____________ Hour meter ____________ VIN _________________________________

▶ **TASK** Exhibit knowledge of symbol identification through demonstration.

**AED
3.5**

Time off_____________

Time on_____________

CDX Tasksheet Number: EO107

Total time_____________

1. **Research common symbols used in hydraulic schematics and how they represent component function and hydraulic system operation. Answer the questions below.**

    a. **Draw the symbol that represents a vented reservoir.**

    b. **Draw the symbol that represents a filter.**

    c. **Draw the symbol that represents an oil cooler.**

    d. **Draw the symbol that represents a three-position, four-way, tandem center directional control valve (DCV).**

e. Draw the symbol that represents a bidirectional motor.

f. Draw the symbol that represents a unidirectional pump.

2. Draw a hydraulic schematic that includes the following components:

    a. Pressurized reservoir

    b. Return filter

    c. Internal combustion prime mover

    d. Pilot operated variable relief valve

    e. Double acting cylinder

    f. Three-position four-way tandem center DCV

    g. Conductors.

3. **Instructor Comments:**

____

____

____

____

Performance Rating

CDX Tasksheet Number: E0107

| 0 | 1 | 2 | 3 | 4 |

Supervisor/instructor signature _______________________________ Date ___________

▶ **TASK** Given a selected schematic, exhibit your knowledge of schematics by using JIC, ISO, and various symbols to identify locations of various components.

AED
3.5

Time off____________________

Time on____________________

Total time____________________

CDX Tasksheet Number: EO108

1. **Locate the components, operation, and power flow through a hydraulic circuit on an instructor-assigned machine. Use machine manufacturer's service manual and the correct hydraulic schematic.**

   a. **Locate the components that make up the machines circuit on the hydraulic schematic, and list below:**

   ____________________   ____________________

   ____________________   ____________________

   ____________________   ____________________

   ____________________   ____________________

   ____________________   ____________________

   ____________________   ____________________

   ____________________   ____________________

   ____________________   ____________________

   b. **Using the schematic trace the flow of fluid through the circuit as various functions are used.**

   i. **Function ____________**

   a. **Flow:**

   ____

   ____

   ____

   ____

   ii. **Function ____________**

   a. **Flow:**

   ____

   ____

   ____

   ____

iii. Function _______________

    a. Flow:

____

____

____

____

4. **Using the schematic and service information locate the circuit component locations on the machine.**

5. **Instructor Comments:**

____

____

____

Performance Rating

CDX Tasksheet Number: E0108

0	1	2	3	4

Supervisor/instructor signature _________________________________ Date _____________

Diagnostic Systems and Component Troubleshooting

Student/Intern information:

Name _________________________________ Date ____________ Class _________________________

Vehicle used for this activity:

Year ______________ Make _______________________ Model _____________________

Odometer ____________ Hour meter __________ VIN _________________________________

Time off_________________

Time on_________________

Total time_________________

Learning Objective/Task	CDX Tasksheet Number	2014 Edition Rev2/12/16 AED Standard
• Exhibit the ability to reason with regard to a specific malfunction.	E0109	3.6
• Exhibit mastering the use of all test equipment including flow meters, pressure gauges, vacuum gauges, and temperature measuring devices, in both the metric and standard scales.	E0110	3.6
• Demonstrate the ability to use schematic diagrams and follow a troubleshooting flow chart using a selected technical manual.	E0111	3.6
• Demonstrate the ability to follow an operational check procedure using a selected technical manual.	E0112	3.6
• Troubleshooting of load-sensing hydraulics.	E0113	3.6
• Demonstrate technical write-up competency.	E0114	3.6

Materials Required

- Machines/simulators with complaints
- Equipment manufacturer's, motor manufacturer's workshop manual including schematic hydraulic diagrams
- Basic mechanics tool set
- Hydraulic test equipment
- Manufacturer-specific tools depending on the concern
- Lifting equipment, if applicable

Some Safety Issues to Consider

- Activities require you to measure hydraulic pressures. Always ensure that the instructor/ supervisor checks test instrument connections prior to connecting power or taking measurements. High fluid pressures can be dangerous; avoid opening hydraulic circuits until system is deenergized.
- Activities may require test driving the equipment on the school grounds, which carries severe risks. Attempt this task only with full permission from your supervisor/instructor, and follow all the guidelines exactly.

- Lifting equipment such as equipment jacks and stands, hoists, and engine hoists are important tools that increase productivity and make the job easier. However, they can also cause severe injury or death if used improperly. Make sure you follow the manufacturer's operation procedures. Also make sure you have your supervisor/instructor's permission to use any particular type of lifting equipment.
- Comply with personal and environmental safety practices associated with clothing; eye protection; hand tools; power equipment; proper ventilation; and the handling, storage, and disposal of chemicals/materials in accordance with federal, state, and local regulations.
- Always wear the correct protective eyewear and clothing and use the appropriate safety equipment, as well as fender covers, seat protectors, and floor mat protectors.
- Make sure you understand and observe all legislative and personal safety procedures when carrying out practical assignments. If you are unsure of what these are, ask your supervisor/instructor.

Student/Intern information:

Name _______________________________ Date ___________ Class _______________________

Vehicle used for this activity:

Year _______________ Make _______________________ Model _______________________

Odometer ___________ Hour meter ___________ VIN _______________________________

▶ TASK Exhibit the ability to reason with regard to a specific malfunction. **AED 3.6**

Time off_______________

Time on_______________

CDX Tasksheet Number: E0109

1. **Using an instructor-assigned machine and its service and troubleshooting information identify the possible causes of low hydraulic fluid level.**

   a. **List all possible causes:**

   ____
   ____
   ____
   ____
   ____
   ____
   ____

Total time_______________

   b. **Other alternative causes:**

   ____
   ____
   ____
   ____
   ____
   ____
   ____

2. **From the list above, determine what actions should be taken to diagnose these causes:**

   ____
   ____
   ____
   ____
   ____
   ____

3. **From the list above, determine what actions should be taken to repair these causes:**

   ____
   ____
   ____
   ____
   ____
   ____

   a. **In your own words list some questions you may ask the customer to aid you in the troubleshooting process.**

   ____
   ____
   ____
   ____

4. Instructor Comments:

5. Using an instructor-assigned machine and its service and troubleshooting information identify the possible causes of inoperative system.

    a. List all possible causes:

    b. Other alternative causes:

6. From the list above, determine what actions should be taken to diagnose these causes:

7. From the list above, determine what actions should be taken to repair these causes:

    a. In your own words list some questions you may ask the customer to aid you in the troubleshooting process.

8. **Instructor Comments:**

____

____

____

9. **Using an instructor-assigned machine and its service and troubleshooting information identify the possible causes of low or erratic system pressure.**

    a. **List all possible causes:**

____

____

____

____

____

____

    b. **Other alternative causes:**

____

____

____

____

____

____

10. **From the list above, determine what actions should be taken to diagnose these causes:**

____

____

____

____

____

____

11. **From the list above, determine what actions should be taken to repair these causes:**

____

____

____

____

____

    a. **In your own words list some questions you may ask the customer to aid you in the troubleshooting process.**

____

____

____

____

____

____

12. **Instructor Comments:**

____

____

____

13. **Using an instructor-assigned machine and its service and troubleshooting information identify the possible causes of system operates erratically.**

    a. **List all possible causes:**

    b. **Other alternative causes:**

14. **From the list above, determine what actions should be taken to diagnose these causes:**

15. **From the list above, determine what actions should be taken to repair these causes:**

    a. **In your own words list some questions you may ask the customer to aid you in the troubleshooting process.**

16. **Instructor Comments:**

17. **Using an instructor-assigned machine and its service and troubleshooting information identify the possible causes of system operates slowly.**

a. **List all possible causes:**

b. **Other alternative causes:**

18. **From the list above, determine what actions should be taken to diagnose these causes:**

19. **From the list above, determine what actions should be taken to repair these causes:**

a. **In your own words list some questions you may ask the customer to aid you in the troubleshooting process.**

20. **Instructor Comments:**

21. **Using an instructor-assigned machine and its service and troubleshooting information identify the possible causes of system that overheats.**

a. **List all possible causes:**

b. Other alternative causes:

22. From the list above, determine what actions should be taken to diagnose these causes:

23. From the list above, determine what actions should be taken to repair these causes:

    a. In your own words list some questions you may ask the customer to aid you in the troubleshooting process.

24. Instructor Comments:

25. Using an instructor-assigned machine and its service and troubleshooting information identify the possible causes of incorrect actuator movement and leakage (internal and external).

    a. List all possible causes:

    b. Other alternative causes:

---

26. From the list above, determine what actions should be taken to diagnose these causes:

---

27. From the list above, determine what actions should be taken to repair these causes:

---

    a. In your own words list some questions you may ask the customer to aid you in the troubleshooting process.

---

28. Instructor Comments:

---

29. Using an instructor-assigned machine and its service and troubleshooting information identify the possible causes of control valve leakage problems (internal/external).

    a. List all possible causes:

---

    b. Other alternative causes:

____

30. **From the list above, determine what actions should be taken to diagnose these causes:**

____

31. **From the list above, determine what actions should be taken to repair these causes:**

____

    a. In your own words list some questions you may ask the customer to aid you in the troubleshooting process.

____

32. Instructor Comments:

____

▶ TASK Exhibit mastering the use of all test equipment including flow meters, pressure gauges, vacuum gauges, and temperature measuring devices, in both the metric and standard scales.

AED 3.6

Time off_________________

Time on_________________

Total time_________________

CDX Tasksheet Number: EO110

1. **Using an instructor-assigned machine/simulator and its service and troubleshooting information identify the correct procedure for performing system temperature, pressure, flow, and cycle time tests.**

   a. **List the procedure and all safety requirements for performing system temperature, pressure, flow, and cycle time tests:**

   ____

   ____

   ____

   ____

   ____

   ____

   ____

2. **Perform each procedure listed above under instructor supervision.**

   a. **Connect the hydraulic pressure testing equipment into the vehicle's hydraulic circuit as outlined in the manufacturer's workshop manual.**

   b. **Perform the system check; check the hydraulic oil temperature.**

   i. **Is it within specifications? Yes: _____________ No: _____________**

   c. **Perform the system check; check the hydraulic system pressure.**

   i. **Is it within specifications? Yes: _____________ No: _____________**

   d. **Perform the system check; check the hydraulic flow.**

   i. **Is it within specifications? Yes: _____________ No: _____________**

   e. **Perform the system check; check the hydraulic system cycle time test.**

   i. **Is it within specifications? Yes: _____________ No: _____________**

3. **Instructor Comments:**

Performance Rating

CDX Tasksheet Number: EO110

0	1	2	3	4

Supervisor/instructor signature _______________________________ Date _____________

▶ TASK Demonstrate the ability to use schematic diagrams and follow a troubleshooting flow chart using a selected technical manual.

AED 3.6

Time off____________

Time on____________

Total time____________

CDX Tasksheet Number: EO111

> **NOTE:** This tasksheet will require the use of a machine or simulator. Ask your instructor which machine or simulator you are to use.

1. **Using the machine or simulator service manual, hydraulic diagrams, and schematics perform the diagnostic procedures necessary to determine the root cause of system malfunction.**

   a. **Ask your instructor for the customer's complaint and interview the customer.**

      i. **Complaint:**

         ____

         ____

      ii. **Interview questions:**

         ____

         ____

   b. **Verify if the complaint exists.**

   c. **Using the manufacturer's service information troubleshoot and diagnose the system malfunction. List the steps you took to determine the root cause.**

      ____

      ____

      ____

      ____

      ____

      ____

      ____

      ____

      ____

      ____

   d. **List the recommended repair needed to fix the problem.**

      ____

      ____

      ____

      ____

2. Discuss your findings with the instructor.

CDX Tasksheet Number: EO111

<table>
<tr><td>☐</td><td>☐</td><td>☐</td><td>☐</td><td>☐</td></tr>
<tr><td>0</td><td>1</td><td>2</td><td>3</td><td>4</td></tr>
</table>

Supervisor/instructor signature ____ Date _______________

Student/Intern information:

Name _________________________ Date ___________ Class _______________

Vehicle used for this activity:

Year _____________ Make _________________________ Model _______________

Odometer ___________ Hour meter ___________ VIN _______________________

▶ **TASK** Demonstrate the ability to follow an operational check procedure
using a selected technical manual.

AED
3.6

Time off_______________

Time on_______________

Total time_______________

CDX Tasksheet Number: EO112

> **NOTE:** This tasksheet will require the use of a machine or simulator. Ask your
> instructor which machine or simulator you are to use.

1. **Using the machine or simulator service manual, hydraulic diagrams,
   and schematics perform the operational check procedures necessary to
   determine the operation of the system.**

   a. **Ask your instructor for the system that needs inspection.**

      i. **System:**

      ____

      ____

   b. **Using the manufacturer's service information perform the operational
      check procedures necessary to determine the operation of the
      system.**

   c. **List the operational check procedures performed:**

      ____

      ____

      ____

      ____

      ____

      ____

      ____

      ____

      ____

      ____

2. **Discuss your findings with the instructor.**

CDX Tasksheet Number: EO112

0	1	2	3	4

Supervisor/instructor signature ____ Date _______________

<table>
<tr><td>▶ **TASK** Troubleshooting of load-sensing hydraulics.</td><td>**AED** 3.6</td></tr>
</table>

CDX Tasksheet Number: EO113

Time off______________

Time on______________

Total time______________

> **NOTE:** This tasksheet will require the use of a machine or simulator. Ask your instructor which machine or simulator you are to use.

1. **Using the machine or simulator service manual, hydraulic diagrams, and schematics perform the troubleshooting procedures necessary to determine the root cause of system malfunction.**

    a. **Ask your instructor for the customer's complaint and interview the customer.**

      i. **Complaint:**

      ____

      ____

      ii. **Interview questions:**

      ____

      ____

    b. **Verify if the complaint exists.**

    c. **Using the manufacturer's service information troubleshoot and diagnose the system malfunction. List the steps you took to determine the root cause:**

      ____

      ____

      ____

      ____

      ____

      ____

      ____

      ____

      ____

      ____

    d. **List the recommended repair needed to fix the problem:**

      ____

      ____

      ____

      ____

2. Discuss your findings with the instructor.

CDX Tasksheet Number: EO113

| 0 | 1 | 2 | 3 | 4 |

Supervisor/instructor signature _______________________________________ Date _____________

▶ **TASK** Demonstrate technical write-up competency.

AED 3.6

Time off _______________

Time on _______________

CDX Tasksheet Number: EO114

Total time _______________

> **NOTE:** This tasksheet will require you to complete technical write-ups for all troubleshooting and diagnostic tasks in this section.

1. **Using the following scenario, write up the 3 Cs as listed on most repair orders. Assume that the customer authorized the recommended repairs.**

    a. **Concern/complaint:** ____

    ____

    ____

    ____

    ____

    ____

    ____

    b. **Cause:**

    ____

    ____

    ____

    ____

    ____

    ____

    ____

    c. **Correction:**

    ____

    ____

    ____

    ____

    ____

    ____

    ____

2. **Discuss your findings with the instructor.**

Performance Rating

CDX Tasksheet Number: EO114

0	1	2	3	4

Supervisor/instructor signature ____ Date _____________

Characteristics of Oils

Student/Intern information:

Name ___________________________ Date ___________ Class ___________________

Vehicle used for this activity:

Year _____________ Make _________________________ Model ___________________

Odometer ___________ Hour meter ___________ VIN _________________________

Learning Objective/Task	CDX Tasksheet Number	2014 Edition Rev2/12/16 AED Standard
• Understand oils and show familiarity with various fluids and their effects on hydraulic systems.	EO115	3.3

Time off___________

Time on___________

Total time___________

Materials Required

- Machines/simulators with complaints
- Equipment manufacturer's, motor manufacturer's workshop manual including schematic hydraulic diagrams
- Basic mechanics tool set
- Hydraulic test equipment
- Manufacturer-specific tools depending on the concern
- Lifting equipment, if applicable

Some Safety Issues to Consider

- Activities require you to measure hydraulic pressures. Always ensure that the instructor/supervisor checks test instrument connections prior to connecting power or taking measurements. High fluid pressures can be dangerous; avoid opening hydraulic circuits until system is deenergized.
- Activities may require test driving the equipment on the school grounds, which carry severe risks. Attempt this task only with full permission from your supervisor/instructor, and follow all the guidelines exactly.
- Lifting equipment such as equipment jacks and stands, hoists, and engine hoists are important tools that increase productivity and make the job easier. However, they can also cause severe injury or death if used improperly. Make sure you follow the manufacturer's operation procedures. Also make sure you have your supervisor/instructor's permission to use any particular type of lifting equipment.
- Comply with personal and environmental safety practices associated with clothing; eye protection; hand tools; power equipment; proper ventilation; and the handling, storage, and disposal of chemicals/materials in accordance with federal, state, and local regulations.

- Always wear the correct protective eyewear and clothing and use the appropriate safety equipment, as well as fender covers, seat protectors, and floor mat protectors.
- Make sure you understand and observe all legislative and personal safety procedures when carrying out practical assignments. If you are unsure of what these are, ask your supervisor/instructor.

Name _________________________________ Date _____________ Class _____________________________

Vehicle used for this activity:

Year _______________ Make _________________________________ Model _____________________________

Odometer _____________ Hour meter _____________ VIN _______________________________________

▶ **TASK** Understand oils and show familiarity with various fluids and their effects on hydraulic systems.

AED 3.3

Time off_______________

Time on_______________

CDX Tasksheet Number: EO115

1. **Using an instructor-assigned machine and its service and troubleshooting information identify the possible causes of hydraulic system contamination.**

Total time_______________

    a. List all possible causes:

    ____
    ____
    ____
    ____
    ____
    ____

    b. Other alternative causes:

    ____
    ____
    ____
    ____
    ____
    ____

2. **From the list above, determine what actions should be taken to diagnose these causes:**

    ____
    ____
    ____
    ____
    ____
    ____

3. **From the list above, determine what actions should be taken to repair these causes:**

    ____
    ____
    ____
    ____
    ____
    ____

4. Instructor Comments:

____

____

____

5. Using an instructor-assigned machine and its service and troubleshooting information check the hydraulic reservoir fluid level and condition.

    a. Is fluid level at the correct height? Yes: _______ No: _______

6. Check the condition of the fluid. Look for burnt fluid or contamination.

    a. Condition of the fluid? Good: _______ Bad: _______

7. Using the machine service information identify what type(s) of hydraulic fluid is recommended and list below:

____

____

____

____

____

____

____

8. From the list above, answer below:

    a. What type of fluid is recommended for your climate?

____

    b. What type of fluid is recommended for extreme climate(s)?

____

    c. What type of fluid is recommended for rough service conditions?

____

    d. What type of fluid is recommended where fire hazards exist?

____

9. Explain oil viscosity and the effects of temperature on oils of varying viscosity.

____

10. Explain the flash point of oils.

____

____

11. Explain the effects of mixing oil types in the hydraulic system.

____

____

12. Instructor Comments:

____

____

____

Performance Rating

CDX Tasksheet Number: EO115

<table>
<tr><td>☐</td><td>☐</td><td>☐</td><td>☐</td><td>☐</td></tr>
<tr><td>0</td><td>1</td><td>2</td><td>3</td><td>4</td></tr>
</table>

Supervisor/instructor signature _________________________________ Date ___________

Fluid Cleanliness

Student/Intern information:

Name _________________________________ Date ____________ Class ___________________

Vehicle used for this activity:

Year _____________ Make ______________________ Model ___________________

Odometer ____________ Hour meter ___________ VIN _______________________________

<table>
<tr><td rowspan="2">Learning Objective/Task</td><td>CDX
Tasksheet
Number</td><td>2014
Edition
Rev2/12/16
AED
Standard</td></tr>
<tr></tr>
<tr><td>• Understand ISO cleanliness code principles.</td><td>E0116</td><td>3.3</td></tr>
</table>

Time off___________

Time on___________

Total time___________

Materials Required

- Oil sample kits
- Oil sample reports
- Machines/simulators with complaints
- Equipment manufacturer's, motor manufacturer's workshop manual including schematic hydraulic diagrams
- Basic mechanics tool set
- Hydraulic test equipment
- Manufacturer-specific tools depending on the concern
- Lifting equipment, if applicable

Some Safety Issues to Consider

- Activities require you to measure hydraulic pressures. Always ensure that the instructor/supervisor checks test instrument connections prior to connecting power or taking measurements. High fluid pressures can be dangerous; avoid opening hydraulic circuits until system is deenergized.
- Activities may require test driving the equipment on the school grounds, which carry severe risks. Attempt this task only with full permission from your supervisor/instructor, and follow all the guidelines exactly.
- Lifting equipment such as equipment jacks and stands, hoists, and engine hoists are important tools that increase productivity and make the job easier. However, they can also cause severe injury or death if used improperly. Make sure you follow the manufacturer's operation procedures. Also make sure you have your supervisor/instructor's permission to use any particular type of lifting equipment.
- Comply with personal and environmental safety practices associated with clothing; eye protection; hand tools; power equipment; proper ventilation; and the handling, storage, and disposal of chemicals/materials in accordance with federal, state, and local regulations.

- Always wear the correct protective eyewear and clothing and use the appropriate safety equipment, as well as fender covers, seat protectors, and floor mat protectors.
- Make sure you understand and observe all legislative and personal safety procedures when carrying out practical assignments. If you are unsure of what these are, ask your supervisor/instructor.

AED
3.3

Time off________________

Time on________________

CDX Tasksheet Number: EO116

1. **Using an instructor-assigned machine follow the procedures to draw an oil sample from the hydraulic system.**

    a. **Follow the procedures to clean the area before sample is drawn.**

    b. **Follow the procedures to ensure system is at temperature before sample is drawn.**

    c. **Follow the procedures to draw sample.**

    d. **Follow the procedures to fill out paperwork for sample and package for shipping.**

Total time________________

2. **Send sample in to be analyzed.**

3. **Using an oil sample analysis report, identify key elements found in oil analysis and the types of failures related to each. Identify key indicators on a fluid analysis report that illustrate the questions below:**

    a. **The proper fluid type is being used.**

    b. **Fluid types have not been mixed.**

    c. **Indicators of fluid degradation.**

    d. **Trend analysis.**

____

____

____

____

____

____

Section A4: Power Trains

CONTENTS

Basic Principles of Power Trains

Student/Intern information:

Name _________________________________ Date ____________ Class _________________________

Vehicle used for this activity:

Year ______________ Make _________________________ Model _______________________

Odometer ____________ Hour meter ___________ VIN _____________________________________

Learning Objective/Task	CDX Tasksheet Number	2014 Edition Rev2/12/16 AED Standard
• Demonstrate understanding of various types of bearings and proper adjustment procedures.	E0119	4.1
• Identify components of a torque converter and describe the relationship of those components to one another.	E0120	4.1
• Describe the operation of a given torque converter and various stages of operation.	E0121	4.1
• Use OEM manuals/service information to test a torque converter unit and determine if operation is within specifications.	E0122	4.1

Time off____________

Time on____________

Total time____________

Materials Required

- Various bearing types
- Torque converter disassembled or cutaway
- Wheeled and track machines
- Equipment manufacturer's workshop manual
- Hydraulic pressure gauges
- Manufacturer-specific tools depending on the concern
- Machine lifting equipment, if applicable
- Machine lockout/tagout equipment, if applicable

Some Safety Issues to Consider

- Activities require you to test hydraulic pressures. Always ensure that the instructor/supervisor checks test instrument connections prior to connecting power or taking measurements. High fluid pressures can be dangerous; avoid opening hydraulic circuits until system is deenergized.

- Activities may require test driving the equipment on the school grounds, which carries severe risks. Attempt this task only with full permission from your supervisor/instructor, and follow all the guidelines exactly.
- Lifting equipment such as equipment jacks and stands, hoists, and engine hoists are important tools that increase productivity and make the job easier. However, they can also cause severe injury or death if used improperly. Make sure you follow the manufacturer's operation procedures. Also make sure you have your supervisor/instructor's permission to use any particular type of lifting equipment.
- Comply with personal and environmental safety practices associated with clothing; eye protection; hand tools; power equipment; proper ventilation; and the handling, storage, and disposal of chemicals/materials in accordance with federal, state, and local regulations.
- Always wear the correct protective eyewear and clothing and use the appropriate safety equipment, as well as fender covers, seat protectors, and floor mat protectors.
- Make sure you understand and observe all legislative and personal safety procedures when carrying out practical assignments. If you are unsure of what these are, ask your supervisor/instructor.

▶ **TASK** Demonstrate understanding of various types of bearings and proper adjustment procedures.

AED
4.1

Time off____________

Time on____________

Total time____________

CDX Tasksheet Number: E0119

1. **Research the various types of bearings used in heavy equipment and identify their applications.**

    a. **Identify various bearing types that have been labeled and displayed by your instructor.**

    b. **List their applications in the table below:**

Bearing Type	Application
1.	
2.	
3.	
4.	
5.	
6.	
7.	

2. **What types of bearings used in heavy equipment applications require preload adjustment?**

    ____

    ____

    ____

    ____

3. **Using an instructor-assigned machine and its service information, demonstrate how to adjust bearing preload.**

    a. **List the steps for bearing preload adjustment below:**

    ____

    ____

    ____

    b. **Adjust preload following outlined steps under instructor supervision.**

4. Instructor Comments:

____

____

____

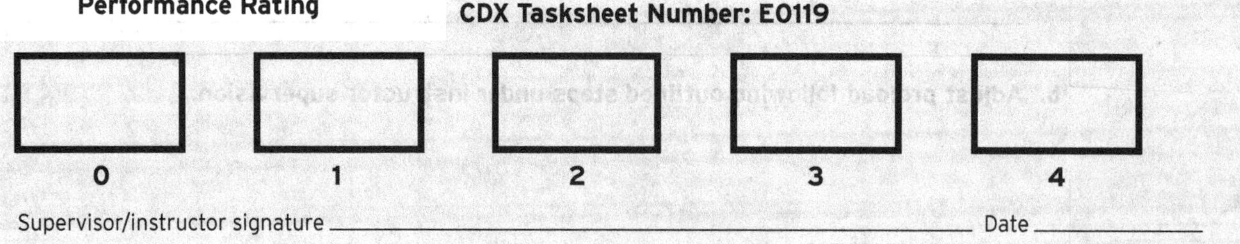

Student/Intern information:

Name _________________________________ Date _____________ Class _________________________

Vehicle used for this activity:

Year _______________ Make _________________________ Model _________________________

Odometer _____________ Hour meter _____________ VIN _________________________________

▶ **TASK** Identify components of a torque converter and describe the relationship of those components to one another.

AED 4.1

Time off_________________

Time on_________________

Total time_________________

CDX Tasksheet Number: E0120

1. **Research the components that make up torque converters used in heavy equipment. List the components and their function below:**

Component	Function

2. **Describe the relationship of the torque converter components to one another (how they interact with each other).**

____
____
____
____
____
____.

3. **Using an instructor-assigned torque converter, identify and label each of the components.**

4. Instructor Comments:

____
____
____.

© 2019 Jones & Bartlett Learning, LLC, an Ascend Learning Company

Student/Intern information:

Name ___________________________________ Date ____________ Class _______________________

Vehicle used for this activity:

Year _______________ Make ___________________________ Model _______________________

Odometer ____________ Hour meter ____________ VIN _________________________________

▶ **TASK** Describe the operation of a given torque converter and various stages of operation.

AED
4.1

Time off____________

Time on____________

Total time____________

CDX Tasksheet Number: EO121

1. **In the table below, describe the operation of a given torque converter and various stages of operation.**

Stage of operation	Operation
Vortex flow	
Stall	
Torque multiplication	
Lock-up clutches	
Rotary flow	
Cooler flow	

2. **Instructor Comments:**

____
____
____
____

Student/Intern information:

Name ______________________________ Date ____________ Class ____________________

Vehicle used for this activity:

Year ______________ Make ______________________ Model ____________________

Odometer ____________ Hour meter ____________ VIN ________________________

▶ **TASK** Use OEM manuals/service information to test a torque converter
unit and determine if operation is within specifications.

AED
4.1

Time off____________

Time on____________

CDX Tasksheet Number: E0122

Total time____________

1. **Using an instructor-assigned machine and its service information, test a
    torque converter unit and determine if operation is within specifications.**

    a. **Using the machines service information, look up the procedures to
        test the torque converter. List the procedures below:**

        ____
        ____
        ____
        ____
        ____
        ____
        ____

    b. **For each test, list the manufacturer's specifications below:**

        ____
        ____
        ____
        ____
        ____
        ____
        ____

    c. **Perform the tests following manufacturer's procedures under
        instructor supervision and list the results below.**

        ____
        ____
        ____
        ____
        ____
        ____

    d. **Did all tests fall within specifications? Yes ________ No ________**

    e. **If no, list test failures below.**

        ____
        ____
        ____

2. **Instructor Comments:**

Performance Rating

CDX Tasksheet Number: EO122

☐	☐	☐	☐	☐
0	1	2	3	4

Supervisor/instructor signature _________________________________ Date _____________

Theory and Principles of Manual Transmissions

Student/Intern information:

Name _____________________________ Date ___________ Class _______________

Vehicle used for this activity:

Year _____________ Make _____________________ Model _______________

Odometer __________ Hour meter __________ VIN _______________

Time off___________

Time on___________

Total time___________

Learning Objective/Task	CDX Tasksheet Number	2014 Edition Rev2/12/16 AED Standard
• Exhibit your understanding of "sliding gear" transmissions by identifying components, explaining operation, and demonstrating power flow through all gear sets.	EO123	4.1
• Exhibit your understanding of "collar shift" transmissions by identifying components, explaining operation, and demonstrating power flow through all gear sets.	EO124	4.1
• Exhibit your understanding of "synchromesh" transmissions by identifying components, explaining operation, and demonstrating power flow through all gear sets.	EO125	4.1
• Identify shifting control components and explain their operation.	EO126	4.1
• Demonstrate ability to perform adjustments to transmissions as instructed in the OEM service manual/information.	EO127	4.1

Materials Required

- Various manual transmissions
- Wheeled and track machines
- Equipment manufacturer's workshop manual
- Basic tool set
- Manufacturer-specific tools depending on the concern
- Machine lifting equipment, if applicable
- Machine lockout/tagout equipment, if applicable

Some Safety Issues to Consider

- Activities may require test driving the equipment on the school grounds, which carries severe risks. Attempt this task only with full permission from your supervisor/instructor, and follow all the guidelines exactly.

- Lifting equipment such as equipment jacks and stands, hoists, and engine hoists are important tools that increase productivity and make the job easier. However, they can also cause severe injury or death if used improperly. Make sure you follow the manufacturer's operation procedures. Also make sure you have your supervisor/instructor's permission to use any particular type of lifting equipment.
- Comply with personal and environmental safety practices associated with clothing; eye protection; hand tools; power equipment; proper ventilation; and the handling, storage, and disposal of chemicals/materials in accordance with federal, state, and local regulations.
- Always wear the correct protective eyewear and clothing and use the appropriate safety equipment, as well as fender covers, seat protectors, and floor mat protectors.
- Make sure you understand and observe all legislative and personal safety procedures when carrying out practical assignments. If you are unsure of what these are, ask your supervisor/instructor.

Student/Intern information:

Name _________________________________ Date ____________ Class _________________________

Vehicle used for this activity:

Year ______________ Make _________________________ Model _____________________________

Odometer ____________ Hour meter ____________ VIN _________________________________

▶ TASK Exhibit your understanding of "sliding gear" transmissions by identifying components, explaining operation, and demonstrating power flow through all gear sets.

AED
4.1

Time off_______________

Time on_______________

Total time_______________

CDX Tasksheet Number: E0123

1. **Research sliding gear transmissions used in heavy equipment and identify their components.**

    a. **Identify various components that have been labeled and displayed by your instructor.**

    b. **List their function in the operation of the transmission in the table below:**

Component	Function
1.	
2.	
3.	
4.	
5.	
6.	
7.	
8.	
9.	
10.	
11.	
12.	
13.	
14.	
15.	
16.	
17.	
18.	
19.	
20.	

2. **Instructor Comments:**

3. **Using an instructor-assigned transmission and its service information, demonstrate how transmission operates.**

   a. **In your own words describe operation in neutral.**

   b. **In your own words describe operation in reverse.**

   c. **In your own words describe operation in forward.**

4. **Using an instructor-assigned transmission and its service information, demonstrate how power flows through the transmission through all gear (speed) ratios.**

   a. **Power flow in neutral.**
   b. **Power flow in reverse.**
   c. **Power flow in 1st gear.**
   d. **Power flow in 2nd gear.**
   e. **Power flow in 3rd gear.**
   f. **Power flow in 4th gear.**
   g. **Power flow in 5th gear.**

5. **Instructor Comments:**

Performance Rating

CDX Tasksheet Number: EO123

0	1	2	3	4

Supervisor/instructor signature ___________________________________ Date _____________

Student/Intern information:

Name _________________________________ Date _____________ Class _________________________

Vehicle used for this activity:

Year _______________ Make _________________________ Model _____________________________

Odometer _____________ Hour meter _____________ VIN _________________________________

▶ TASK Exhibit your understanding of "Collar shift" transmissions by identifying components, explaining operation, and demonstrating power flow through all gear sets.

AED
4.1

Time off___________________

Time on____________________

Total time__________________

CDX Tasksheet Number: E0124

1. **Research collar shift transmissions used in heavy equipment and identify their components.**

     a. **Identify various components that have been labeled and displayed by your instructor.**

     b. **List their function in the operation of the transmission in the table below:**

Component	Function
1.	
2.	
3.	
4.	
5.	
6.	
7.	
8.	
9.	
10.	
11.	
12.	
13.	
14.	
15.	
16.	
17.	
18.	
19.	
20.	

2. **Instructor Comments:**

3. **Using an instructor-assigned transmission and its service information, demonstrate how transmission operates.**

   a. **In your own words describe operation in neutral.**

   b. **In your own words describe operation in reverse.**

   c. **In your own words describe operation in forward.**

4. **Using an instructor-assigned transmission and its service information, demonstrate how power flows through the transmission through all gear (speed) ratios.**

   a. **Power flow in neutral.**
   b. **Power flow in reverse.**
   c. **Power flow in 1st gear.**
   d. **Power flow in 2nd gear.**
   e. **Power flow in 3rd gear.**
   f. **Power flow in 4th gear.**
   g. **Power flow in 5th gear.**

5. **Instructor Comments:**

Performance Rating

CDX Tasksheet Number: E0124

0	1	2	3	4

Supervisor/instructor signature ____ Date _____________

▶ TASK Exhibit your understanding of "Syncromesh " transmissions by identifying components, explaining operation, and demonstrating power flow through all gear sets.

AED
4.1

Time off_______________

Time on_______________

Total time_______________

CDX Tasksheet Number: E0125

1. **Research Syncromesh transmissions used in heavy equipment and identify their components.**

    a. **Identify various components that have been labeled and displayed by your instructor.**
    b. **List their function in the operation of the transmission in the table below:**

Component	Function
1.	
2.	
3.	
4.	
5.	
6.	
7.	
8.	
9.	
10.	
11.	
12.	
13.	
14.	
15.	
16.	
17.	
18.	
19.	
20.	

2. **Instructor Comments:**

____

____

____

3. **Using an instructor-assigned transmission and its service information, demonstrate how transmission operates.**

   a. **In your own words describe operation in neutral.**

   ____

   ____

   ____

   b. **In your own words describe operation in reverse.**

   ____

   ____

   ____

   c. **In your own words describe operation in forward.**

   ____

   ____

   ____

4. **Using an instructor-assigned transmission and its service information, demonstrate how power flows through the transmission through all gear (speed) ratios.**

   a. **Power flow in neutral.**
   b. **Power flow in reverse.**
   c. **Power flow in 1st gear.**
   d. **Power flow in 2nd gear.**
   e. **Power flow in 3rd gear.**
   f. **Power flow in 4th gear.**
   g. **Power flow in 5th gear.**

5. **Instructor Comments:**

____

____

____

Performance Rating

CDX Tasksheet Number: E0125

0	1	2	3	4

Supervisor/instructor signature ___________________________ Date __________

Student/Intern information:

Name ________________________________ Date ____________ Class ________________________________

Vehicle used for this activity:

Year ______________ Make ________________________________ Model ________________________________

Odometer ____________ Hour meter ____________ VIN ________________________________

▶ TASK Identify shifting control components and explain their operation.

AED 4.1

Time off____________

Time on____________

CDX Tasksheet Number: EO126

Total time____________

1. **Research the components that make up shifting control components used in heavy equipment. List the components and their function below:**

Component	Function
1.	
2.	
3.	
4.	
5.	
6.	

2. **Using an instructor-assigned transmission and its service information, remove the shift bar housing and completely disassemble.**

   a. **Using the service manual, inspect the components for wear. List any worn parts below:**

   ____

   ____

   ____

   b. **Using the service manual, reassemble shift bar housing and reinstall on transmission.**

3. **Instructor Comments:**

Performance Rating

CDX Tasksheet Number: EO126

☐	☐	☐	☐	☐
0	1	2	3	4

Supervisor/instructor signature ____ Date _____________

Student/Intern information:

Name _________________________________ Date _____________ Class _________________________

Vehicle used for this activity:

Year _______________ Make _____________________________ Model _____________________________

Odometer _____________ Hour meter _____________ VIN _________________________________

▶ TASK Demonstrate ability to perform adjustments to transmissions as instructed in the OEM service manual/information.

AED 4.1

Time off____________

Time on____________

CDX Tasksheet Number: EO127

Total time____________

1. **Using an instructor-assigned machine, research the procedure and specifications for inspecting, adjusting, and installing shift linkages in the appropriate service information.**

   a. **List or print off and attach to this sheet the procedure for adjusting the shift linkages, if applicable:**

   ____

   ____

   ____

2. **Following the specified procedure, inspect the shift linkage components. List your observations for each component:**

   ____

   ____

   ____

3. **Using an instructor-assigned machine, research the procedure and specifications for inspecting, adjusting, endplay, preload, backlash in the appropriate service information.**

   a. **List or print off and attach to this sheet the procedure:**

   ____

   ____

   ____

4. **Following the specified procedure, inspect, and adjust, endplay, preload, backlash. List your observations for each component:**

   ____

   ____

   ____

5. Instructor Comments:

Theory and Principles of Powershift Transmissions

Student/Intern information:

Name _________________________________ Date ____________ Class _________________________

Vehicle used for this activity:

Year ______________ Make ________________________ Model _________________________

Odometer ____________ Hour meter ____________ VIN ___________________________________

Learning Objective/Task	CDX Tasksheet Number	2014 Edition Rev2/12/16 AED Standard
• Demonstrate your understanding of the operation of powershift transmissions by explaining which clutches and/or brakes are engaged, and which planetary gear sets are being used during a specific gear selection.	E0128	4.1
• Explain the differences, advantages, and disadvantages of planetary and countershaft transmissions.	E0129	4.1

Time off________________

Time on________________

Total time________________

Materials Required

- Various powershift transmissions
- Wheeled and track machines
- Equipment manufacturer's workshop manual
- Basic tool set
- Manufacturer-specific tools depending on the concern
- Machine lifting equipment, if applicable
- Machine lockout/tagout equipment, if applicable

Some Safety Issues to Consider

- Activities may require test driving the equipment on the school grounds, which carry severe risks. Attempt this task only with full permission from your supervisor/instructor, and follow all the guidelines exactly.
- Lifting equipment such as equipment jacks and stands, hoists, and engine hoists are important tools that increase productivity and make the job easier. However, they can also cause severe injury or death if used improperly. Make sure you follow the manufacturer's operation procedures. Also make sure you have your supervisor/instructor's permission to use any particular type of lifting equipment.
- Comply with personal and environmental safety practices associated with clothing; eye protection; hand tools; power equipment; proper ventilation; and the handling, storage, and disposal of chemicals/materials in accordance with federal, state, and local regulations.

- Always wear the correct protective eyewear and clothing and use the appropriate safety equipment, as well as fender covers, seat protectors, and floor mat protectors.
- Make sure you understand and observe all legislative and personal safety procedures when carrying out practical assignments. If you are unsure of what these are, ask your supervisor/instructor.

▶ **TASK** Demonstrate your understanding of the operation of powershift transmissions by explaining which clutches and/or brakes are engaged, and which planetary gear sets are being used during a specific gear selection.

AED
4.1

Time off__________________

Time on__________________

Total time__________________

CDX Tasksheet Number: E0128

1. **Research powershift transmissions used in heavy equipment and identify their components.**

    a. **Identify various components that have been labeled and displayed by your instructor.**

    b. **List their function in the operation of the transmission in the table below:**

Component	Function
1.	
2.	
3.	
4.	
5.	
6.	
7.	
8.	
9.	
10.	
11.	
12.	
13.	
14.	
15.	
16.	
17.	
18.	
19.	
20.	

2. Instructor Comments:

____

____

____

3. Using an instructor-assigned transmission and its service information, demonstrate how the transmission operates.

   a. In your own words describe operation in neutral.

____

____

____

   b. In your own words describe operation in reverse.

____

____

____

   c. In your own words describe operation in forward.

____

____

____

4. Using an instructor-assigned transmission and its service information, demonstrate how power flows through the transmission (including which clutches and or brakes are engaged, and which planetary sets are being used) through all gear (speed) ratios.

   a. Power flow in neutral.

   b. Power flow in reverse.

   c. Power flow in 1st gear.

   d. Power flow in 2nd gear.

   e. Power flow in 3rd gear.

   f. Power flow in 4th gear.

   g. Power flow in 5th gear.

   h. Power flow in 6th gear.

5. Instructor Comments:

____

____

____

Performance Rating

| 0 | 1 | 2 | 3 | 4 |

Supervisor/instructor signature ____ Date ____________

Student/Intern information:

Name _________________________________ Date _____________ Class _________________________

Vehicle used for this activity:

Year _______________ Make _________________________ Model _________________________

Odometer _____________ Hour meter ___________ VIN _________________________________

▶ TASK Explain the differences, advantages, and disadvantages of planetary and countershaft transmissions.

AED
4.1

Time off_________________

Time on_________________

Total time_________________

CDX Tasksheet Number: E0129

1. **In the table below, list the advantages and disadvantages of countershaft transmissions:**

Advantages	Disadvantages

2. **In the table below, list the advantages and disadvantages of planetary transmissions:**

Advantages	Disadvantages

3. **In your own words describe the heavy equipment applications where countershaft transmissions would be better suited.**

____

____

____

____

4. **In your own words describe the heavy equipment applications where planetary transmissions would be better suited.**

____

____

____

____

5. Instructor Comments:

____

____

____

____

Theory and Principles of Clutches

Student/Intern information:

Name _______________________________ Date ____________ Class _______________________

Vehicle used for this activity:

Year _______________ Make ___________________________ Model ___________________________

Odometer _____________ Hour meter ____________ VIN _________________________________

Learning Objective/Task	CDX Tasksheet Number	2014 Edition Rev2/12/16 AED Standard
• Use service manual/information to test and/or troubleshoot a powershift transmission (on-highway truck transmissions do not qualify), and verify if it is within OEM specifications.	EO130	4.1
• Identify all components in a single- and multiple-disc and plate-type clutch, including flywheel, pilot and release bearings, disc and pressure plate parts, and power train input shaft. Also, explain differences and benefits of solid- and button-type clutches.	EO131	4.1
• Explain operation of a selected clutch.	EO132	4.1
• Demonstrate knowledge and operation of single- and multiple-disc clutches by explaining the relationship of the clutch components to each other and their roles in the transfer of power.	EO133	4.1
• Describe the relationship of the number of discs, types of discs (wet or dry), and type of clutch material to the transfer of torque and horsepower to the ground.	EO134	4.1
• Explain the operation of magnetic clutches and name various applications.	EO135	4.1

Materials Required

- Machines equipped with various clutch systems
- Selection of various clutch assemblies
- Equipment manufacturer's workshop manual
- Basic tool set
- Manufacturer-specific tools depending on the concern
- Machine lifting equipment, if applicable
- Machine lockout/tagout equipment, if applicable

Some Safety Issues to Consider

- Activities may require test driving the equipment on the school grounds, which carries severe risks. Attempt this task only with full permission from your supervisor/instructor, and follow all the guidelines exactly.
- Lifting equipment such as equipment jacks and stands, hoists, and engine hoists are important tools that increase productivity and make the job easier. However, they can also cause severe injury or death if used improperly. Make sure you follow the manufacturer's operation procedures. Also make sure you have your supervisor/instructor's permission to use any particular type of lifting equipment.
- Comply with personal and environmental safety practices associated with clothing; eye protection; hand tools; power equipment; proper ventilation; and the handling, storage, and disposal of chemicals/materials in accordance with federal, state, and local regulations.
- Always wear the correct protective eyewear and clothing and use the appropriate safety equipment, as well as fender covers, seat protectors, and floor mat protectors.
- Make sure you understand and observe all legislative and personal safety procedures when carrying out practical assignments. If you are unsure of what these are, ask your supervisor/instructor.

▶ TASK Use service manual/information to test and/or troubleshoot a powershift transmission (on-highway truck transmissions do not qualify), and verify if it is within OEM specifications.

AED
4.1

Time off_______________

Time on_______________

Total time_______________

CDX Tasksheet Number: EO130

1. **Using an instructor-assigned machine and its service testing and troubleshooting information, identify the possible causes of powershift transmission hydraulic oil-related problems.**

   a. **List all possible causes:**

   ____
   ____
   ____

   b. **Other alternative causes:**

   ____
   ____
   ____

2. **From the list above, determine what actions should be taken to diagnose these causes:**

   ____
   ____
   ____

3. **From the list above, determine what actions should be taken to repair these causes:**

   ____
   ____
   ____

   a. **In your own words list some questions you may ask the customer to aid you in the troubleshooting process.**

   ____
   ____
   ____

4. Instructor Comments:

____

____

____

5. Using an instructor-assigned machine and its service testing and troubleshooting information, identify the possible causes of powershift transmission overheating problems.

   a. List all possible causes:

____

____

____

   b. Other alternative causes:

____

____

____

6. From the list above, determine what actions should be taken to diagnose these causes:

____

____

____

7. From the list above, determine what actions should be taken to repair these causes:

____

____

____

   a. In your own words list some questions you may ask the customer to aid you in the troubleshooting process.

____

____

____

8. Instructor Comments:

____

____

____

9. Using an instructor-assigned machine and its service testing and troubleshooting information, identify the possible causes of powershift transmission clutch slippage problems.

   a. List all possible causes:

____

____

____

   b. Other alternative causes:

____

____

____

10. **From the list above, determine what actions should be taken to diagnose these causes:**

____

____

____

11. **From the list above, determine what actions should be taken to repair these causes:**

____

____

____

    a. **In your own words list some questions you may ask the customer to aid you in the troubleshooting process.**

____

____

____

12. **Instructor Comments:**

____

____

____

13. **Using an instructor-assigned machine and its service testing and troubleshooting information, identify the possible causes of powershift transmission abnormal transmission noise problems.**

    a. **List all possible causes:**

____

____

____

    b. **Other alternative causes:**

____

____

____

14. **From the list above, determine what actions should be taken to diagnose these causes:**

____

____

____

15. **From the list above, determine what actions should be taken to repair these causes:**

____

____

____

a. In your own words list some questions you may ask the customer to aid you in the troubleshooting process.

____

____

____

16. Instructor Comments:

____

____

____

17. Using an instructor-assigned machine and its service testing and troubleshooting information, identify the possible causes of powershift transmission vibration problems.

a. List all possible causes:

____

____

____

b. Other alternative causes:

____

____

____

18. From the list above, determine what actions should be taken to diagnose these causes:

____

____

____

19. From the list above, determine what actions should be taken to repair these causes:

____

____

____

a. In your own words list some questions you may ask the customer to aid you in the troubleshooting process.

____

____

____

20. Instructor Comments:

____

____

____

21. **Using an instructor-assigned machine and its service testing and troubleshooting information, identify the possible causes of powershift transmission oil leak problems.**

    a. **List all possible causes:**

____

____

____

    b. **Other alternative causes:**

____

____

____

22. **From the list above, determine what actions should be taken to diagnose these causes:**

____

____

____

23. **From the list above, determine what actions should be taken to repair these causes:**

____

____

____

    a. **In your own words list some questions you may ask the customer to aid you in the troubleshooting process.**

____

____

____

24. **Instructor Comments:**

____

____

____

25. **Using an instructor-assigned machine and its service testing and troubleshooting information, identify the possible causes of powershift transmission gear selection, shifting, and directional change problems.**

    a. **List all possible causes:**

____

____

____

    b. **Other alternative causes:**

____

____

____

26. From the list above, determine what actions should be taken to diagnose these causes:

____

____

____

27. From the list above, determine what actions should be taken to repair these causes:

____

____

____

    a. In your own words list some questions you may ask the customer to aid you in the troubleshooting process.

____

____

____

28. Instructor Comments:

____

____

____

> **Note:** This tasksheet will require the use of a machine. Ask your instructor which machine or simulator you are to use.

29. Using the machine service manual testing and troubleshooting information, perform the troubleshooting procedures necessary to determine the root cause of system malfunction.

    a. Ask your instructor for the customer's complaint and interview the customer.

        i. Complaint:

____

____

        ii. Interview questions:

____

____

    b. Verify the complaint exists.

    c. Using the manufacturer's service information, troubleshoot and diagnose the system malfunction. List the steps you took to determine the root cause.

____

____

____

d. List the recommended repair needed to fix the problem.

____

____

____

30. Discuss your findings with the instructor.

Performance Rating

CDX Tasksheet Number: E0130

0	1	2	3	4

Supervisor/instructor signature _______________________________ Date _____________

Student/Intern information:

Name _________________________________ Date ____________ Class _______________________

Vehicle used for this activity:

Year _______________ Make ______________________________ Model ______________________

Odometer ____________ Hour meter ____________ VIN ___________________________________

▶ TASK Identify all components in a single- and multiple-disc and plate-type clutch, including flywheel, pilot and release bearings, disc and pressure plate parts, and power train input shaft. Also, explain the differences and benefits of solid- and button-type clutches.

AED 4.1

Time off_______________

Time on_______________

Total time_______________

CDX Tasksheet Number: EO131

1. **Research the various clutch assemblies used in heavy equipment and identify their components.**

    a. **Identify various components that have been labeled and displayed by your instructor.**

    b. **List their function in the operation of the clutch assembly in the table below:**

Component	Function
Single-disc clutch components	
1.	
2.	
3.	
4.	
5.	
6.	
7.	
8.	
9.	
10.	
11.	
Multiple-disc clutch components	
12.	
13.	
14.	
15.	
16.	
17.	

18.	
19.	
20.	
21.	
22.	

2. **Identify a button-type clutch disc and explain its benefits below.**

____
____
____

3. **Identify a solid-type clutch disc and explain its benefits below.**

____
____
____

4. **Identify a push-type clutch assembly and describe its function below.**

____
____
____

5. **Identify a pull-type clutch assembly and describe its function below.**

____
____
____

6. **What are the differences between a coil spring clutch, diaphragm spring clutch, and an angle spring clutch? Identify each type to your instructor.**

    a.   ____**coil spring clutch**

    b.   ____**diaphragm spring clutch**

    c.   ____**angle spring clutch**

7. **Discuss your findings with the instructor.**

Performance Rating

CDX Tasksheet Number: EO131

| 0 | 1 | 2 | 3 | 4 |

Supervisor/instructor signature ____ Date ____________

Name ___________________________ Date ___________ Class ___________________________

Vehicle used for this activity:

Year _____________ Make ___________________________ Model ___________________________

Odometer ___________ Hour meter ___________ VIN ___________________________

▶ TASK Explain operation of a selected clutch.

AED
4.1

Time off___________

Time on___________

CDX Tasksheet Number: EO132

1. **Using an instructor-assigned clutch assembly, explain the operation and power flow through engagement and disengagement.**

   a. **Operation and power flow disengaged.**
   b. **Operation and power flow engaged.**

Total time___________

2. **Instructor Comments:**

____
____
____

Performance Rating

CDX Tasksheet Number: EO132

0	**1**	**2**	**3**	**4**

Supervisor/instructor signature ____ Date ___________

Student/Intern information:

Name _________________________________ Date _____________ Class _________________________

Vehicle used for this activity:

Year _______________ Make _____________________________ Model _____________________________

Odometer _____________ Hour meter _____________ VIN _________________________________

▶ **TASK** Demonstrate knowledge and operation of single- and multiple-disc
clutches by explaining the relationship of the clutch components
to each other and their roles in the transfer of power.

AED
4.1

Time off_______________

Time on_______________

CDX Tasksheet Number: E0133

Total time_______________

1. **Using an instructor-assigned single- and multiple-disc clutches, explain the relationship of the clutch components to each other.**

   a. **Single-disc clutch.**
   b. **Multiple-disc clutch.**

2. **Using an instructor-assigned single- and multiple-disc clutches, explain their roles in the transfer of power.**

   a. **Single-disc clutch.**
   b. **Multiple-disc clutch.**

3. **Instructor Comments:**

   ____
   ____
   ____

Performance Rating

CDX Tasksheet Number: E0133

| 0 | 1 | 2 | 3 | 4 |

Supervisor/instructor signature ____ Date _____________

Student/Intern information:

Name ___________________________ Date ___________ Class _______________________

Vehicle used for this activity:

Year _______________ Make _______________________ Model ___________________

Odometer ____________ Hour meter ____________ VIN _________________________

▶ TASK Describe the relationship of the number of discs, types of discs (wet or dry), and type of clutch material to the transfer of torque and horsepower to the ground.

AED 4.1

Time off_____________

Time on_____________

Total time_____________

CDX Tasksheet Number: EO134

1. **Using an instructor-assigned wet or dry clutch discs, identify and describe the types of material used in different applications allowing for maximum torque transfer.**

   ____

   ____

   ____

2. **Instructor Comments:**

   ____

   ____

   ____

Performance Rating

CDX Tasksheet Number: EO134

0	1	2	3	4

Supervisor/instructor signature ___________________________________ Date ___________

Student/Intern information:

Name _________________________ Date _________ Class _________________

Vehicle used for this activity:

Year _____________ Make _____________________ Model _______________

Odometer ___________ Hour meter ___________ VIN _____________________

▶ TASK Explain the operation of magnetic clutches and name various applications.

AED
4.1

Time off___________

Time on___________

Total time___________

CDX Tasksheet Number: EO135

1. **Research the various magnetic clutch assemblies used in heavy equipment and identify their components.**

   a. **Identify various components that have been labeled and displayed by your instructor.**

   b. **List their function in the table below:**

Component	Function

2. **Given a magnetic clutch, explain its operation to your instructor.**

3. **Instructor Comments:**

   ____
   ____
   ____

4. **List the applications that magnetics clutches are used in on heavy equipment:**

   ____
   ____
   ____

5. **Instructor Comments:**

____

____

____

Theory and Principles of Electronic-Controlled Transmissions

Student/Intern information:

Name _________________________ Date _____________ Class _________________________

Vehicle used for this activity:

Year _______________ Make _______________________ Model _________________________

Odometer ____________ Hour meter ___________ VIN _________________________________

Learning Objective/Task	CDX Tasksheet Number	2014 Edition Rev2/12/16 AED Standard
• Exhibit knowledge of electronic control systems by identifying components used on a specific unit.	E0136	4.1
• Demonstrate understanding of a specific unit's operation by explaining the functions of all components and their relationships to one another.	E0137	4.1

Time off_______________

Time on_______________

Total time_______________

Materials Required

- Machines equipped with electronic-controlled transmissions systems
- Selection of various electronic-controlled transmissions system components
- Equipment manufacturer's workshop manual
- Basic tool set
- Manufacturer-specific tools depending on the concern
- Machine lifting equipment, if applicable
- Machine lockout/tagout equipment, if applicable

Some Safety Issues to Consider

- Activities may require test driving the equipment on the school grounds, which carries severe risks. Attempt this task only with full permission from your supervisor/instructor, and follow all the guidelines exactly.
- Lifting equipment such as equipment jacks and stands, hoists, and engine hoists are important tools that increase productivity and make the job easier. However, they can also cause severe injury or death if used improperly. Make sure you follow the manufacturer's operation procedures. Also make sure you have your supervisor/instructor's permission to use any particular type of lifting equipment.
- Comply with personal and environmental safety practices associated with clothing; eye protection; hand tools; power equipment; proper ventilation; and the handling, storage, and disposal of chemicals/materials in accordance with federal, state, and local regulations.

- Always wear the correct protective eyewear and clothing and use the appropriate safety equipment, as well as fender covers, seat protectors, and floor mat protectors.
- Make sure you understand and observe all legislative and personal safety procedures when carrying out practical assignments. If you are unsure of what these are, ask your supervisor/instructor.

Student/Intern information:

Name ________________________________ Date ____________ Class ________________________

Vehicle used for this activity:

Year ______________ Make ________________________ Model ________________________

Odometer ____________ Hour meter ____________ VIN ________________________

▶ TASK Exhibit knowledge of electronic control systems by identifying components used on a specific unit.

AED 4.1

Time off ________________

Time on ________________

Total time ________________

CDX Tasksheet Number: EO136

1. **Research electronic over hydraulic control transmission systems used in heavy equipment and identify their components.**

   a. **Identify various components that have been labeled and displayed by your instructor.**

Component Name	Label Number
Transmission Control Unit (TCU)	
Engine Control Unit (ECU)	
Accelerator Position Sensor	
Transmission Input Speed Sensor	
Transmission Turbine Speed Sensor	
Transmission Output Speed Sensor	
Transmission Shift Control	
Transmission Control Module	
Transmission Pressure Switch	
Transmission Temperature Sensor	
Transmission Oil Level Sensor	
Transmission Solenoid Valves	
Transmission Retarder Control	
Diagnostic Link Connector	

Component Name	Label Number
Transmission Control Unit (TCU)	
Engine Control Unit (ECU)	
Accelerator Position Sensor	
Transmission Input Speed Sensor	
Transmission Turbine Speed Sensor	

Transmission Output Speed Sensor	
Transmission Shift Control	
Transmission Control Module	
Transmission Pressure Switch	
Transmission Temperature Sensor	
Transmission Oil Level Sensor	
Transmission Solenoid Valves	
Transmission Retarder Control	
Diagnostic Link Connector	

    b. Review with instructor.

2. **Research electronic over air control transmission systems used in heavy equipment and identify their components.**

    a. Identify various components that have been labeled and displayed by your instructor.

Component Name	Label Number
Transmission Control Unit (TCU)	
Engine Control Unit (ECU)	
Transmission Input Speed Sensor	
Transmission Output Speed Sensor	
Driver (operator) Interface (electronic shifter)	
Start Enable Relay	
MEIIR (momentary engine ignition interrupt relay)	
Shift Motors	
Shift Position Sensors	
Electronically Controlled Range Valve Assembly	
Electronically Controlled Splitter Valve Assembly	
Inertia Brake	

3. **Discuss your findings with the instructor.**

____
____
____

Performance Rating   **CDX Tasksheet Number: E0136**

  0   1   2   3   4

Supervisor/instructor signature ________________________________ Date ______________

▶ **TASK** Demonstrate understanding of a specific unit's operation by explaining the functions of all components and their relationships to one another.

AED
4.1

Time off____________

Time on____________

Total time____________

CDX Tasksheet Number: EO137

1. **Using an instructor-assigned machine or components, explain the function of the electronic over hydraulic control transmission system components and their operational relationship to system operation.**

Component Name	Function and Operational Relationship to System
Transmission Control Unit (TCU)	
Engine Control Unit (ECU)	
Accelerator Position Sensor	
Transmission Input Speed Sensor	
Transmission Turbine Speed Sensor	
Transmission Output Speed Sensor	
Transmission Shift Control	
Transmission Control Module	
Transmission Pressure Switch	
Transmission Temperature Sensor	
Transmission Oil Level Sensor	
Transmission Solenoid Valves	
Transmission Retarder Control	
Diagnostic Link Connector	

2. **Instructor Comments:**

____

____

____

3. **Using an instructor-assigned machine or components, explain the function of the electronic over air control transmission systems components and their operational relationship to system operation.**

Component Name	Function and Operational Relationship to System
Transmission Control Unit (TCU)	
Engine Control Unit (ECU)	
Transmission Input Speed Sensor	
Transmission Output Speed Sensor	
Driver (operator) Interface (electronic shifter)	
Start Enable Relay	
MEIIR (momentary engine ignition interrupt relay)	
Shift Motors	
Shift Position Sensors	
Electronically Controlled Range Valve Assembly	
Electronically Controlled Splitter Valve Assembly	
Inertia Brake	

4. **Instructor Comments:**

____

____

____

Performance Rating

CDX Tasksheet Number: E0137

0	**1**	**2**	**3**	**4**

Supervisor/instructor signature _________________________________ Date _____________

Theory and Principles of Hydrostatic Transmissions

Student/Intern information:

Name _________________________________ Date ____________ Class _________________________

Vehicle used for this activity:

Year _______________ Make _______________________________ Model _____________________________

Odometer ____________ Hour meter ____________ VIN _____________________________________

Learning Objective/Task	CDX Tasksheet Number	2014 Edition Rev2/12/16 AED Standard
• Demonstrate understanding of theory and principles of hydrostatic systems by explaining, in writing, how a basic hydrostatic system functions.	E0138	4.1
• Exhibit knowledge of hydrostatic transmission operation by explaining the flow of fluids through the charge circuit, pump, motor, control and loop circuits.	E0139	4.1
• Explain the differences between fixed and variable pumps and motors, and the effects of their various combinations.	E0140	4.1

Time off_______________

Time on_______________

Total time_______________

Materials Required

- Machines equipped with hydrostatic transmissions systems
- Various hydrostatic transmission components
- Equipment manufacturer's workshop manual
- Basic tool set
- Manufacturer-specific tools depending on the concern
- Machine lifting equipment, if applicable
- Machine lockout/tagout equipment, if applicable

Some Safety Issues to Consider

- Activities may require test driving the equipment on the school grounds, which carries severe risks. Attempt this task only with full permission from your supervisor/instructor, and follow all the guidelines exactly.
- Lifting equipment such as equipment jacks and stands, hoists, and engine hoists are important tools that increase productivity and make the job easier. However, they can also cause severe injury or death if used improperly. Make sure you follow the manufacturer's operation procedures. Also make sure you have your supervisor/instructor's permission to use any particular type of lifting equipment.
- Comply with personal and environmental safety practices associated with clothing; eye protection; hand tools; power equipment; proper ventilation; and the handling, storage, and disposal of chemicals/materials in accordance with federal, state, and local regulations.

- Always wear the correct protective eyewear and clothing and use the appropriate safety equipment, as well as fender covers, seat protectors, and floor mat protectors.
- Make sure you understand and observe all legislative and personal safety procedures when carrying out practical assignments. If you are unsure of what these are, ask your supervisor/instructor.

Student/Intern information:

Name _________________________________ Date ____________ Class _____________________

Vehicle used for this activity:

Year _______________ Make _________________________ Model _____________________

Odometer ____________ Hour meter ___________ VIN _________________________________

▶ TASK Demonstrate understanding of theory and principles of hydrostatic systems by explaining, in writing, how a basic hydrostatic system functions.

AED 4.1

Time off_________________

Time on_________________

Total time_______________

CDX Tasksheet Number: EO138

1. **Research the theory and principles of hydrostatic systems used in heavy equipment and identify their components.**

    a. **Identify various components that have been labeled and displayed by your instructor.**

    b. **List their function in the operation of the hydrostatic system in the table below:**

Component	Function
Open-loop hydrostatic components	
1.	
2.	
3.	
4.	
5.	
6.	

7.	
8.	
Closed-loop hydrostatic components	
9.	
10.	
11.	
12.	
13.	
14.	
15.	

2. **Answer the questions about hydrostatic systems below:**

   a. **In your own words describe the major differences between open- and closed-loop hydrostatic systems.**

   ____

   ____

   ____

   b. **What is the purpose of the charge circuit in a hydrostatic system?**

   ____

   ____

   ____

c. What is the purpose of the lubrication circuit in a hydrostatic system?

____

____

____

d. Describe the purpose of the charge circuit in a hydrostatic system.

____

____

____

3. Discuss your findings with the instructor.

____

____

____

Performance Rating

CDX Tasksheet Number: E0138

0	1	2	3	4

Supervisor/instructor signature _______________________________________ Date _______________

Student/Intern information:

Name _________________________ Date ___________ Class _____________________

Vehicle used for this activity:

Year _____________ Make _____________________ Model _____________________

Odometer ___________ Hour meter ___________ VIN _____________________

▶ TASK Exhibit knowledge of hydrostatic transmission operation by explaining the flow of fluids through the charge circuit, pump, motor, control and loop circuits.

AED 4.1

Time off_____________

Time on_____________

Total time_____________

CDX Tasksheet Number: EO139

1. **Locate the components, operation, and power flow through a hydrostatic system on an instructor-assigned machine. Use machine manufacturer's service manual and the correct hydraulic schematic.**

   a. **Locate the components that make up the machines circuits on the hydraulic schematic, and list below:**

   ____
   ____
   ____
   ____
   ____
   ____

   b. **Using the schematic, trace the flow of fluid through the circuit as various functions are used.**

   i. **Function _____________**

   a. **Flow:**

   ____
   ____
   ____
   ____

   ii. **Function _____________**

   a. **Flow:**

   ____
   ____
   ____
   ____

   iii. **Function _____________**

   a. **Flow:**

   ____
   ____
   ____
   ____

2. **Using the schematic and service information, locate the circuit component
    locations on the machine.**

3. **Instructor Comments:**

____

____

____

CDX Tasksheet Number: EO139

| 0 | 1 | 2 | 3 | 4 |

Supervisor/instructor signature ____ Date ______________

Student/Intern information:

Name _________________________________ Date _____________ Class _____________________

Vehicle used for this activity:

Year _______________ Make _________________________ Model _______________________

Odometer ____________ Hour meter ____________ VIN _________________________________

▶ TASK Explain the differences between fixed and variable pumps and motors, and the effects of their various combinations.

AED 4.1

Time off_______________

Time on_______________

Total time_______________

CDX Tasksheet Number: E0140

1. **Explain the various combinations of pumps and motors, and their effects when used in hydrostatic transmission applications.**

   a. **Explain fixed-fixed combinations of pumps and motors and their effects.**

   ____

   ____

   ____

   ____

   b. **Explain variable-fixed combinations of pumps and motors and their effects.**

   ____

   ____

   ____

   ____

   c. **Explain fixed-variable combinations of pumps and motors and their effects.**

   ____

   ____

   ____

   ____

   d. **Explain variable-variable combinations of pumps and motors and their effects.**

   ____

   ____

   ____

   ____

2. **Instructor Comments:**

____

____

____

Performance Rating   **CDX Tasksheet Number: EO140**

| 0 | 1 | 2 | 3 | 4 |

Supervisor/instructor signature __________________________________ Date ____________

Driveshaft Function and Construction

Student/Intern information:

Name _________________________________ Date _____________ Class _________________________

Vehicle used for this activity:

Year _______________ Make _________________________________ Model _________________________

Odometer _____________ Hour meter _____________ VIN _________________________________

Learning Objective/Task	CDX Tasksheet Number	2014 Edition Rev2/12/16 AED Standard
• Demonstrate knowledge of driveshafts by recognizing components, realizing the effects of driveline angle, and studying why driveline failures occur.	EO141	4.2

Time off_______________

Time on_______________

Total time_______________

Materials Required

- Machines equipped with driveshaft systems
- Various driveshaft components
- Equipment manufacturer's workshop manual
- Basic tool set
- Manufacturer-specific tools depending on the concern
- Machine lifting equipment, if applicable
- Machine lockout/tagout equipment, if applicable

Some Safety Issues to Consider

- Activities may require test driving the equipment on the school grounds, which carries severe risks. Attempt this task only with full permission from your supervisor/instructor, and follow all the guidelines exactly.
- Lifting equipment such as equipment jacks and stands, hoists, and engine hoists are important tools that increase productivity and make the job easier. However, they can also cause severe injury or death if used improperly. Make sure you follow the manufacturer's operation procedures. Also make sure you have your supervisor/instructor's permission to use any particular type of lifting equipment.
- Comply with personal and environmental safety practices associated with clothing; eye protection; hand tools; power equipment; proper ventilation; and the handling, storage, and disposal of chemicals/materials in accordance with federal, state, and local regulations.
- Always wear the correct protective eyewear and clothing and use the appropriate safety equipment, as well as fender covers, seat protectors, and floor mat protectors.
- Make sure you understand and observe all legislative and personal safety procedures when carrying out practical assignments. If you are unsure of what these are, ask your supervisor/instructor.

Name _________________________________ Date _____________ Class _________________________________

Vehicle used for this activity:

Year _______________ Make _______________________ Model _____________________

Odometer _____________ Hour meter _____________ VIN _________________________________

▶ TASK Demonstrate knowledge of driveshafts by recognizing components, realizing the effects of driveline angle, and studying why driveline failures occur.

AED
4.2

Time off____________

Time on____________

Total time____________

CDX Tasksheet Number: EO141

1. **Research the theory and principles of driveshaft systems used in heavy equipment and identify their components.**

    a. **Identify various components that have been labeled and displayed by your instructor.**

    b. **List their function in the operation of the driveshaft system in the table below:**

Component	Function

2. Research the procedure and specifications for inspecting, measuring, and adjusting the driveline balance, phasing, and angles in the appropriate service information.

    a. List any special tools required to perform this task:

    b. List any specific precautions to be taken when performing this task:

c. List any specifications for balance, phasing, runout, and driveline angles:

____

____

____

d. List or print off and attach to this sheet the procedure for inspecting, measuring, and adjusting the driveline balance, phasing, and angles:

e. Check the drive shaft for phasing. List your observation(s):

____

____

____

f. Measure drive shaft runout: _______________ in/mm

g. Measure driveline angles.

    i. Front: _______________ degrees

    ii. Center (if specified): _______________ degrees

    iii. Rear: _______________ degrees

h. Check the drive shaft for balance. List your observation(s):

____

____

____

3. Discuss your findings with the instructor.

____

____

____

4. Research the common types of failures in driveshaft systems and identify various failures using instructor assigned parts.

Failed Component	Failure Type
1.	
2.	
3.	
4.	
5.	

6.	
7.	
8.	
9.	
10.	
11.	
12.	
13.	
14.	
15.	

Performance Rating

CDX Tasksheet Number: EO141

0	1	2	3	4

Supervisor/instructor signature _________________________________ Date _____________

Theory and Principles of Differentials

Student/Intern information:

Name _______________________________ Date ____________ Class _________________________

Vehicle used for this activity:

Year ______________ Make ________________________________ Model ____________________________

Odometer ____________ Hour meter ____________ VIN ___________________________________

<table>
<tr><td rowspan="2">Learning Objective/Task</td><td>CDX
Tasksheet
Number</td><td>2014
Edition
Rev2/12/16
AED
Standard</td></tr>
<tr></tr>
<tr><td>• Exhibit understanding of basic differential operation by identifying the components and explaining how pinion, ring, and bevel gears operate in relationship to each other.</td><td>E0142</td><td>4.2</td></tr>
<tr><td>• Identify each type of differential locking device and explain in detail how each one operates.</td><td>E0143</td><td>4.2</td></tr>
<tr><td>• Given a specific component and proper manuals/information, perform all adjustments on a differential with a new ring and pinion, and also perform all adjustments with original ring and pinion but with new bearings.</td><td>E0144</td><td>4.2</td></tr>
<tr><td>• Identify the most common root causes of failure with differentials.</td><td>E0145</td><td>4.2</td></tr>
</table>

Time off__________

Time on__________

Total time__________

Materials Required

- Machines equipped with differential systems
- Various differential components
- Equipment manufacturer's workshop manual
- Basic tool set
- Manufacturer-specific tools depending on the concern
- Machine lifting equipment, if applicable
- Machine lockout/tagout equipment, if applicable

Some Safety Issues to Consider

- Activities may require test driving the equipment on the school grounds, which carries severe risks. Attempt this task only with full permission from your supervisor/instructor, and follow all the guidelines exactly.
- Lifting equipment such as equipment jacks and stands, hoists, and engine hoists are important tools that increase productivity and make the job easier. However, they can also cause severe injury or death if used improperly. Make sure you follow the manufacturer's operation procedures. Also make sure you have your supervisor/instructor's permission to use any particular type of lifting equipment.

- Comply with personal and environmental safety practices associated with clothing; eye protection; hand tools; power equipment; proper ventilation; and the handling, storage, and disposal of chemicals/materials in accordance with federal, state, and local regulations.
- Always wear the correct protective eyewear and clothing and use the appropriate safety equipment, as well as fender covers, seat protectors, and floor mat protectors.
- Make sure you understand and observe all legislative and personal safety procedures when carrying out practical assignments. If you are unsure of what these are, ask your supervisor/instructor.

▶ TASK Exhibit understanding of basic differential operation by identifying the components and explaining how pinion, ring, and bevel gears operate in relationship to each other.

AED 4.2

Time off_________________

Time on_________________

Total time_________________

CDX Tasksheet Number: EO142

1. **Research the theory and differential systems used in heavy equipment and identify their components.**

    a. **Identify various components that have been labeled and displayed by your instructor.**

    b. **List their function in the operation of the differential system in the table below:**

Component	Function
1.	
2.	
3.	
4.	
5.	
6.	
7.	

8.	
9.	
10.	
11.	
12.	
13.	
14.	
15.	

2. Given a differential cutaway, describe how power flows through the differential starting with the input shaft and ending at the axle shafts.

    a. Power flow driving straight:

____
____
____
____
____
____
____

    b. Power flow driving through a turn:

____
____
____
____
____
____
____

3. **Given a differential with power divider cutaway, describe how power flows through the differential and power divider starting with the input shaft and ending at the axle shafts and through shaft.**

    a. **Power flow driving straight:**

    b. **Power flow driving through a turn:**

    c. **Power flow driving front and rear axle at different speeds:**

4. **Discuss your findings with the instructor.**

Performance Rating

CDX Tasksheet Number: E0142

| 0 | 1 | 2 | 3 | 4 |

Supervisor/instructor signature ____ Date ______________

Student/Intern information:

Name _____________________________ Date _____________ Class _____________________

Vehicle used for this activity:

Year _____________ Make _____________________ Model _____________________

Odometer _____________ Hour meter _____________ VIN _____________________

▶ TASK Identify each type of differential locking device and explain in detail how each one operates.

AED 4.2

Time off_____________

Time on_____________

Total time_____________

CDX Tasksheet Number: E0143

1. **Research locking differential systems used in heavy equipment and identify their components and operation.**

   a. **Identify various components that have been labeled and displayed by your instructor.**

   b. **List their function in the operation of the locking differential system in the table below:**

Component	Function
Mechanical locking differentials	
1.	
2.	
3.	
4.	
5.	
6.	

7.	
8.	
Hydraulic locking differentials	
9.	
10.	
11.	
12.	
13.	
14.	
15.	
16.	
17.	
18.	

Automatic no-spin locking differentials	
19.	
20.	
21.	
22.	
23.	
24.	
25.	

2. Given a locking differential cutaway, describe how power flows through the differential starting with the input shaft and ending at the axle shafts.

    a. Power flow driving straight:

    b. Power flow driving through a turn:

   c. Power flow driving on split-coefficient surface:

____
____
____
____
____
____
____
____

3. **Discuss your findings with the instructor.**

____
____
____

Name _______________________________ Date ____________ Class _______________________________

Vehicle used for this activity:

Year _______________ Make _______________________________ Model _______________________________

Odometer _____________ Hour meter _____________ VIN _______________________________

▶ TASK Given a specific component and proper manuals/information, perform all adjustments on a differential with a new ring and pinion, and also perform all adjustments with original ring and pinion but with new bearings.

AED 4.2

Time off________________

Time on________________

Total time________________

CDX Tasksheet Number: EO144

1. **Research the procedure and specifications for disassembling, inspecting, measuring, and replacing the differential gears and associated parts.**

   a. **List any special tools required to perform this task:**

   ____
   ____
   ____

   b. **List any specific precautions when performing this task:**

   ____
   ____
   ____

   c. **List any specifications for this task:**

   ____
   ____
   ____

   d. **List or print off and attach to this sheet the procedure for performing the task.**

2. **Following the specified procedure, disassemble the differential pinion gears and associated parts.**

3. **Inspect and measure the following parts. List your observations for each:**

   a. **Drive pinion:**

   ____
   ____
   ____

   b. **Ring gear:**

   ____
   ____
   ____

   c. **Spacers and sleeves:**

   ____
   ____
   ____

d. Bearings:

____

____

____

e. Differential pinion gears:

____

____

____

f. Shafts:

____

____

____

g. Differential side gears:

____

____

____

h. Side bearings:

____

____

____

i. Thrust washers:

____

____

____

j. Case:

____

____

____

4. Follow manufacturer's procedures to reassemble differential assembly.

   a. Demonstrate how to set bearing preload of ring gear and pinion assemblies.

____

____

____

   b. Demonstrate how to set backlash of ring gear and pinion assemblies.

____

____

____

   c. Demonstrate how to check tooth contact pattern of ring gear and pinion assemblies.

____

____

____

____

____

____

▶ **TASK** Identify the most common root causes of failure with differentials.

AED
4.2

Time off________________

Time on________________

CDX Tasksheet Number: EO145

1. **Research the common types of failures of differential systems and identify various failures using instructor-assigned parts.**

Total time________________

Failed Component	Failure Type
1.	
2.	
3.	
4.	
5.	
6.	
7.	
8.	

9.	
10.	
11.	
12.	
13.	
14.	
15.	

Theory and Principles of Final Drives

Name _________________________ Date __________ Class _______________________

Vehicle used for this activity:

Year _____________ Make _______________________ Model _______________________

Odometer __________ Hour meter __________ VIN _______________________________

Learning Objective/Task	CDX Tasksheet Number	2014 Edition Rev2/12/16 AED Standard
• Exhibit knowledge of final drives by identifying the different types, and the components that make up final drives.	E0146	4.2
• Perform adjustments according to OEM standards.	E0147	4.2

Time off____________

Time on____________

Total time____________

Materials Required

- Machines equipped with final drive systems
- Various final drive components
- Equipment manufacturer's workshop manual
- Basic tool set
- Manufacturer-specific tools depending on the concern
- Machine lifting equipment, if applicable
- Machine lockout/tagout equipment, if applicable

Some Safety Issues to Consider

- Activities may require test driving the equipment on the school grounds, which carries severe risks. Attempt this task only with full permission from your supervisor/instructor, and follow all the guidelines exactly.
- Lifting equipment such as equipment jacks and stands, hoists, and engine hoists are important tools that increase productivity and make the job easier. However, they can also cause severe injury or death if used improperly. Make sure you follow the manufacturer's operation procedures. Also make sure you have your supervisor/instructor's permission to use any particular type of lifting equipment.
- Comply with personal and environmental safety practices associated with clothing; eye protection; hand tools; power equipment; proper ventilation; and the handling, storage, and disposal of chemicals/materials in accordance with federal, state, and local regulations.
- Always wear the correct protective eyewear and clothing and use the appropriate safety equipment, as well as fender covers, seat protectors, and floor mat protectors.
- Make sure you understand and observe all legislative and personal safety procedures when carrying out practical assignments. If you are unsure of what these are, ask your supervisor/instructor.

Student/Intern information:

Name ___________________________ Date ___________ Class ___________________________

Vehicle used for this activity:

Year _______________ Make _______________________ Model ___________________________

Odometer _____________ Hour meter _____________ VIN ___________________________

▶ TASK Exhibit knowledge of final drives by identifying the different types, and the components that make up final drives.

AED 4.2

Time off___________

Time on___________

Total time___________

CDX Tasksheet Number: E0146

1. **Research the theory and function of final drive systems used in heavy equipment and identify their components.**

   a. **Identify various components that have been labeled and displayed by your instructor.**

   b. **List their function in the operation of the final drive system in the table below:**

Component	Function
Rigid axle: Full-floating	

Rigid axle: Semi-floating	
Flexible axle shaft	

Pinion/bull gear	
Inboard planetary	

Outboard planetary	
Double reduction planetary	

2. **Given a final drive cutaway, describe how power flows through the final drive starting with the input shaft and ending at the axle shafts.**

____

____

____

____

____
____
____
____
____
____
____
____
____
____
____

3. Discuss your findings with the instructor.

____
____
____

Performance Rating

CDX Tasksheet Number: E0146

0	1	2	3	4

Supervisor/instructor signature _______________________________________ Date _______________

Student/Intern information:

Name ______________________________ Date ____________ Class ____________________________

Vehicle used for this activity:

Year _______________ Make ____________________________ Model ____________________________

Odometer _____________ Hour meter _____________ VIN ____________________________

▶ **TASK** Perform adjustments according to OEM standards.   **AED 4.2**

CDX Tasksheet Number: EO147

Time off______________

Time on______________

Total time______________

1. **Ask your instructor to assign you a machine equipped with a final drive.**

2. **Research the procedure and specifications for performing adjustments on final drives and associated parts.**

    a. **List any special tools required to perform this task:**

    ____
    ____
    ____

    b. **List any specific precautions to be taken when performing this task:**

    ____
    ____
    ____

    c. **List any specifications for this task:**

    ____
    ____
    ____

    d. **List or print off and attach to this sheet the procedure for performing the task.**

3. **Following the specified procedure, perform adjustments required.**

4. **Discuss your findings with the instructor.**

    ____
    ____
    ____

Performance Rating   **CDX Tasksheet Number: EO147**

0	1	2	3	4

Supervisor/instructor signature ____________________________________ Date ____________

Fundamental Theory of Hydraulic and Pneumatic Braking Systems

Student/Intern information:

Name _______________________________ Date _____________ Class _______________________

Vehicle used for this activity:

Year _______________ Make _______________________________ Model _______________________

Odometer _____________ Hour meter _____________ VIN _______________________________

Learning Objective/Task	CDX Tasksheet Number	2014 Edition Rev2/12/16 AED Standard
• Fundamental theory, adjustments, and repair of hydraulic braking systems used primarily in mobile construction equipment.	E0148	4.3
• Fundamental theory, adjustments, and repair of pneumatic braking systems used primarily in mobile construction equipment.	E0149	4.3

Time off_______________

Time on_______________

Total time_______________

Materials Required

- Machines equipped with hydraulic brake systems
- Machines equipped with pneumatic brake systems
- Various final hydraulic brake components
- Various pneumatic brake components
- Equipment manufacturer's workshop manual
- Basic tool set
- Manufacturer-specific tools depending on the concern
- Machine lifting equipment, if applicable
- Machine lockout/tagout equipment, if applicable

Some Safety Issues to Consider

- Activities may require test driving the equipment on the school grounds, which carries severe risks. Attempt this task only with full permission from your supervisor/instructor, and follow all the guidelines exactly.
- Lifting equipment such as equipment jacks and stands, hoists, and engine hoists are important tools that increase productivity and make the job easier. However, they can also cause severe injury or death if used improperly. Make sure you follow the manufacturer's operation procedures. Also make sure you have your supervisor/instructor's permission to use any particular type of lifting equipment.
- Comply with personal and environmental safety practices associated with clothing; eye protection; hand tools; power equipment; proper ventilation; and the handling, storage, and disposal of chemicals/materials in accordance with federal, state, and local regulations.

- Always wear the correct protective eyewear and clothing and use the appropriate safety equipment, as well as fender covers, seat protectors, and floor mat protectors.
- Make sure you understand and observe all legislative and personal safety procedures when carrying out practical assignments. If you are unsure of what these are, ask your supervisor/instructor.

Student/Intern information:

Name ___________________________ Date ___________ Class ___________________

Vehicle used for this activity:

Year ______________ Make _______________________ Model ___________________

Odometer ____________ Hour meter ___________ VIN ___________________________

▶ TASK Fundamental theory, adjustments, and repair of hydraulic braking systems used primarily in mobile construction equipment.

AED 4.3

Time off___________

Time on___________

Total time___________

CDX Tasksheet Number: E0148

1. **Research the theory of hydraulic brake systems used in heavy equipment and identify their components.**

   a. **Identify various components that have been labeled and displayed by your instructor.**

   b. **List their function in the operation of the hydraulic brake system in the table below:**

Component	Function

2. **Inspect and test master cylinder for internal/external leaks and damage.**

    a. **Check the brake master cylinder fluid for correct quantity and test its suitability.**

        i. Locate the brake master cylinder.

        ii. Clean the top of the cylinder.

        iii. Open the master cylinder cap and check the fluid level.

        At the specified level: _____________ Requires top off: _____________

    b. **Collect a sample of the current brake fluid.**

        i. Test the sample for contamination. Satisfactory: _____________ Requires changing: _____________

    c. **Refill with correct fluid type.**

        i. Fluid type: _____________

        ii. Recommendations:

        ____

        ____

        ____

3. **Referencing the manufacturer's workshop manual, list the recommended steps to check for internal and external leakage of the master cylinder.**

____

____

____

4. **Discuss your findings with the instructor.**

____

____

____

5. **Inspect hydraulic system brake lines, flexible hoses, and fittings for leaks and damage.**

    a. **Inspect the brake lines, fittings, flexible hoses, and valves for leaks and damage.**

        i. Inspect the condition and security of all brake lines, both solid and flexible.

        Meets manufacturer's specifications:

        Yes: _____________ Requires maintenance: _____________

        ii. If they require maintenance, list the areas and recommendations

        ____

        ____

        ____

        ____

    b. Inspect the brake valves for signs of external leakage.

      i. Meets manufacturer's specifications:

        Yes: _____________   Requires maintenance: _____________

      ii. If they require maintenance, list the areas and recommendations.

        ____

        ____

        ____

        ____

6. Discuss your findings with the instructor.

    ____

    ____

    ____

7. Inspect and test metering (hold-off).

    a. Following the procedures outlined in the manufacturer's workshop manual, inspect and test metering (hold-off).

      i. Meets manufacturer's specifications:

        Yes: _____________   Requires maintenance: _____________

      ii. If they require maintenance, list the areas and recommendations.

        ____

        ____

        ____

        ____

8. Inspect and test load sensing/proportioning.

    a. Following the procedures outlined in the manufacturer's workshop manual, inspect and test load sensing/proportioning.

      i. Meets manufacturer's specifications:

        Yes: _____________   Requires maintenance: _____________

      ii. If they require maintenance, list the areas and recommendations.

        ____

        ____

        ____

        ____

9. Inspect and test proportioning and combination valves.

    a. Following the procedures outlined in the manufacturer's workshop manual, inspect and test proportioning and combination valves.

      i. Meets manufacturer's specifications:

        Yes: _____________   Requires maintenance: _____________

    ii. **If they require maintenance, list the areas and recommendations.**

10. **Check brake master cylinder fluid for correct quantity and test suitability.**

    a. **Clean top of master cylinder.**

    b. **Check fluid level of master cylinder.**

    c. **Following the procedures outlined in the manufacturer's workshop manual, drain the old brake fluid and flush the braking system.**

    d. **Following the procedures outlined in the manufacturer's workshop manual, refill braking circuit master cylinder.**

    e. **Determine proper fluid type and quantity:**

        i. **Type** _______________________________

        ii. **Quantity** _______________________________

    f. **Fill to capacity and follow manufacturer's procedure to bleed brake system.**

11. **Follow manufacturer's procedures to perform a foundation brake job.**

    a. **Remove all components required.**

    b. **Follow manufacturers cleaning procedures.**

    c. **Inspect all components and list failed parts.**

    d. **Reassemble according to manufacturer's recommendations.**

    e. **Adjust brakes according to manufacturer's recommendations.**

    f. **Instructor's Comments:**

12. **Referencing manufacturer's service information, list the procedure for inspecting and checking parking brake operation, inspecting parking brake application and holding device, and note all safety precautions listed by the manufacturer.**

13. **Following the procedures outlined in the manufacturer's workshop manual, check parking brake operation and inspect parking brake application and holding device.**

____
____
____
____
____

14. **Discuss your findings with the instructor.**

____
____
____

Performance Rating

CDX Tasksheet Number: EO148

0	1	2	3	4

Supervisor/instructor signature _________________________________ Date __________

▶ TASK Fundamental theory, adjustments, and repair of pneumatic braking systems used primarily in mobile construction equipment.

AED 4.3

Time off___________

Time on___________

Total time___________

CDX Tasksheet Number: EO149

1. **Research the theory of pneumatic brake systems used in heavy equipment and identify their components.**

   a. **Identify various components that have been labeled and displayed by your instructor.**

   b. **List their function in the operation of the pneumatic brake system in the table below:**

Component	Function

2. **Check air system build-up time; determine needed action.**

> **NOTE:** The air compressor is responsible for air pressure buildup. If the system itself is leak free, any delay in buildup may be a direct result of worn internal compressor parts.

3. **Check the time for air pressure to complete build up.**

    a. **Reference the manufacturer's workshop manual for the correct procedure for checking the air pressure build timing.**

    b. **Block and/or hold the machine by means other than air brakes during these tests.**

    c. **Drain the air reservoirs; ensure that a container(s) is placed under the drain cocks to collect any contaminants, then close the drain cocks.**

    d. **Pressure gauge reading: _______________ psi (should be 0 psi)**

    e. **Start time of air build: _______________ minutes _______________ seconds**

    f. **Start the engine and build up system pressure.**

    g. **Check the pressure in the buildup system.**

        i. **Record the pressure when the "low-pressure warning device" goes out:**

        Pressure gauge reading: _______________ psi

        ii. **Time to build up to shut off low-pressure warning device:**

        Minute's _______________ seconds _______________

    h. **Check the pressure in the buildup system.**

        i. **Record the pressure when the "governor cut-out" pressure is reached.**

        Pressure gauge reading: _______________ psi

        ii. **Time to build up to shut off low-pressure warning device:**

        Minutes _______________ seconds _______________

    i. **Record the total time from the start of pumping to reach the governor cut-out pressure:**

        Minutes _______________ seconds _______________

    j. **Did the air pressure build-up time operate within specifications?**

        Yes: _______________ No: _______________

Recommendations:

____

____

____

4. Referencing the manufacturer's workshop manual, list the recommended steps to check for internal and external leakage of the pneumatic brake system.

____

____

____

5. Discuss your findings with the instructor.

____

____

____

6. Drain air reservoir/tanks; check for oil, water, and foreign material; determine needed action.

   • **Engine "OFF" checks**

7. Drain the air reservoir using an appropriate drain container to capture any discharged fluid.

   a. With the drain container under each air reservoir tank, open the drain cock and collect all waste fluid.

8. Check for any contamination:

   a. Is there any oil present in the reservoir residue?
   Yes: _______________ No: _______________

   b. Is there any water present in the reservoir residue?
   Yes: _______________ No: _______________

   c. Is there any foreign material present in the reservoir residue?
   Yes: _______________ No: _______________

9. Determine what action is required to rectify any faults within the system.

   a. List all the actions that are required to identify the source of the contamination:

   ____

   ____

   ____

   ____

   ____

   b. List all the actions that are required to rectify any identified fault(s):

   ____

   ____

   ____

   ____

   ____

10. Discuss your findings with the instructor.

____

____

____

11. **Inspect air compressor drive gear, belts, and coupling; adjust or replace as needed.**

    a. **Condition of drive gear:**

    Good: _______ **Needs adjustment:** _______ **Needs replacing:** _______

    b. **Condition of coupling (if applicable):**

    Good: _______ **Needs adjustment:** _______ **Needs replacing:** _______

    c. **Condition and tightness of belt:**

    Good: _______ **Needs adjustment:** _______ **Needs replacing:** _______

12. **List adjustment procedures as per manufacturer workbook manual.**

____

____

____

13. **Discuss your findings with the instructor.**

____

____

____

14. **Inspect the air compressor inlet connections and piping.**

    a. **Condition of inlet connectors:**

    Good: _______ **Needs replacing:** _______

    b. **Condition of air compressor piping from air compressor to reservoirs:**

    Good: _______ **Needs replacing:** _______

    c. **Inspect the air compressor coolant lines and connections.**

    Good: _______ **Needs replacing:** _______

    d. **Inspect the air compressor mounting bracket(s).**

    Good: _______ **Needs replacing:** _______

15. **Discuss your findings with the instructor.**

____

____

____

Engine "OFF" checks

16. **With reference to the workshop manual, inspect the governor and unloader valve for any signs of malfunction.**

    No visible signs of problems: _______ **Visible signs of problem(s):** _______

    If visible signs are present, list them:

____

____

____

17. **Reference the manufacturer's workshop manual and record the recommended "governor cut-out pressure":**

Specification: _________________ psi

Engine "ON" checks

18. **Testing the governor operation.**

    a. **Block and/or hold the machine by means other than air brakes during these tests.**

    b. **Drain the vehicle reservoirs. Ensure that a container(s) is placed under the drain cocks to collect any contaminants, then close the drain cocks.**

    c. **Start the engine and build up system pressure.**

    d. **As pressure in the system builds up, there should be no pressure reading on the test gauge. You will need to time this procedure.**

    e. **Record the air pressure when the compressor is cut out by the governor.**

       Air gauge reading when the governor cut out: _________________ psi/kpa

       Time taken: _________________ minutes

    f. **Are these measurements within the manufacturer's specifications:**
       Yes: _________________ No: _________________

19. **Discuss your findings with the instructor.**

    ____

    ____

    ____

20. **Carry out a visual inspection of the air compressor air filter, lines, hoses, and fittings.**

    a. **With reference to the workshop manual, inspect, clean, or replace the air cleaner filter. Filter required:**
       Yes: _________________ No: _________________

    b. **Cleaning and refitting required:**
       Yes: _________________ No: _________________

21. **With reference to the workshop manual, inspect the machine's air system lines and hoses.**

    Reusable: Yes: _________________ No: _________________

> NOTE: When working with the air system, always wear safety glasses as a rupture in the system could cause injury to the eyes.

22. **Build air system up to manufacturer specifications.**

    Manufacture specifications: _________________ psi

23. **Check for air leaks at all connections by using a soap and water solution in a spray bottle. Leaks Present: Yes: _______________ No: _______________**

24. **Test the air tank relief valve by setting the governor air pressure above pop off limit. This can also be done utilizing shop air. When pop off limit is reached, the valve should open up and vent the excess air pressure to prevent air system damage. Some relief valves have a ring or cable to pull to manually make the valve work.**

    Working Condition: Good: _______________ Bad: _______________

25. **Test the one-way check valve by removing it and blowing air in the direction of flow through the supply port, then reverse air pressure to the delivery port. Air should flow through the supply port but not through the delivery port in the reverse direction.**

    Working Condition: Good: _______________ Bad: _______________

26. **Manual and automatic drain valves are used for removing moisture from the system. Manual valves have a pull cord to manually activate the valve to manually remove the moisture. Automatic valves operate when the air dryer pops off when air pressure in the system is fully charged. Automatic valves can be tested by unseating the wire stem valve with a sideways motion utilizing a finger or a plastic probe.**

    Working Condition: Good: _______________ Bad: _______________

27. **Discuss your findings with the instructor.**

    ____

    ____

    ____

28. **Inspect and test air system valves following manufacturer's procedures.**

    a. **Using manufacturer's service information, identify the valves used in the air system.**

    b. **Locate each valve and inspect and test valve operation.**

Valve Name	Inspection Procedure	Test Procedure

29. **Follow manufacturer's procedures to perform a foundation brake job.**

    a. **Remove all components required.**

    b. **Follow manufacturers cleaning procedures.**

    c. **Inspect all components and list failed parts.**

    d. **Reassemble according to manufacturer's recommendations.**

    e. **Adjust brakes according to manufacturer's recommendations.**

30. **Instructor's Comments:**

31. **Referring manufacturer's service information, list the procedure for inspecting and checking parking brake operation, inspecting parking brake application and holding device, and note all safety precautions listed by the manufacturer.**

32. **Following the procedures outlined in the manufacturer's workshop manual, check parking brake operation and inspect parking brake application and holding device.**

____
____
____
____
____

33. **Discuss your findings with the instructor.**

____
____
____
____

Performance Rating

CDX Tasksheet Number: E0149

0	1	2	3	4

Supervisor/instructor signature _______________________________ Date _____________

Failure Analysis

Learning Objective/Task	CDX Tasksheet Number	2014 Edition Rev2/12/16 AED Standard
• Demonstrate the ability to follow reference information, test, and determine if unit is within specifications for a hydraulic/ hydrostatic trainer or equipment with a hydrostatic drive using service manuals/information/software; demonstrate the ability to follow a diagnostic troubleshooting chart for a specific system.	EO150	4.3

Time off____________

Time on____________

Total time____________

Materials Required

- Machines equipped with final drive systems
- Various final drive components
- Equipment manufacturer's workshop manual
- Basic tool set
- Manufacturer-specific tools depending on the concern
- Machine lifting equipment, if applicable
- Machine lockout/tagout equipment, if applicable

Some Safety Issues to Consider

- Activities may require test driving the equipment on the school grounds, which carry severe risks. Attempt this task only with full permission from your supervisor/instructor, and follow all the guidelines exactly.
- Lifting equipment such as equipment jacks and stands, hoists, and engine hoists are important tools that increase productivity and make the job easier. However, they can also cause severe injury or death if used improperly. Make sure you follow the manufacturer's operation procedures. Also make sure you have your supervisor/instructor's permission to use any particular type of lifting equipment.
- Comply with personal and environmental safety practices associated with clothing; eye protection; hand tools; power equipment; proper ventilation; and the handling, storage, and disposal of chemicals/materials in accordance with federal, state, and local regulations.

- Always wear the correct protective eyewear and clothing and use the appropriate safety equipment, as well as fender covers, seat protectors, and floor mat protectors.
- Make sure you understand and observe all legislative and personal safety procedures when carrying out practical assignments. If you are unsure of what these are, ask your supervisor/instructor.

Name _________________________________ Date ____________ Class _________________________

Vehicle used for this activity:

Year _______________ Make _______________________ Model _________________________

Odometer ____________ Hour meter ____________ VIN _________________________________

▶ **TASK** Demonstrate the ability to follow reference information, test, and determine if unit is within specifications for a hydraulic/hydrostatic trainer or equipment with a hydrostatic drive using service manuals/information/software; demonstrate the ability to follow a diagnostic troubleshooting chart for a specific system.

AED
4.3

Time off_____________

Time on_____________

Total time_____________

CDX Tasksheet Number: EO150

> **NOTE:** This tasksheet will require the use of a machine or simulator. Ask your instructor which machine or simulator you are to use.

1. **Using the machine or simulator service manual, test and determine if unit is within specifications; perform the troubleshooting procedures necessary to determine the root cause of system malfunction.**

   a. **Ask your instructor for the customer's complaint and interview the customer.**

      i. **Complaint:**

      ____

      ____

      ii. **Interview questions:**

      ____

      ____

   b. **Verify if the complaint exists.**

   c. **Using the manufacturer's service information, troubleshoot and diagnose the system malfunction. List the steps you took to determine the root cause:**

   ____
   ____
   ____
   ____
   ____
   ____
   ____
   ____
   ____
   ____
   ____

d. List the recommended repair needed to fix the problem:

____
____
____
____
____

2. Discuss your findings with the instructor.

Section A5:
Diesel Engines

CONTENTS

Maintenance Practices

Student/Intern information:

Name _________________________________ Date ___________ Class _________________________

Vehicle used for this activity:

Year _______________ Make _____________________________ Model _____________________

Odometer ____________ Hour meter ___________ VIN _____________________________________

Learning Objective/Task	CDX Tasksheet Number	2014 Edition Rev2/12/16 AED Standard
• Inspect fuel, oil, Diesel Exhaust Fluid (DEF), and coolant levels, and the condition.	E0151	5.4
• Identify engine fuel, oil, coolant, air, and other leaks; determine needed action.	E0152	5.4
• Listen for engine noises; determine needed action.	E0153	5.4
• Observe engine exhaust smoke color and quantity; determine needed action	E0154	5.4

Time off_______________

Time on_______________

Total time_______________

Materials Required

- Machines or simulators with possible engine concern
- Machine manufacturer's workshop manual
- Machine lifting equipment, if applicable
- Manufacturer-specific tools depending on the concern

Some Safety Issues to Consider

- Activities may require test driving the equipment on the school grounds, which carry severe risks. Attempt this task only with full permission from your supervisor/instructor, and follow all the guidelines exactly.
- All practices and procedures must be performed according to current mandates, standards, and regulations.
- Comply with personal and environmental safety practices associated with clothing; eye protection; hand tools; power equipment; proper ventilation; and the handling, storage, and disposal of chemicals/materials in accordance with federal, state, and local regulations.
- Always wear the correct protective eyewear and clothing and use the appropriate safety equipment, as well as fender covers, seat protectors, and floor mat protectors.
- Lifting equipment such as machine jacks and stands, machine hoists, and engine hoists are important tools that increase productivity and make the job easier. However, they can also cause severe injury or death if used improperly. Make sure you follow the manufacturer's operation procedures. Also make sure you have your supervisor/instructor's permission to use any particular type of lifting equipment.

Performance Standard

0—No exposure: No information or practice provided during the program; complete training required

1—Exposure only: General information provided with no practice time; close supervision needed; additional training required

2—Limited practice: Has practiced job during training program; additional training required to develop skill

3—Moderately skilled: Has performed job independently during training program; limited additional training may be required

4—Skilled: Can perform job independently with no additional training

Student/Intern information:

Name _________________________________ Date _____________ Class _____________________________

Vehicle used for this activity:

Year _______________ Make _________________________________ Model _____________________________

Odometer _____________ Hour meter _____________ VIN ____

▶ TASK Inspect fuel, oil, Diesel Exhaust Fluid (DEF), and coolant levels, and the condition.

AED 5.4

Time off_____________________

Time on_____________________

Total time_____________________

CDX Tasksheet Number: EO151

1. **Inspect fuel, oil, and coolant levels, and the condition.**

2. **Refer to the appropriate manufacturer's workshop manual.**

    a. **Identify the engine oil type and quantity.**

        i. **Recommended type of engine oil:** _____________________________

        ii. **Recommended engine oil quantity:** _____________________________

    b. **Identify the engine coolant type and quantity:**

        i. **Recommended type of engine coolant:** _____________________________

        ii. **Recommended engine coolant quantity:** _____________________________

3. **Using a clean glass container, drain off a sample of the diesel fuel from the fuel tank(s).**

    a. **Ensure there is no fuel spillage.**

    b. **Inspect the sample for any contaminations (e.g., water). Note: In case of poor running conditions or catastrophic failure, it may be necessary to send the sample out to an independent lab to check the fuel.**

        i. **Within manufacturer's specifications: Yes:** _________ **No:** _________

        ii. **If no, list the problems and your recommendation(s) for rectification:**

        ____

        ____

        ____

        ____

        ____

        ____

    c. **If directed by your instructor, arrange for the fuel sample to be tested in accordance with industry standards.**

4. **Check the engine oil level as outlined in the manufacturer's workshop manual.**

    a. **Within manufacturer's specifications: Yes:** _________ **No:** _________

b. If no, list the problems and your recommendation(s) for rectification:

    c. Using a clean glass container, drain off a sample of the diesel engine oil.

    d. If directed by your instructor, arrange for the engine oil sample to be tested in accordance with industry standards.

5. Using a clean glass container, drain off a sample of the engine coolant from the radiator.

    a. Ensure there is no coolant spillage.

    b. Check the coolant levels:

        i. Within manufacturer's specifications: Yes: _______ No: _______

        ii. If no, list the problems and your recommendation(s) for rectification:

    c. Inspect the sample for any contaminations (e.g., oil).

        i. Within manufacturer's specifications: Yes: _______ No: _______

        ii. If no, list the problems and your recommendation(s) for rectification:

    d. If directed by your instructor, arrange for the coolant sample to be tested in accordance with industry standards.

6. Discuss the findings with your instructor.

Performance Rating

CDX Tasksheet Number: EO151

| 0 | 1 | 2 | 3 | 4 |

Supervisor/instructor signature _______________________________________ Date ____________

▶ TASK Identify engine fuel, oil, coolant, air, and other leaks; determine needed action.

AED 5.4

Time off____________

Time on____________

Total time____________

CDX Tasksheet Number: EO152

1. **Refer to the manufacturer's workshop manual for the correct procedure for checking and identifying the main causes of engine fuel leaks.**

    a. **List all possible causes for engine fuel leaks:**

    ____
    ____
    ____
    ____
    ____
    ____

2. **Refer to the manufacturer's workshop manual for the correct procedure for checking and identifying the main causes of engine oil leaks.**

    a. **List all possible causes for engine oil leaks:**

    ____
    ____
    ____
    ____
    ____
    ____

3. **Refer to the manufacturer's workshop manual for the correct procedure for checking and identifying the main causes of engine coolant leaks.**

    a. **List all possible causes for engine coolant leaks:**

    ____
    ____
    ____
    ____
    ____
    ____

4. **Refer to the manufacturer's workshop manual for the correct procedure for checking and identifying the main causes of air and other leaks.**

    a. **List all possible causes of air intake and other air filter-related problems and leaks:**

    ____
    ____
    ____
    ____
    ____
    ____

5. Discuss the findings with your instructor.

Performance Rating

CDX Tasksheet Number: EO152

0	1	2	3	4

Supervisor/instructor signature ___________________________ Date ___________

Student/Intern information:

Name _________________________ Date ___________ Class _________________________

Vehicle used for this activity:

Year _______________ Make _____________________________ Model _____________________

Odometer _____________ Hour meter ___________ VIN _________________________________

▶ TASK Listen for engine noises; determine needed action.

AED 5.4

CDX Tasksheet Number: EO153

1. **Start the engine.**

2. **Check engine starting/operation for any unusual noises and vibrations.**

    a. **List any usual noises from the engine (if any):**

    ____

    ____

    ____

    ____

    ____

    b. **Are there any vibrations coming from the engine?**
       Yes: ________ No: ________

        i. **If yes, record where they are coming from:**

        ____

        ____

        ____

        ____

        ____

        ____

3. **Shut the engine down.**

4. **Referencing the appropriate manufacturer's workshop manual, record your recommendations to identify the source of the abnormal engine noises and vibrations:**

    ____

    ____

    ____

    ____

    ____

Time off_____________

Time on_____________

Total time_____________

5. **Discuss the findings with your instructor.**

____

____

____

Performance Rating

CDX Tasksheet Number: EO153

0	**1**	**2**	**3**	**4**

Supervisor/instructor signature _________________________________ Date _____________

▶ **TASK** Observe engine exhaust smoke color and quantity; determine needed action.

AED 5.4

Time off _______________

Time on _______________

Total time _______________

CDX Tasksheet Number: EO154

1. **Start the engine.**

2. **Check the exhaust smoke.**

   a. **With the engine running:**

      i. **Record the idle RPM:** _______________________________

      ii. **Record the governed RPM:** _______________________________

      iii. **Record the color of the exhaust smoke from the stack:**

      ____

      iv. **Record the quantity of smoke coming from the exhaust (e.g., short bus then cleared; longer bust, then cleared; large amount after start up and at idle):** _______________________________

3. **Shut the engine down.**

4. **Referencing the appropriate manufacturer's workshop manual, record your recommendations to rectify the source of the abnormal exhaust smoke.**

   ____

   ____

   ____

   ____

   ____

5. **Discuss the findings with your instructor.**

   ____

   ____

   ____

Performance Rating

CDX Tasksheet Number: EO154

0	1	2	3	4

Supervisor/instructor signature ____ Date _______________

General System Diagnostics

Student/Intern information:

Name _________________________________ Date _____________ Class _________________________

Vehicle used for this activity:

Year _______________ Make _________________________________ Model _________________________

Odometer _____________ Hour meter _____________ VIN _________________________________

Learning Objective/Task	CDX Tasksheet Number	2014 Edition Rev2/12/16 AED Standard
• Check engine not cranking, cranks but fails to start, hard starting, and starts but does not continue to run problems; determine needed action.	E0155	5.4
• Identify engine surging, rough operation, misfiring, low power, slow deceleration, slow acceleration, and shutdown problems; determine needed action.	E0156	5.4
• Identify engine vibration problems.	E0157	5.4
• Check and record electronic diagnostic codes.	E0158	5.4

Materials Required

- Machine with possible engine concern
- Machine manufacturer's workshop manual
- Manufacturer-specific tools depending on the concern
- Machine lifting equipment, if applicable

Some Safety Issues to Consider

- Diagnosis of this fault may require test driving the machine on the school grounds or on a hoist, both of which carry severe risks. Attempt this task only with full permission from your supervisor/instructor and follow all the guidelines exactly.
- Caution: If you are working in an area where there could be "brake dust" present (may contain asbestos, which has been determined to cause cancer when inhaled or ingested), ensure you wear and use all OSHA-approved asbestos protective/removal equipment.
- Lifting equipment such as machine jacks and stands, machine hoists, and engine hoists are important tools that increase productivity and make the job easier. However, they can also cause severe injury or death if used improperly. Make sure you follow the manufacturer's operation procedures. Also make sure you have your supervisor/instructor's permission to use any particular type of lifting equipment.
- Comply with personal and environmental safety practices associated with clothing; eye protection; hand tools; power equipment; proper ventilation; and the handling, storage, and disposal of chemicals/materials in accordance with federal, state, and local regulations.

- Always wear the correct protective eyewear and clothing and use the appropriate safety equipment, as well as fender covers, seat protectors, and floor mat protectors.
- Make sure you understand and observe all legislative and personal safety procedures when carrying out practical assignments. If you are unsure of what these are, ask your supervisor/instructor.

Performance Standard

0—No exposure: No information or practice provided during the program; complete training required

1—Exposure only: General information provided with no practice time; close supervision needed; additional training required

2—Limited practice: Has practiced job during training program; additional training required to develop skill

3—Moderately skilled: Has performed job independently during training program; limited additional training may be required

4—Skilled: Can perform job independently with no additional training

Student/Intern information:

Name ________________________________ Date ____________ Class ________________________

Vehicle used for this activity:

Year ______________ Make ____________________________ Model __________________________

Odometer ____________ Hour meter ____________ VIN ________________________________

▶ TASK Check engine not cranking, cranks but fails to start, hard starting, and starts but does not continue to run problems; determine needed action.

AED 5.4

Time off________________

Time on________________

Total time________________

CDX Tasksheet Number: EO155

1. **Refer to the manufacturer's workshop manual for the common causes for a "no crank" situation.**

   a. **List all possible causes for "no crank" (the engine will not turn over using the starter motor):**

   ____

   ____

   ____

   ____

   ____

   b. **Determine what action will be required for "no crank":**

   ____

   ____

   ____

   ____

   ____

2. **Refer to the manufacturer's workshop manual for the common causes for a "cranks but fails to start" situation.**

   a. **List all possible causes for "cranks but fails to start" (the engine turns over by the starter motor but fails to run):**

   ____

   ____

   ____

   ____

   ____

   b. **Determine what action will be required for "cranks but fails to start":**

   ____

   ____

   ____

   ____

   ____

3. Refer to the manufacturer's workshop manual for the common causes for a "hard starting" situation (the engine will run but takes longer than usual to start).

    a. List all possible causes for "hard starting":

____
____
____
____
____

    b. Determine what action will be required for "hard starting":

____
____
____
____
____

4. Refer to the manufacturer's workshop manual for the common causes for a "starts but does not continue to run problem" situation.

    a. List all possible causes for "starts but does not continue to run problem":

____
____
____
____
____

    b. Determine what action will be required for "starts but does not continue to run problem":

____
____
____
____
____

5. Discuss the findings with your instructor.

____
____
____

Performance Rating

CDX Tasksheet Number: EO155

0	1	2	3	4

Supervisor/instructor signature ____ Date ______________

Name _________________________ Date __________ Class _______________________

Vehicle used for this activity:

Year _______________ Make _________________________ Model _______________________

Odometer ____________ Hour meter ____________ VIN _______________________

▶ TASK Identify engine surging, rough operation, misfiring, low power, slow deceleration, slow acceleration, and shutdown problems; determine needed action.

AED 5.4

Time off_______________

Time on_______________

Total time_______________

CDX Tasksheet Number: E0156

1. **Refer to the manufacturer's workshop manual for the common causes for a "surging" situation.**

   a. **List all possible causes of "surging" (uneven or rolling idle):**

   ____
   ____
   ____
   ____
   ____

   b. **Determine what action will be required of "surging":**

   ____
   ____
   ____
   ____
   ____

2. **Refer to the manufacturer's workshop manual for the common causes for an engine's "rough operation" situation.**

   a. **List all possible causes for an engine's "rough operation":**

   ____
   ____
   ____
   ____
   ____

   b. **Determine what action will be required for an engine's "rough operation":**

   ____
   ____
   ____
   ____
   ____

3. **Refer to the manufacturer's workshop manual for the common causes for an engine's "misfiring" situation.**

   a. **List all possible causes for an engine's "misfiring":**

   ____
   ____

____
____
____
____

  b. Determine what action will be required for an engine's "misfiring":

____
____
____
____
____

4. **Refer to the manufacturer's workshop manual for the common causes for an engine's "low power" situation.**

  a. List all possible causes for an engine's "low power" condition:

____
____
____
____
____

  b. Determine what action will be required for an engine's "low power" condition:

____
____
____
____
____

5. **Refer to the manufacturer's workshop manual for the common causes for an engine's "slow deceleration and/or slow acceleration" situation.**

  a. List all possible causes for an engine's "slow deceleration and/or slow acceleration" situation:

____
____
____
____
____

  b. Determine what action will be required for an engine's "slow deceleration and/or slow acceleration" situation:

____
____
____
____
____

6. **Refer to the manufacturer's workshop manual for the common causes for an engine's "shutdown" problems.**

  a. List all possible causes for an engine's "shutdown" problems:

____
____
____
____

b. Determine what action will be required for an engine's "shutdown" problems:

____

____

____

____

____

7. Discuss the findings with your instructor.

____

____

____

▶ **TASK** Identify engine vibration problems.

AED 5.4

Time off________

CDX Tasksheet Number: EO157

Time on________

1. **Refer to the manufacturer's workshop manual for the common causes for an engine's possible "vibration" problems.**

    a. **List all possible causes for an engine's possible "vibration" problems:**

Total time________

    ____
    ____
    ____
    ____
    ____

    b. **Determine what action will be required for an engine's possible "vibration" problems:**

    ____
    ____
    ____
    ____
    ____

2. **Discuss the findings with your instructor.**

    ____
    ____
    ____

Performance Rating

CDX Tasksheet Number: EO157

0	1	2	3	4

Supervisor/instructor signature _________________________________ Date ___________

Student/Intern information:

Name _________________________________ Date _____________ Class _________________________

Vehicle used for this activity:

Year _______________ Make _________________________________ Model _____________________

Odometer _____________ Hour meter _____________ VIN _________________________________

▶ TASK Check and record electronic diagnostic codes.

AED 5.4

Time off_________________

Time on_________________

CDX Tasksheet Number: EO158

Total time_______________

ENGINE "ON" or "OFF"

1. **Check the operation of all accessories.**

   a. **Check all accessory devices fitted and ensure they are activating and working in accordance with the manufacturer's specifications.**

      i. **Are they operating as described in the manufacturer's workshop manual? Yes: _________ No: _________**

      ii. **If no, list the problems and your recommendation(s):**

      ____

      ____

      ____

      ____

      ____

2. **Using diagnostic tool or on-board diagnostic system, extract the engine monitoring information.**

   a. **Are there any fault codes listed in the diagnostic tester? Yes: _________ No: _________**

      i. **If yes, list the code(s):**

      ____

      ____

      ____

      ____

      ____

3. **Following the diagnostic tool's instructions and operating manual, carry out a full monitoring of the engine system.**

   a. **Record the information in the table below:**

Test	Description	Result
1		
2		
3		
4		
5		
6		

7		
8		
9		
10		
11		
12		
13		
14		
15		
16		
17		
18		
19		
20		
21		
22		
23		
24		
25		

    b. Any other readings current:

____

____

____

____

____

4. Refer to the machine manufacturer's specifications.

    a. Are all the results within the manufacturer's specifications: Yes: _________ No: _________

    b. If no, list the results that are outside the manufacturer's specifications:

____

____

____

____

5. Discuss your findings with your instructor.

6. If directed by your instructor:

    a. Clear any stored codes.

    b. Investigate and rectify where possible any of the "out of specifications" recorded.

7. **Discuss the results with your instructor.**

Engine Subsystems: Cylinder Heads and Valve Train Disassembly

Student/Intern information:

Name _________________________________ Date ____________ Class _________________________

Vehicle used for this activity:

Year ______________ Make ___________________________ Model _____________________________

Odometer ____________ Hour meter ____________ VIN _______________________________________

<table>
<tr><td rowspan="2">Learning Objective/Task</td><td>CDX Tasksheet Number</td><td>2014 Edition Rev2/12/16 AED Standard</td></tr>
<tr><td></td><td></td></tr>
<tr><td>• Inspect cylinder head for cracks/damage; check mating surfaces for warpage; check condition of passages; inspect core/expansion and gallery plugs; determine needed action.</td><td>E0159</td><td>5.6</td></tr>
<tr><td>• Disassemble head and inspect valves, guides, seats, springs, retainers, rotators, locks, and seals; determine needed action.</td><td>E0160</td><td>5.6</td></tr>
<tr><td>• Measure valve head height relative to deck and valve face-to-seat contact; determine needed action.</td><td>E0161</td><td>5.6</td></tr>
<tr><td>• Inspect injector sleeves and seals; measure injector tip or nozzle protrusion; determine needed action.</td><td>E0162</td><td>5.6</td></tr>
<tr><td>• Inspect valve train components; determine needed action.</td><td>E0163</td><td>5.6</td></tr>
</table>

Time off________________

Time on________________

Total time________________

Materials Required

- Machine with possible engine concern
- Machine manufacturer's workshop manual
- Manufacturer-specific tools depending on the concern
- Machine lifting equipment, if applicable

Some Safety Issues to Consider

- Diagnosis of this fault may require test driving the machine on the school grounds or on a hoist, both of which carry severe risks. Attempt this task only with full permission from your supervisor/instructor and follow all the guidelines exactly.
- Caution: If you are working in an area where there could be "brake dust" present (may contain asbestos, which has been determined to cause cancer when inhaled or ingested), ensure you wear and use all OSHA-approved asbestos protective/removal equipment.
- Lifting equipment such as machine jacks and stands, machine hoists, and engine hoists are important tools that increase productivity and make the job easier. However, they can also cause severe injury or death if used improperly. Make sure you follow the manufacturer's operation procedures. Also make sure you have your supervisor/instructor's permission to use any particular type of lifting equipment.

- Comply with personal and environmental safety practices associated with clothing; eye protection; hand tools; power equipment; proper ventilation; and the handling, storage, and disposal of chemicals/materials in accordance with federal, state, and local regulations.
- Always wear the correct protective eyewear and clothing and use the appropriate safety equipment, as well as fender covers, seat protectors, and floor mat protectors.
- Make sure you understand and observe all legislative and personal safety procedures when carrying out practical assignments. If you are unsure of what these are, ask your supervisor/instructor.

Performance Standard

0–No exposure: No information or practice provided during the program; complete training required

1–Exposure only: General information provided with no practice time; close supervision needed; additional training required

2–Limited practice: Has practiced job during training program; additional training required to develop skill

3–Moderately skilled: Has performed job independently during training program; limited additional training may be required

4–Skilled: Can perform job independently with no additional training

▶ TASK Inspect cylinder head for cracks/damage; check mating surfaces for warpage; check condition of passages; inspect core/expansion and gallery plugs; determine needed action.

AED
5.6

Time off_______________

Time on_______________

Total time_______________

CDX Tasksheet Number: EO159

1. **Determine the type of crack detention process(es) that your workshop utilizes.**

    a. **Magnetic particle inspection: Yes: _________ No: _________**

    b. **Penetrating dyes: Yes: _________ No: _________**

    c. **Pressure testing: Yes: _________ No: _________**

    d. **Vacuum testing: Yes: _________ No: _________**

    e. **Ultrasonic testing: Yes: _________ No: _________**

    f. **None of the above: Yes: _________**

    g. **If none of the above, name the method your workshop uses: _______________________________**

    h. **Outsource testing and repairs: Yes: _________ No: _________**

2. **Refer to the manufacturer's workshop manual; list the procedure and all safety precautions that must be observed when carrying out an inspection of a cylinder head for cracks/damage.**

    a. **List the steps involved in inspecting a cylinder head for cracks/ damage:**

    ____

    ____

    ____

    ____

    ____

    b. **Determine what safety precautions must be observed when inspecting a cylinder head for cracks/damage:**

    ____

    ____

    ____

    ____

    ____

3. Discuss these procedures and safety precautions with your instructor. Determine what method of testing will be carried out:

____

____

____

____

____

4. If directed by your instructor, commence crack testing the cylinder head. Follow the procedures listed above and refer to the manufacturer's workshop manual.

    a. Meets the manufacturer's specifications:
    Yes: _________ No: _________

____

____

____

____

____

    b. If no, list the areas of cracking and your recommendations for any rectifications:

____

____

____

____

____

5. Refer to the manufacturer's workshop manual; list the procedure for checking for warpage of the cylinder head mating surfaces:

    a. List the steps involved in checking the cylinder head for warpage:

____

____

____

____

____

    b. Determine what safety precautions must be observed when checking the cylinder head for warpage:

____

____

____

____

____

6. Following the procedures listed above, and while referencing the manufacturer's workshop manual, check for any warpage of the cylinder head mating surfaces.

    a. Meets the manufacturer's specifications:
    Yes: _________ No: _________

    b. If no, list your recommendations for any rectifications:

    ____

    ____

    ____

    ____

    ____

7. **Referring to the manufacturer's workshop manual, check the condition of passages and inspect the core/expansion and gallery plugs.**

    a. Meets the manufacturer's specifications:
    Yes: _________ No: _________

    b. If no, list the areas of concerns and your recommendations for any rectifications:

    ____

    ____

    ____

    ____

    ____

8. **Discuss the results with your instructor.**

    ____

    ____

    ____

Performance Rating

CDX Tasksheet Number: EO159

0	1	2	3	4

Supervisor/instructor signature ____ Date _______________

Student/Intern information:

Name _________________________________ Date ____________ Class _________________________

Vehicle used for this activity:

Year _______________ Make __________________________ Model _________________________

Odometer ____________ Hour meter ___________ VIN _________________________________

▶ TASK Disassemble head and inspect valves, guides, seats, springs, retainers, rotators, locks, and seals; determine needed action.

AED 5.6

Time off______________

Time on______________

Total time______________

CDX Tasksheet Number: E0160

1. **Referring to the appropriate manufacturer's workshop manual, dismantle the cylinder head using the recommended special tools. When removing the components, store all the nuts and bolts in order in storage trays.**

    a. **As you dismantle each valve assembly, keep them in order so that they can be evaluated as a unit. Lay out the components so as to identify their original position.**

    b. **Remove and inspect the retainers.**

       i. **Meets the manufacturer's specifications: Yes: ________ No: ________**

    c. **Remove and inspect the rotators.**

       i. **Meets the manufacturer's specifications: Yes: ________ No: ________**

    d. **Remove and inspect the springs for twists, distortions, and nicks.**

       i. **Meets the manufacturer's specifications: Yes: ________ No: ________**

    e. **Remove and inspect the seals.**

       i. **Meets the manufacturer's specifications: Yes: ________ No: ________**

    f. **Remove and inspect the valves.**

       i. **Meets the manufacturer's specifications: Yes: ________ No: ________**

    g. **Inspect the valve seats.**

       i. **Meets the manufacturer's specifications: Yes: ________ No: ________**

    h. **If no to any of the above, list the areas of concerns and your recommendations for any rectifications:**

    ____

    ____

    ____

    ____

    ____

2. Discuss the results with your instructor.

____

____

____

▶ TASK Measure valve head height relative to deck and valve face-to-seat contact; determine needed action.

AED 5.6

Time off _______________

Time on _______________

Total time _______________

CDX Tasksheet Number: EO161

1. **Referring to the appropriate manufacturer's workshop manual, measure valve head height relative to the deck and valve face-to-seat contact using the recommended special tools. Record your findings in the tables below.**

Cylinder	Valve Head Height Relative to Deck					
	Valve 1	Valve 2	Valve 3	Valve 4	Valve 5	Valve 6
1						
2						
3						
4						
5						
6						
7						
8						
9						
10						

Cylinder	Valve Face-to-Seat Contact (Use coding below)					
	Valve 1	Valve 2	Valve 3	Valve 4	Valve 5	Valve 6
1						
2						
3						
4						
5						
6						
7						
8						
9						
10						

   a. **List the areas of concern and your recommendations for any rectifications:**

____

____

____

____

____

2. **Discuss the results with your instructor.**

____

____

____

Performance Rating

CDX Tasksheet Number: EO161

0	1	2	3	4

Supervisor/instructor signature ___________________________ Date ____________

Name ___________________________ Date ____________ Class ___________________________

Vehicle used for this activity:

Year _______________ Make _______________________ Model ___________________________

Odometer ____________ Hour meter ____________ VIN ___________________________

▶ **TASK** Inspect injector sleeves and seals; measure injector tip or nozzle protrusion; determine needed action.

AED 5.6

Time off______________

Time on______________

Total time______________

CDX Tasksheet Number: E0162

1. **Referring to the appropriate manufacturer's workshop manual, inspect injector sleeves and seals.**

    a. **Meets the manufacturer's specifications:**
       Yes: _________ No: _________

    b. **If no, list the areas of concern and your recommendations for any rectifications:**

    ____
    ____
    ____
    ____

2. **Referring to the appropriate manufacturer's workshop manual and using the recommended special tools, measure the injector tip or nozzle protrusion and record your findings in the table below.**

Cylinder	Injector Tip or Nozzle Protrusion
1	
2	
3	
4	
5	
6	
7	
8	
9	
10	

    a. **List the areas of concern and your recommendations for any rectifications:**

3. **Discuss the results with your instructor.**

____

____

____

Name _________________________________ Date _____________ Class _______________________

Vehicle used for this activity:

Year _______________ Make _________________________ Model _____________________________

Odometer _____________ Hour meter _____________ VIN _________________________________

▶ TASK Inspect valve train components; determine needed action.

**AED
5.6**

Time off_________________

Time on_________________

CDX Tasksheet Number: EO163

1. **Referring to the appropriate manufacturer's workshop manual, inspect the injector sleeves and seals. Inspect and report on all the components that are applicable to your task in the table below.**

Total time_________________

Component	Serviceable	Repairable	Unserviceable
Camshaft/lobes			
Cam followers			
Bucket tappets			
Adjusting shims			
Rockers			
Cam rollers			
Cam gear(s)			
Cam retaining caps			
Timing belt/chain			
Rocker shaft(s)			

2. **Valve inspection and assessment:**

Cylinder	Valve Spring Height (inches or mm) and Tension (ft-lb or N·m)							
	Valve 1	Valve 2	Valve 3	Valve 4	Valve 5	Valve 6	Valve 7	Valve 8
	Height	Tension	Height	Tension	Height	Tension	Height	Tension
1								
2								
3								
4								
5								
6								
7								
8								
9								
10								

a. **Meets the manufacturer's specifications:**
   Yes: _________ **No:** _________

b. **If no, list the areas of concern and your recommendations for any rectifications:**

____
____
____
____
____

3. **Discuss the results with your instructor.**

____
____
____

Engine Subsystems: Cylinder Heads and Valve Train Assembly

Student/Intern information:

Name _________________________ Date _________ Class _________________

Vehicle used for this activity:

Year _____________ Make _________________________ Model _________________

Odometer _____________ Hour meter _____________ VIN _________________________

Learning Objective/Task	CDX Tasksheet Number	2014 Edition Rev2/12/16 AED Standard
• Reassemble cylinder head.	E0164	5.6
• Inspect, measure, and replace/reinstall overhead camshaft; measure/adjust end play and backlash.	E0165	5.6
• Inspect electronic wiring harness and brackets for wear, bending, cracks, and looseness; determine needed action.	E0166	5.6
• Adjust valve bridges (crossheads); adjust valve clearances and injector settings.	E0167	5.6

Time off_________

Time on_________

Total time_________

Materials Required

- Machine with possible engine concern
- Machine manufacturer's workshop manual
- Manufacturer-specific tools depending on the concern
- Machine lifting equipment, if applicable

Some Safety Issues to Consider

- Diagnosis of this fault may require test driving the machine on the school grounds or on a hoist, both of which carry severe risks. Attempt this task only with full permission from your supervisor/instructor and follow all the guidelines exactly.
- Caution: If you are working in an area where there could be "brake dust" present (may contain asbestos, which has been determined to cause cancer when inhaled or ingested), ensure you wear and use all OSHA-approved asbestos protective/removal equipment.
- Lifting equipment such as machine jacks and stands, machine hoists, and engine hoists are important tools that increase productivity and make the job easier. However, they can also cause severe injury or death if used improperly. Make sure you follow the manufacturer's operation procedures. Also make sure you have your supervisor/instructor's permission to use any particular type of lifting equipment.
- Comply with personal and environmental safety practices associated with clothing; eye protection; hand tools; power equipment; proper ventilation; and the handling, storage, and disposal of chemicals/materials in accordance with federal, state, and local regulations.

- Always wear the correct protective eyewear and clothing and use the appropriate safety equipment, as well as fender covers, seat protectors, and floor mat protectors.
- Make sure you understand and observe all legislative and personal safety procedures when carrying out practical assignments. If you are unsure of what these are, ask your supervisor/instructor.

Performance Standard

0–No exposure: No information or practice provided during the program; complete training required

1–Exposure only: General information provided with no practice time; close supervision needed; additional training required

2–Limited practice: Has practiced job during training program; additional training required to develop skill

3–Moderately skilled: Has performed job independently during training program; limited additional training may be required

4–Skilled: Can perform job independently with no additional training

---

▶ TASK Reassemble cylinder.　　　　　　　　　　　　　　　　　　　**AED 5.6**

CDX Tasksheet Number: E0164

1. Ensure all cylinder head and associated parts have been cleaned and dried.

2. Secure the cylinder head to the bench to ensure it does not fall and is safe to work on.

3. Have the valves been ground in accordance with the manufacturer's specifications?

    a. Meets the manufacturer's specifications:
    Yes: ________ No: ________

    b. If no, discuss with your instructor if he or she requires this action to be undertaken.

    c. If yes, list the steps involved in servicing the valves:

    ____

    ____

    ____

4. Have the valve seats been ground in accordance with the manufacturer's specifications?

    a. Meets the manufacturer's specifications:
    Yes: ________ No: ________

    b. If no, discuss with your instructor if he or she requires this action to be undertaken.

    c. If yes, list the steps involved in servicing the valve seats:

    ____

    ____

    ____

5. Have the valves been lapped into the valve seats in accordance with the manufacturer's specifications?

    a. Meets the manufacturer's specifications:
    Yes: ________ No: ________

    b. If no, discuss with your instructor if he or she requires this action to be undertaken.

Time off___________

Time on___________

Total time___________

c. If yes, list the steps involved in lapping the valves faces onto the valve seats:

____

____

____

6. Reference the manufacturer's workshop manual; list the procedure and all safety precautions that must be observed when reassembling the cylinder head.

    a. List the steps involved in reassembling the cylinder head:

    ____

    ____

    ____

    b. Determine what safety precautions must be observed when reassembling the cylinder head:

    ____

    ____

    ____

7. Discuss these procedures and safety precautions with your instructor.

8. If directed by your instructor, commence reassembling the cylinder head.

    a. Source all the special tooling required as recommended by the manufacturer and your instructor.

    b. Source all necessary spare parts and repair kit(s) from your stores area or spare parts department.

9. Following the procedures listed above, and referring to the manufacturer's workshop manual, reassemble the cylinder head.

    a. Insert the valves into their relative seats as per their reconditioning results (or original position, depending on the level of servicing).

    b. Fit the valve oil seal(s) if applicable.

    c. Fit the new valve springs and shims (or original valve springs and to the original position, depending on the level of servicing).

    d. Fit the rotators and valve keepers as required.

10. Following the procedures listed above, and referring to the manufacturer's workshop manual, measure the assembled height of all the valves assembled in the cylinder head.

    a. Measure valve head height relative to the deck and valve face-to-seat contact using the recommended special tools and compare the heights before and after in the table below:

Cylinder	Valve Head Height Relative to the Deck					
	Valve 1	Valve 2	Valve 3	Valve 4	Valve 5	Valve 6
1						
2						

3						
4						
5						
6						
7						
8						
9						
10						

   i. **Meets the manufacturer's specifications:**
      Yes: _________ No: _______

   ii. **If no, list your recommendations for any rectifications:**

____
____
____
____
____

b. **Measure the assembled valve head height relative to the deck using the recommended special tools. Record your findings in the table below:**

Cylinder	Valve Head Height Relative to the Deck					
	Valve 1	**Valve 2**	**Valve 3**	**Valve 4**	**Valve 5**	**Valve 6**
1						
2						
3						
4						
5						
6						
7						
8						
9						
10						

   i. **Meets the manufacturer's specifications:**
      Yes: _________ No: _______

   ii. **If no, list your recommendations for any rectifications:**

____
____
____
____
____

11. Following the procedures listed above, and referring to the manufacturer's workshop manual, reassemble the injector sleeves and injectors into the cylinder head.

   a. Insert the injector sleeves and new seals into their position relative as per their reconditioning results (or original position, depending on the level of servicing).

   b. Fit the injectors into their original position.

   c. Assemble the injector clamps or securing devices and tighten to the recommended manufacturer's tension specifications.

12. Referring to the appropriate manufacturer's workshop manual and using the recommended special tools, measure the injector tip or nozzle protrusion.

   a. Measure the injector tip or nozzle protrusion using the recommended special tools, and compare the heights before and after in the table below:

Cylinder	Injector Tip or Nozzle Protrusion
1	
2	
3	
4	
5	
6	
7	
8	
9	
10	

   i. Meets the manufacturer's specifications:
   Yes: _________ No: _________

   ii. If no, list your recommendations for any rectifications:

   ____
   ____
   ____
   ____
   ____

13. **If your workshop has pressure testing capabilities, pressure test the newly installed valves for correct sealing under operating pressures. Refer to your instructor for directions and assistance to carry this operation out, if applicable.**

14. **Have your instructor inspect your assembled cylinder head to this point and discuss the findings with him or her.**

____

____

____

Performance Rating

CDX Tasksheet Number: E0164

0	1	2	3	4

Supervisor/instructor signature _________________________________ Date _____________

Student/Intern information:

Name _________________________________ Date _____________ Class _________________________________

Vehicle used for this activity:

Year _________________ Make _________________________ Model _________________________

Odometer _____________ Hour meter _____________ VIN _________________________________

▶ **TASK** Inspect, measure, and replace/reinstall overhead camshaft; measure/adjust end play and backlash.

AED 5.6

Time off_________________

Time on_________________

Total time_________________

CDX Tasksheet Number: EO165

1. **Follow the procedures listed in the manufacturer's workshop manual. Reassemble the camshaft into the cylinder head (applicable only to overhead mounted camshaft(s)). Continue if this is an OHC assembly.**

    a. **Check the camshaft bearings; replace if necessary.**

    b. **Locate the camshaft into the cylinder head.**

    c. **Fit cam mounting brackets and torque to the manufacturer's specifications.**

        i. **Manufacturer's specified torque:** _______________ **ft-lb (N·m)**

        ii. **Actual torque:** _______________ **ft-lb (N·m)**

2. **Following the procedures listed in the manufacturer's workshop manual, use the recommended special tools to measure/adjust end play and backlash (applicable only to overhead mounted camshaft(s)). Continue if this is an OHC assembly.**

    a. **Manufacturer's specification - end play:** _______________ **in/mm**

    b. **Actual end play measurement:** _______________ **in/mm**

    c. **Manufacturer's specification - backlash:** _______________ **in/mm**

    d. **Actual backlash measurement:** _______________ **in/mm**

3. **Have your instructor inspect your assembled cylinder head to this point and discuss the findings with him or her.**

    ____

    ____

    ____

Performance Rating

CDX Tasksheet Number: EO165

0	1	2	3	4

Supervisor/instructor signature ____ Date _____________

Student/Intern information:

Name _______________________________ Date _____________ Class _______________________

Vehicle used for this activity:

Year _______________ Make _________________________ Model ___________________________

Odometer ____________ Hour meter ____________ VIN _________________________________

▶ TASK Inspect electronic wiring harness and brackets for wear, bending, cracks, and looseness; determine needed action.

AED 5.6

Time off________________

Time on________________

CDX Tasksheet Number: E0166

1. **Follow the procedures listed in the manufacturer's workshop manual. Inspect the electronic wiring harness and brackets for wear, bending, cracks, and looseness.**

Total time________________

   a. **Meets the manufacturer's specifications:**
      Yes: _________ No: _________

   b. **If no, list the areas of concerns and your recommendations for any rectifications:**

   ____
   ____
   ____
   ____
   ____
   ____

2. **Discuss your findings with your instructor.**

   ____
   ____
   ____

Performance Rating

CDX Tasksheet Number: E0166

0	1	2	3	4

Supervisor/instructor signature ____ Date ______________

▶ TASK Adjust valve bridges (crossheads); adjust valve clearances and injector settings.

AED
5.6

Time off________________

Time on________________

Total time________________

CDX Tasksheet Number: EO167

1. **Refer the manufacturer's workshop manual; list the procedure to adjust valve bridges (crossheads).**

   a. **List the steps involved to adjust valve bridges (crossheads):**

   ____
   ____
   ____
   ____
   ____

   b. **Determine what safety precautions must be observed when adjusting the valve bridges (crossheads):**

   ____
   ____
   ____
   ____
   ____

2. **Following the procedures listed above, and referring to the manufacturer's workshop manual, adjust the valve bridges (crossheads).**

   a. **Meets the manufacturer's specifications:**
      Yes: _________ No: _________

   b. **If no, list your recommendations for any rectifications:**

   ____
   ____
   ____
   ____
   ____

3. **Refer the manufacturer's workshop manual; list the procedure to adjust the valve injector settings.**

   a. **List the steps involved to adjust the valve injector settings:**

   ____
   ____
   ____
   ____
   ____

b. Determine what safety precautions must be observed when adjusting the settings:

____

____

____

____

____

4. **Following the procedures listed above, and referring to the manufacturer's workshop manual, adjust the valve injector settings.**

    a. Meets the manufacturer's specifications:
    Yes: _________ No: _________

    b. If no, list your recommendations for any rectifications:

____

____

____

____

____

5. **Discuss your findings with your instructor.**

____

____

____

Performance Rating

CDX Tasksheet Number: EO167

☐	☐	☐	☐	☐
0	1	2	3	4

Supervisor/instructor signature ____ Date _____________

Engine Subsystems: Engine Block Disassembly and Inspection

Student/Intern information:

Name ___________________________ Date __________ Class ___________________

Vehicle used for this activity:

Year ____________ Make ____________________ Model ____________________

Odometer __________ Hour meter __________ VIN ____________________

Learning Objective/Task	CDX Tasksheet Number	2014 Edition Rev2/12/16 AED Standard
• Perform crankcase pressure test; determine needed action.	E0168	5.8
• Remove, inspect, service, and install pans, covers, gaskets, seals, wear rings, and crankcase ventilation components.	E0169	5.6
• Disassemble, clean, and inspect engine block for cracks/ damage; measure mating surfaces for warpage; check condition of passages, core/expansion, and gallery plugs; inspect threaded holes, studs, dowel pins, and bolts for serviceability; determine needed action.	E0170	5.6
• Inspect cylinder sleeve counter bore and lower bore; check bore distortion; determine needed action.	E0171	5.6
• Clean, inspect, and measure cylinder walls or liners for wear and damage; determine needed action.	E0172	5.6

Materials Required

- Machine with possible engine concern
- Machine manufacturer's workshop manual
- Manufacturer-specific tools depending on the concern
- Machine lifting equipment, if applicable

Some Safety Issues to Consider

- Diagnosis of this fault may require test driving the machine on the school grounds or on a hoist, both of which carry severe risks. Attempt this task only with full permission from your supervisor/instructor and follow all the guidelines exactly.
- Caution: If you are working in an area where there could be "brake dust" present (may contain asbestos, which has been determined to cause cancer when inhaled or ingested), ensure you wear and use all OSHA-approved asbestos protective/removal equipment.
- Lifting equipment such as machine jacks and stands, machine hoists, and engine hoists are important tools that increase productivity and make the job easier. However, they can also cause severe injury or death if used improperly. Make sure you follow the manufacturer's

operation procedures. Also make sure you have your supervisor/instructor's permission to use any particular type of lifting equipment.

- Comply with personal and environmental safety practices associated with clothing; eye protection; hand tools; power equipment; proper ventilation; and the handling, storage, and disposal of chemicals/materials in accordance with federal, state, and local regulations.
- Always wear the correct protective eyewear and clothing and use the appropriate safety equipment, as well as fender covers, seat protectors, and floor mat protectors.
- Make sure you understand and observe all legislative and personal safety procedures when carrying out practical assignments. If you are unsure of what these are, ask your supervisor/instructor.

Performance Standard

0–No exposure: No information or practice provided during the program; complete training required

1–Exposure only: General information provided with no practice time; close supervision needed; additional training required

2–Limited practice: Has practiced job during training program; additional training required to develop skill

3–Moderately skilled: Has performed job independently during training program; limited additional training may be required

4–Skilled: Can perform job independently with no additional training

Student/Intern information:

Name _________________________________ Date _____________ Class _________________________

Vehicle used for this activity:

Year _______________ Make _____________________________ Model _________________________

Odometer _____________ Hour meter _____________ VIN _________________________________

▶ TASK Perform crankcase pressure test; determine needed action. **AED 5.8**

Time off_______________

CDX Tasksheet Number: E0168

Time on_______________

1. **Refer to the manufacturer's workshop manual; list the procedure and all safety precautions that must be observed when performing a crankcase pressure test.**

   Total time_______________

   a. **List the steps involved when performing a crankcase pressure test:**

   ____
   ____
   ____
   ____
   ____
   ____

   b. **Determine what safety precautions must be observed when performing a crankcase pressure test:**

   ____
   ____
   ____
   ____
   ____
   ____

2. **Discuss these procedures and safety precautions with your instructor.**

3. **If directed by your instructor, perform a crankcase pressure test following the procedures listed above and while referring to the manufacturer's workshop manual.**

   a. **Meets the manufacturer's specifications:**
      Yes: _________ No: _________

   b. **If no, list your recommendations for any rectifications:**

   ____
   ____
   ____
   ____
   ____

4. Discuss the findings with your instructor.

Student/Intern information:

Name _______________________________________ Date _____________ Class _________________________

Vehicle used for this activity:

Year _______________ Make _____________________________ Model _______________________

Odometer _____________ Hour meter _____________ VIN _________________________________

▶ TASK Remove, inspect, service, and install pans, covers, gaskets, seals, wear rings, and crankcase ventilation components.

AED 5.6

Time off_______________

Time on_______________

Total time_______________

CDX Tasksheet Number: E0169

1. **Refer to the manufacturer's workshop manual; remove, inspect, service, and install pans, covers, gaskets, seals, and wear rings.**

    a. **Meets the manufacturer's specifications:**
       Yes: _________ **No:** _________

    b. **If no, list your recommendations for any rectifications:**

    ____

    ____

    ____

    ____

    ____

2. **Refer to the manufacturer's workshop manual; remove, inspect, service, and install the crankcase ventilation components.**

    a. **Meets the manufacturer's specifications:**
       Yes: _________ **No:** _________

    b. **If no, list your recommendations for any rectifications:**

    ____

    ____

    ____

    ____

    ____

3. **Discuss the findings with your instructor.**

Performance Rating

CDX Tasksheet Number: E0169

0	1	2	3	4

Supervisor/instructor signature ____ Date _____________

Name _________________________________ Date ____________ Class ___________________

Vehicle used for this activity:

Year _______________ Make _________________________________ Model _________________

Odometer ____________ Hour meter ____________ VIN _______________________________

▶ TASK Disassemble, clean, and inspect engine block for cracks/damage; measure mating surfaces for warpage; check condition of passages, core/expansion and gallery plugs; inspect threaded holes, studs, dowel pins, and bolts for serviceability; determine needed action.

AED
5.6

Time off______________

Time on______________

Total time______________

CDX Tasksheet Number: E0170

1. **Refer to the manufacturer's workshop manual; list the procedure and all safety precautions that must be observed when disassembling an engine assembly.**

   a. **List the steps involved in disassembling the engine block:**

   ____

   ____

   ____

   ____

   ____

   ____

   b. **Determine what safety precautions must be observed during the disassembly of the engine block:**

   ____

   ____

   ____

   ____

   ____

2. **Discuss these procedures and safety precautions with your instructor.**

3. **If directed by your instructor, mount the engine block into an engine rollover stand.**

4. **Following the removal procedures listed above, and referring to the manufacturer's workshop manual, disassemble the engine block.**

   a. **If the removed engine block is dirty and possibly has an oily coating in the designated cleaning bay, clean the engine.**

   b. **Using a suitable sized container, drain the oil from the engine and dispose of it in accordance with your shop's procedure and in compliance with all government and environmental protect legislation.**

   c. **If there is any coolant in the engine, use a suitable sized container, drain the remaining coolant from the engine, and dispose of it in accordance with your shop's procedure and in compliance with all government and environmental protect legislation.**

5. Commence the dismantling procedure.

    a. When removing the components, store all the nuts and bolts in order in storage trays.

    b. Store all removed components on oil collection trays to prevent oil spillage on the workshop floor (safety requirement).

    c. Remove all the necessary components as indicated by the manufacturer in their manufacturer's workshop manual.

6. Before removing the connecting rod caps, mark their location and position.

7. Check to see if there is a wear lip at the top of the cylinder that could catch the piston rings: Yes: _________ No: _________

    a. If yes, list your recommendations for any rectifications:

____

____

____

____

____

8. Discuss these recommendations with your instructor.

9. If directed by your instructor, remove the wear lip as described in the manufacturer's workshop manual. Follow all procedures and safety requirements.

    a. Remove the piston assemblies and store in a safe location.

    b. Remove the flywheel assembly and store in a safe location.

    c. Remove oil pump and store in an oil collection container.

10. Before removing the crankshaft main bearing caps, mark their location and position.

    a. Loosen and remove the main bearing caps.

    b. Carefully lift the crankshaft out and store in a safe location.

    c. Collect the thrust bearings and store.

    d. Replace the main bearing caps into their original position and screw the cap bolts back finger tight.

11. Referring to the manufacturer's workshop manual, complete the dismantling of the engine block including wet sleeve cylinder liners (if applicable). Ensure that you number their location and store components in a safe location.

12. While observing all lifting safety precautions, transport the disassembled block and disassembled components to the designated cleaning bay.

    a. Clean the block and components.

    b. Dry the engine block and components.

    c. After cleaning and drying the engine block, remount onto an engine rollover stand and secure.

    d. Lay out the components in an orderly manner to assist in reassembly.

13. Refer to the manufacturer's workshop manual; list the procedure for measure mating surfaces for warpage.

    a. List the steps involved to measure mating surfaces for warpage:

____

____

____

____

____

    b. Determine what safety precautions must be observed when measuring mating surfaces for warpage:

____

____

____

____

____

14. Following the procedures listed above, and referring to the manufacturer's workshop manual, measure mating surfaces for warpage.

    a. Meets the manufacturer's specifications:
    Yes: _________ No: _________

    b. If no, list your recommendations for any rectifications:

____

____

____

____

____

15. Referring to the manufacturer's workshop manual, check the condition of passages, core/expansion, and gallery plugs; inspect threaded holes, studs, dowel pins, and bolts for serviceability.

    a. Meets the manufacturer's specifications:
    Yes: _________ No: _________

    b. If no, list the areas of concern and your recommendations for any rectifications:

____

____

____

____

____

16. Discuss the findings with your instructor.

Student/Intern information:

Name _________________________ Date __________ Class _________________________

Vehicle used for this activity:

Year _____________ Make _____________________ Model _________________________

Odometer __________ Hour meter __________ VIN _________________________

▶ **TASK** Inspect cylinder sleeve counter bore and lower bore;
check bore distortion; determine needed action.

AED
5.6

Time off_____________

Time on_____________

Total time_____________

CDX Tasksheet Number: EO171

1. **Referring to the manufacturer's workshop manual, inspect the cylinder sleeve counterbore and lower bore; check bore distortion.**

   a. **Measure counter in multiple spots for concentricity and depth according to manufacturer's specification.**

      I. **Record the measurements:** _________________________

   b. **Inspect counter bore for cracking and gaulling from loose liner condition.**

      I. **Meets manufacturer's specifications? Yes:** _______ **No:** _______

   c. **Inspect lower counter bore for pitting or water damage due to improper pH or alkalinity levels.**

      I. **Meets manufacturer's specifications? Yes:** _______ **No:** _______

2. **Discuss the findings with your instructor.**

Performance Rating   **CDX Tasksheet Number: EO171**

0	1	2	3	4

Supervisor/instructor signature _________________________________ Date _____________

Student/Intern information:

Name _______________________________ Date ____________ Class _________________________

Vehicle used for this activity:

Year _______________ Make _________________________________ Model _______________________

Odometer _____________ Hour meter ____________ VIN _______________________________________

▶ TASK Clean, inspect, and measure cylinder walls or liners for wear and damage; determine needed action.

AED 5.6

Time off_______________

Time on_______________

Total time_______________

CDX Tasksheet Number: E0172

1. **Referring to the manufacturer's workshop manual, inspect and measure cylinder walls or liners for wear and damage.**

   a. **Specifications:**

      I. **Cylinder diameter:** _______________ in/mm

      II. **Cylinder taper:** _______________ in/mm

      III. **Cylinder ovality:** _______________ in/mm

      IV. **Minimum wall thickness:** _______________ in/mm

      V. **Cylinder wall construction:** _______________

   b. **Carry out a visual inspection of the cylinders.**

      I. **Meets the manufacturer's specifications:**
      Yes: _________ **No:** _________

      II. **If no, list the areas of concern and your recommendations for any rectifications:**

      ____

      ____

      ____

      ____

      ____

      ____

   c. **Referring to the manufacturer's workshop manual and using the correct recommended tools, measure and record the reading for each engine's cylinders. Record your findings in the table below:**

Cylinder	Diameter (Top)	Diameter @ 90° (top)	Diameter (Bottom)	Diameter @ 90° (Bottom)	Cylinder Taper	Cylinder Ovality
1						
2						
3						
4						
5						
6						

7					
8					
9					
10					
11					
12					

    i. **Meets the manufacturer's specifications:**

       Yes: _________ No: _________

    ii. **If no, list the areas of concern and your recommendations for any rectifications:**

      ____

      ____

      ____

      ____

      ____

2. Discuss your findings with your instructor.

Performance Rating

CDX Tasksheet Number: EO172

0	1	2	3	4

Supervisor/instructor signature ________________________________ Date __________

Engine Subsystems: Engine Block Assembly

Student/Intern information:

Name ________________________________ Date ____________ Class ________________________

Vehicle used for this activity:

Year ______________ Make ____________________________ Model ____________________

Odometer ____________ Hour meter ____________ VIN ________________________________

Learning Objective/Task	CDX Tasksheet Number	2014 Edition Rev2/12/16 AED Standard
• Replace/reinstall cylinder liners and seals; check and adjust liner height (protrusion).	EO173	5.6
• Inspect in-block camshaft bearings for wear and damage; determine needed action.	EO174	5.6
• Inspect, measure, and replace/reinstall in-block camshaft; measure/adjust end play.	EO175	5.6
• Clean and inspect crankshaft for surface cracks and journal damage; check condition of oil passages; check passage plugs; measure journal diameter; determine needed action.	EO176	5.6
• Inspect main bearings for wear patterns and damage; replace as needed; check bearing clearances; check and correct crankshaft end play.	EO177	5.6
• Inspect, install, and time gear train; measure gear backlash; determine needed action.	EO178	5.6

Materials Required

- Machine with possible engine concern
- Machine manufacturer's workshop manual
- Manufacturer-specific tools depending on the concern
- Machine lifting equipment, if applicable

Some Safety Issues to Consider

- Diagnosis of this fault may require test driving the machine on the school grounds or on a hoist, both of which carry severe risks. Attempt this task only with full permission from your supervisor/instructor and follow all the guidelines exactly.
- Caution: If you are working in an area where there could be "brake dust" present (may contain asbestos, which has been determined to cause cancer when inhaled or ingested), ensure you wear and use all OSHA-approved asbestos protective/removal equipment.

- Lifting equipment such as machine jacks and stands, machine hoists, and engine hoists are important tools that increase productivity and make the job easier. However, they can also cause severe injury or death if used improperly. Make sure you follow the manufacturer's operation procedures. Also make sure you have your supervisor/instructor's permission to use any particular type of lifting equipment.
- Comply with personal and environmental safety practices associated with clothing; eye protection; hand tools; power equipment; proper ventilation; and the handling, storage, and disposal of chemicals/materials in accordance with federal, state, and local regulations.
- Always wear the correct protective eyewear and clothing and use the appropriate safety equipment, as well as fender covers, seat protectors, and floor mat protectors.
- Make sure you understand and observe all legislative and personal safety procedures when carrying out practical assignments. If you are unsure of what these are, ask your supervisor/instructor.

Performance Standard

0–No exposure: No information or practice provided during the program; complete training required

1–Exposure only: General information provided with no practice time; close supervision needed; additional training required

2–Limited practice: Has practiced job during training program; additional training required to develop skill

3–Moderately skilled: Has performed job independently during training program; limited additional training may be required

4–Skilled: Can perform job independently with no additional training

▶ **TASK** Replace/reinstall cylinder liners and seals; check and adjust liner height (protrusion).

AED
5.6

Time off____________________

Time on____________________

Total time____________________

CDX Tasksheet Number: EO173

1. **Refer to the manufacturer's workshop manual. List the procedure and all safety precautions that must be observed when you replace/reinstall the cylinder liners and seals and check and adjust the liner height (protrusion).**

   a. **List the steps involved when you replace/reinstall the cylinder liners and seals and check and adjust the liner height (protrusion):**

   ____
   ____
   ____
   ____
   ____

   b. **Determine what safety precautions must be observed when you replace/reinstall the cylinder liners and seals and check and adjust the liner height (protrusion):**

   ____
   ____
   ____
   ____
   ____

2. **Discuss these procedures and safety precautions with your instructor.**

3. **If directed by your instructor, replace/reinstall the cylinder liners and seals.**

4. **Following the procedures listed above, and referring to the manufacturer's workshop manual, check and adjust the liner height (protrusion).**

   a. **Meets the manufacturer's specifications:**
   Yes: ________ No: ________

   b. **If no, list your recommendations for any rectifications:**

   ____
   ____
   ____
   ____
   ____

5. **Record your measurements in the table below:**

Cylinder	Protrusion Measurement (in/mm)
1	
2	
3	
4	
5	
6	
7	
8	
9	
10	
11	
12	

6. **Discuss the findings with your instructor.**

Student/Intern information:

Name _____________________________ Date ___________ Class _____________________

Vehicle used for this activity:

Year _____________ Make ___________________________ Model _____________________

Odometer ____________ Hour meter ___________ VIN _______________________________

▶ **TASK** Inspect in-block camshaft bearings for wear and damage; determine needed action.

AED
5.6

Time off____________________

Time on____________________

Total time__________________

CDX Tasksheet Number: EO174

1. **Refer to the manufacturer's workshop manual and inspect the in-block camshaft bearings for wear and damage.**

    a. **Meets the manufacturer's specifications:**
       Yes: _____________ No: _____________

    b. **If no, list your recommendations for any rectifications:**

    ____

    ____

    ____

    ____

    ____

2. **Referring to the manufacturer's workshop manual, measure the in-block camshaft bearings for wear.**

    a. **Specifications:**

       i.   **Bearing diameter: _____________ in/mm**
       ii. **Bearing taper: _____________ in/mm**
       iii. **Bearing ovality: _____________ in/mm**
       iv. **Minimum wall thickness: _____________ in/mm**

3. **Using the appropriate measuring instrument(s), measure the bearings and record your findings in the table below:**

Journal	Diameter Front	Diameter @ 90° Front	Diameter Back	Diameter @ 90° Back	Bearing Taper	Bearing Ovality
1						
2						
3						
4						
5						
6						
7						
8						

a. **Meets the manufacturer's specifications:**
   Yes: _______________ No: _______________

b. **If no, list your recommendations for any rectifications:**

____

____

____

____

____

4. **Discuss the findings with your instructor.**

Performance Rating

CDX Tasksheet Number: EO174

0	1	2	3	4

Supervisor/instructor signature ____ Date _______________

Name ___________________________ Date ____________ Class ___________________________

Vehicle used for this activity:

Year ______________ Make ___________________________ Model ___________________________

Odometer ____________ Hour meter ____________ VIN ___________________________

▶ **TASK** Inspect, measure, and replace/reinstall in-block camshaft; measure/adjust end play.

AED
5.6

Time off_______________

Time on_______________

Total time_______________

CDX Tasksheet Number: EO175

1. **Refer to the manufacturer's workshop manual. List the procedure and all safety precautions that must be observed when you inspect, measure, and replace/reinstall the in-block camshaft and measure/adjust the end play.**

    a. **List the steps involved to inspect, measure, and replace/reinstall the in-block camshaft and to measure/adjust the end play:**

    ____
    ____
    ____
    ____
    ____
    ____

    b. **Determine what safety precautions must be observed as you inspect, measure, and replace/reinstall the in-block camshaft and measure/ adjust the end play:**

    ____
    ____
    ____
    ____
    ____

2. **Discuss these procedures and safety precautions with your instructor.**

3. **If directed by your instructor, inspect, measure, and replace/reinstall the in-block camshaft and measure/adjust the end play.**

4. **Referring to the manufacturer's workshop manual, measure the in-block camshaft bearings for wear.**

    a. **Specifications:**
       i. **Camshaft end play:** ____________ **in/mm**
    b. **Actual:**
       i. **Camshaft end play:** ____________ **in/mm**

    c. **Meets the manufacturer's specifications:**
       Yes: _______________ No: _______________

    d. **If no, list your recommendations for any rectifications:**

    ____

    ____

    ____

    ____

    ____

5. Discuss the findings with your instructor.

Performance Rating   **CDX Tasksheet Number: EO175**

0	1	2	3	4

Supervisor/instructor signature ____ Date _______________

Name ________________________________ Date ____________ Class ________________________

Vehicle used for this activity:

Year ______________ Make ________________________________ Model ________________________

Odometer ____________ Hour meter ____________ VIN ________________________________

▶ TASK Clean and inspect crankshaft for surface cracks and journal damage; check condition of oil passages; check passage plugs; measure journal diameter; determine needed action.

AED 5.6

Time off____________

Time on____________

Total time____________

CDX Tasksheet Number: EO176

1. **Refer to the manufacturer's workshop manual. List the procedure and all safety precautions that must be observed when you clean and inspect the crankshaft for surface cracks and journal damage, check the condition of the oil passages, check the passage plugs, and measure the journal diameter.**

    a. **List the steps involved to clean and inspect the crankshaft for surface cracks and journal damage, to check the condition of oil passages, to check the passage plugs, and to measure the journal diameter:**

    ____

    ____

    ____

    ____

    ____

    b. **Determine what safety precautions must be observed as you clean and inspect the crankshaft for surface cracks and journal damage, check the condition of the oil passages, check the passage plugs, and measure the journal diameter:**

    ____

    ____

    ____

    ____

    ____

2. **Following the procedures listed above, and referring to the manufacturer's workshop manual, clean and inspect the crankshaft for surface cracks and journal damage.**

    a. **Meets the manufacturer's specifications:**
       Yes: ______________ **No:** ______________

    b. **If no, list your recommendations for any rectifications:**

    ____

    ____

    ____

    ____

    ____

3. Following the procedures listed above, and referring to the manufacturer's workshop manual, check the condition of the oil passages.

    a. Meets the manufacturer's specifications:
       Yes: _______________ No: _______________

    b. If no, list your recommendations for any rectifications:

       ____

       ____

       ____

       ____

       ____

4. Following the procedures listed above, and referring to the manufacturer's workshop manual, check the passage plugs.

    a. Meets the manufacturer's specifications:
       Yes: _______________ No: _______________

    b. If no, list your recommendations for any rectifications:

       ____

       ____

       ____

       ____

       ____

5. While referring to the manufacturer's workshop manual, measure the crankshaft journal diameter.

    a. Specifications:
       i.   Bearing diameter: _______________ in/mm
       ii.   Bearing taper: _______________ in/mm
       iii. Bearing ovality: _______________ in/mm
       iv. Minimum wall thickness: _______________ in/mm

6. While using the appropriate measuring instrument(s), measure the bearings and record your findings in the table below:

Journal	Diameter Front	Diameter @ 90°	Diameter Back	Diameter @ 90°	Bearing Taper	Bearing Ovality
1						
2						
3						
4						
5						
6						
7						
8						
9						
10						

a. **Meets the manufacturer's specifications:**
   Yes: _______________ **No:** _______________

b. **If no, list your recommendations for any rectifications:**

____

____

____

____

____

7. **Discuss the findings with your instructor.**

Name _________________________________ Date _____________ Class _________________________

Vehicle used for this activity:

Year _______________ Make _______________________________ Model _____________________________

Odometer _______________ Hour meter _____________ VIN _________________________________

▶ **TASK** Inspect main bearings for wear patterns and damage; replace as needed; check bearing clearances; check and correct crankshaft end play.

AED
5.6

Time off________________

Time on________________

Total time________________

CDX Tasksheet Number: EO177

1. **Refer to the manufacturer's workshop manual. List the procedure and all safety precautions that must be observed when you inspect the main bearings for wear patterns and damage and replace as needed, check the bearing clearances, and check and correct the crankshaft end play.**

   a. **List the steps involved to inspect the main bearings for wear patterns and damage and replace as needed, to check the bearing clearances, and to check and correct the crankshaft end play:**

   ____
   ____
   ____
   ____
   ____
   ____
   ____

   b. **Determine what safety precautions must be observed as you inspect the main bearings for wear patterns and damage and replace as needed, check the bearing clearances, and check and correct the crankshaft end play:**

   ____
   ____
   ____
   ____
   ____
   ____
   ____

2. **Following the procedures listed above, and referring to the manufacturer's workshop manual, inspect the main bearings for wear patterns and damage.**

   a. **Meets the manufacturer's specifications:**
      Yes: _______________ No: _______________

   b. **If no, list your recommendations for any rectifications:**

   ____
   ____
   ____
   ____
   ____
   ____

3. Following the procedures listed above, and referring to the manufacturer's workshop manual, check the bearing clearances and record your findings in the table below:

Cylinder	Bearing Clearances (in/mm)
1	
2	
3	
4	
5	
6	
7	
8	
9	
10	
11	
12	

   a. **Meets the manufacturer's specifications:**

   Yes: _______________ No: _______________

   b. **If no, list your recommendations for any rectifications:**

____

____

____

____

____

____

4. **Referring to the manufacturer's workshop manual, install the crankshaft and torque main bearing caps.**

   a. **Specifications:**

   Crankshaft main bearing cap torque: _______________ ft-lb (N·m)

   b. **Actual:**

   Crankshaft main bearing cap torque: _______________ ft-lb (N·m)

5. **Referring to the manufacturer's workshop manual, measure the in-block camshaft bearings for wear.**

   a. **Specifications:**

   Crankshaft end play: _______________ in/mm

   b. **Actual:**

   Crankshaft end play: _______________ in/mm

   c. **Meets the manufacturer's specifications:**

   Yes: _______________ No: _______________

    d. If no, list your recommendations for any rectifications:

____
____
____
____
____
____

6. Discuss the findings with your instructor.

Performance Rating

CDX Tasksheet Number: E0177

0	1	2	3	4

Supervisor/instructor signature ____ Date _____________

▶ **TASK** Inspect, install, and time gear train; measure gear backlash; determine needed action.

AED 5.6

Time off_____________

Time on_____________

Total time_____________

CDX Tasksheet Number: EO178

1. **Referring to the manufacturer's workshop manual, inspect, install, and time the gear train.**

2. **Referring to the manufacturer's workshop manual, measure the gear backlash.**

   a. **Specifications:**
      i. **Gear backlash: _______________ in/mm**

   b. **Actual:**
      i. **Gear backlash: _______________ in/mm**

   c. **Meets the manufacturer's specifications:**
      Yes: _____________ No: _____________

   d. **If no, list your recommendations for any rectifications:**

      ____

      ____

      ____

      ____

      ____

      ____

      ____

3. **Discuss the findings with your instructor.**

Engine Subsystems: Piston and Connecting Rod Inspection and Assembly

Student/Intern information:

Name _________________________ Date ___________ Class _________________________

Vehicle used for this activity:

Year ______________ Make _________________________ Model _________________________

Odometer ___________ Hour meter ___________ VIN _________________________

Time off_________

Time on_________

Total time_________

Learning Objective/Task	CDX Tasksheet Number	2014 Edition Rev2/12/16 AED Standard
• Inspect connecting rod and bearings for wear patterns; measure pistons, pins, retainers, and bushings; perform needed action.	E0179	5.6
• Determine piston-to-cylinder wall clearance; check ring-to-groove fit and end gap; install rings on pistons.	E0180	5.6
• Assemble pistons and connecting rods; install in block; install rod bearings and check clearances.	E0181	5.6

Materials Required

- Machine with possible engine concern
- Machine manufacturer's workshop manual
- Manufacturer-specific tools depending on the concern
- Machine lifting equipment, if applicable

Some Safety Issues to Consider

- Diagnosis of this fault may require test driving the machine on the school grounds or on a hoist, both of which carry severe risks. Attempt this task only with full permission from your supervisor/instructor and follow all the guidelines exactly.
- Caution: If you are working in an area where there could be "brake dust" present (may contain asbestos, which has been determined to cause cancer when inhaled or ingested), ensure you wear and use all OSHA-approved asbestos protective/removal equipment.
- Lifting equipment such as machine jacks and stands, machine hoists, and engine hoists are important tools that increase productivity and make the job easier. However, they can also cause severe injury or death if used improperly. Make sure you follow the manufacturer's operation procedures. Also make sure you have your supervisor/instructor's permission to use any particular type of lifting equipment.
- Comply with personal and environmental safety practices associated with clothing; eye protection; hand tools; power equipment; proper ventilation; and the handling, storage, and disposal of chemicals/materials in accordance with federal, state, and local regulations.
- Always wear the correct protective eyewear and clothing and use the appropriate safety equipment, as well as fender covers, seat protectors, and floor mat protectors.

- Make sure you understand and observe all legislative and personal safety procedures when carrying out practical assignments. If you are unsure of what these are, ask your supervisor/instructor.

Performance Standard

0–No exposure: No information or practice provided during the program; complete training required

1–Exposure only: General information provided with no practice time; close supervision needed; additional training required

2–Limited practice: Has practiced job during training program; additional training required to develop skill

3–Moderately skilled: Has performed job independently during training program; limited additional training may be required

4–Skilled: Can perform job independently with no additional training

Student/Intern information:

Name _________________________________ Date ____________ Class _____________________

Vehicle used for this activity:

Year _______________ Make ___________________________________ Model _____________________

Odometer _____________ Hour meter _____________ VIN _________________________________

▶ TASK Inspect connecting rod and bearings for wear patterns; measure pistons, pins, retainers, and bushings; perform needed action.

**AED
5.6**

Time off_______________

Time on_______________

Total time_______________

CDX Tasksheet Number: EO179

1. **Refer to the manufacturer's workshop manual; inspect connecting rod and bearings for wear patterns:**

   a. **Meets the manufacturer's specifications:**
      Yes: _________ No: _________

   b. **If no, list your recommendations for any rectifications:**

   ____
   ____
   ____
   ____
   ____

2. **Refer to the manufacturer's workshop manual; measure pistons, pins, and retainers, using the appropriate measuring instrument(s).**

Piston	Measurement-in/mm					
	Diameter Top	Diameter @ 90° Top	Diameter Bottom	Diameter @ 90° Bottom	Piston Ovality/ Taper	Piston Pin (outside diameter)

a. **Meets the manufacturer's specifications:**
Yes: _________ No: _________

b. **If no, list your recommendations for any rectifications:**

____

____

____

____

____

3. **Discuss the findings with your instructor.**

Performance Rating

CDX Tasksheet Number: EO179

0	1	2	3	4
☐	☐	☐	☐	☐

Supervisor/instructor signature _________________________________ Date _____________

Student/Intern information:

Name _________________________________ Date _____________ Class _____________________

Vehicle used for this activity:

Year _______________ Make _________________________________ Model _____________________

Odometer _____________ Hour meter _____________ VIN _________________________________

▶ **TASK** Determine piston-to-cylinder wall clearance; check ring-to-groove
fit and end gap; install rings on pistons.

AED
5.6

Time off___________________

Time on___________________

Total time___________________

CDX Tasksheet Number: EO180

1. **Refer to the manufacturer's workshop manual; determine
the piston-to-cylinder wall clearance for all pistons and cylinders.**

    a. **Piston #1:** _________ **in/mm**

    b. **Piston #2:** _________ **in/mm**

    Meets manufacturer's specifications:
    Yes:_________ No:_________

    If no, list the recommendations for any rectifications:

    ____

    ____

    ____

    ____

    ____

2. **Refer to the manufacturer's workshop manual; check ring-to-groove fit and
end gap:**
End gap specification: _________ **in/mm**

Piston	Measurement-in/mm		
	Compression Ring 1	**Compression Ring 2**	**Oil Control Ring**
1			
2			
3			
4			
5			
6			
7			
8			
9			
10			
11			
12			

a. **Meets the manufacturer's specifications:**
   Yes: _________ **No:** _________

b. **If no, list your recommendations for any rectifications:**

____

____

____

____

____

3. **Refer to the manufacturer's workshop manual; install rings on pistons.**

4. **Discuss the findings with your instructor.**

Performance Rating **CDX Tasksheet Number: EO180**

0	1	2	3	4
☐	☐	☐	☐	☐

Supervisor/instructor signature ___________________________________ Date ___________

▶ TASK Assemble pistons and connecting rods; install in block; install rod bearings and check clearances.

AED 5.6

Time off________________

Time on________________

Total time________________

CDX Tasksheet Number: EO181

1. **Refer to the manufacturer's workshop manual; assemble pistons and connecting rods.**

   a. **Meets the manufacturer's specifications:**
      Yes: __________ No: __________

   b. **If no, list your recommendations for any rectifications:**

      ____

      ____

      ____

      ____

      ____

2. **Refer to the manufacturer's workshop manual; assemble pistons and connecting rods:**

   - **Oil the piston assembly.**

   - **Oil the cylinder.**

   - **As each piston assembly is being installed, remove the big end cap as the piston assembly is about to be installed.**

   - **Using an appropriate piston ring installation tool.**
     - **Install each piston assembly;**
     - **Install connecting rod bearing and refit connecting rod cap and bolts.**

3. **Refer to manufacturer's workshop manual; list the procedure when you check clearances for connecting rod bearing:**

   ____

   ____

   ____

   ____

   ____

4. **Refer to the manufacturer's workshop manual; check clearances for connecting rod bearing.**

    a. **Meets the manufacturer's specifications:**
       Yes: _________ No: _________

    b. **If no, list your recommendations for any rectifications:**

       ____

       ____

       ____

       ____

       ____

5. **Discuss the findings with your instructor.**

Performance Rating

CDX Tasksheet Number: EO181

0	1	2	3	4
☐	☐	☐	☐	☐

Supervisor/instructor signature ___________________________________ Date ____________

Engine Subsystems: Crankshaft Sub-component Inspection and Assembly

---

Student/Intern information:

Name _________________________________ Date ____________ Class _________________________

Vehicle used for this activity:

Year _______________ Make _______________________________ Model _________________________

Odometer _____________ Hour meter _____________ VIN _________________________________

Learning Objective/Task	CDX Tasksheet Number	2014 Edition Rev2/12/16 AED Standard
• Check the condition of piston cooling jets (nozzles); determine needed action.	E0182	5.6
• Inspect crankshaft vibration damper; determine needed action.	E0183	5.6
• Install and align flywheel housing; inspect flywheel housing(s) to transmission housing/engine mating surface(s) and measure flywheel housing face and bore runout; determine needed action.	E0184	5.6
• Inspect flywheel/flexplate (including ring gear) and mounting surfaces for cracks and wear; measure runout; determine needed action.	E0185	5.6

Time off_______________

Time on_______________

Total time_______________

Materials Required

- Machine with possible engine concern
- Machine manufacturer's workshop manual
- Manufacturer-specific tools depending on the concern
- Machine lifting equipment, if applicable

Some Safety Issues to Consider

- Diagnosis of this fault may require test driving the machine on the school grounds or on a hoist, both of which carry severe risks. Attempt this task only with full permission from your supervisor/instructor and follow all the guidelines exactly.
- Caution: If you are working in an area where there could be "brake dust" present (may contain asbestos, which has been determined to cause cancer when inhaled or ingested), ensure you wear and use all OSHA-approved asbestos protective/removal equipment.
- Lifting equipment such as machine jacks and stands, machine hoists, and engine hoists are important tools that increase productivity and make the job easier. However, they can also cause severe injury or death if used improperly. Make sure you follow the manufacturer's operation procedures. Also make sure you have your supervisor/instructor's permission to use any particular type of lifting equipment.

- Comply with personal and environmental safety practices associated with clothing; eye protection; hand tools; power equipment; proper ventilation; and the handling, storage, and disposal of chemicals/materials in accordance with federal, state, and local regulations.
- Always wear the correct protective eyewear and clothing and use the appropriate safety equipment, as well as fender covers, seat protectors, and floor mat protectors.
- Make sure you understand and observe all legislative and personal safety procedures when carrying out practical assignments. If you are unsure of what these are, ask your supervisor/instructor.

Performance Standard

0–No exposure: No information or practice provided during the program; complete training required

1–Exposure only: General information provided with no practice time; close supervision needed; additional training required

2–Limited practice: Has practiced job during training program; additional training required to develop skill

3–Moderately skilled: Has performed job independently during training program; limited additional training may be required

4–Skilled: Can perform job independently with no additional training

Student/Intern information:

Name _________________________ Date ____________ Class _________________________

Vehicle used for this activity:

Year _____________ Make _____________________________ Model _________________________

Odometer ____________ Hour meter ____________ VIN _________________________

▶ TASK Check the condition of piston cooling jets (nozzles); determine needed action.

AED 5.6

Time off_________________

Time on_________________

Total time_________________

CDX Tasksheet Number: E0182

1. **Refer to the manufacturer's workshop manual and check the condition of the piston cooling jets (nozzles).**

    a. **Meets the manufacturer's specifications:**
       Yes: ________ **No:** ________

    b. **If no, list your recommendations for any rectifications:**

       ____
       ____
       ____
       ____
       ____
       ____

2. **Discuss the findings with your instructor.**

Performance Rating　　　　　　　　　**CDX Tasksheet Number: E0182**

0	1	2	3	4

Supervisor/instructor signature _________________________________ Date ____________

Name ___________________________ Date ___________ Class ___________________

Vehicle used for this activity:

Year _____________ Make _________________________ Model ___________________

Odometer ___________ Hour meter ___________ VIN _____________________________

▶ TASK Inspect crankshaft vibration damper; determine needed action. **AED 5.6**

Time off_____________

CDX Tasksheet Number: EO183

Time on_____________

1. **Refer to the manufacturer's workshop manual and inspect the crankshaft vibration damper.**

   a. **Inspect the viscous dampers for dents which could render it unusable.**

   b. **Check for run out using a dial indicator.**

      i. **Record the reading:** _______________

      ii. **Compare the reading to manufacturer's specifications.**

2. **Discuss the findings with your instructor.**

Total time_____________

Performance Rating

CDX Tasksheet Number: EO183

0	1	2	3	4

Supervisor/instructor signature ____ Date ___________

Name _________________________________ Date ____________ Class _________________________

Vehicle used for this activity:

Year _______________ Make ___________________________ Model _________________________

Odometer ____________ Hour meter ____________ VIN _________________________________

▶ TASK Install and align flywheel housing; inspect flywheel housing(s) to transmission housing/engine mating surface(s) and measure flywheel housing face and bore runout; determine needed action.

**AED
5.6**

Time off_______________

Time on_______________

Total time_______________

CDX Tasksheet Number: EO184

1. **Refer to the manufacturer's workshop manual and install and align the flywheel housing.**

   a. **Meets the manufacturer's specifications:**
      Yes: _________ **No:** _________
   b. **If no, list your recommendations for any rectifications:**

   ____
   ____
   ____
   ____
   ____

2. **Refer to the manufacturer's workshop manual and inspect the flywheel housing(s) to the transmission housing/engine mating surface(s).**

   a. **Meets the manufacturer's specifications:**
      Yes: _________ **No:** _________
   b. **If no, list your recommendations for any rectifications:**

   ____
   ____
   ____
   ____
   ____

3. **Referring to the manufacturer's workshop manual, measure the flywheel housing face and bore runout.**

   a. **Specifications:**
      i. **Runout:** _______________in/mm
   b. **Actual:**
      i. **Runout:** _______________in/mm
   c. **Meets the manufacturer's specifications:**
      Yes: _________ **No:** _________
   d. **If no, list your recommendations for any rectifications:**

   ____
   ____
   ____
   ____

4. Discuss the findings with your instructor.

▶ TASK Inspect flywheel/flexplate (including ring gear) and mounting surfaces for cracks and wear; measure runout; determine needed action.

AED
5.6

Time off_________________

Time on_________________

Total time_________________

CDX Tasksheet Number: EO185

1. **Refer to the manufacturer's workshop manual and inspect the flywheel/flexplate (including the ring gear) and mounting surfaces for cracks and wear.**

   a. **Meets the manufacturer's specifications:**
   Yes: _________ No: _________

   b. **If no, list your recommendations for any rectifications:**

   ____
   ____
   ____
   ____
   ____

2. **Referring to the manufacturer's workshop manual, measure the runout using a dial indicator.**

   a. **Specifications:**

   i. **Runout: _______________ in/mm**

   b. **Actual:**

   i. **Runout: _______________ in/mm**

   c. **Meets the manufacturer's specifications:**
   Yes: _________ No: _________

   d. **If no, list your recommendations for any rectifications:**

   ____
   ____
   ____
   ____
   ____

3. **Discuss the findings with your instructor.**

Component Repair/Diagnostics: Lubrication System

Student/Intern information:

Name _________________________ Date __________ Class _________________________

Vehicle used for this activity:

Year _____________ Make _________________________ Model _________________________

Odometer ___________ Hour meter ___________ VIN _________________________

Learning Objective/Task	CDX Tasksheet Number	2014 Edition Rev2/12/16 AED Standard
• Test engine oil pressure and check operation of pressure sensor, gauge, and/or sending unit; test engine oil temperature and check operation of temperature sensor; determine needed action.	E0186	5.8
• Check engine oil level, condition, and consumption; determine needed action.	E0187	5.8
• Inspect and measure oil pump, drives, inlet pipes, and pick-up screens; check drive gear clearances; determine needed action.	E0188	5.5
• Inspect oil pressure regulator valve(s), bypass and pressure relief valve(s), oil thermostat, and filters; determine needed action.	E0189	5.5
• Inspect, clean, and test oil cooler and components; determine needed action.	E0190	5.5
• Inspect turbocharger lubrication systems; determine needed action.	E0191	5.5
• Determine proper lubricant and perform oil and filter change.	E0192	5.5

Materials Required

- Machine with possible engine concern
- Machine manufacturer's workshop manual
- Manufacturer-specific tools depending on the concern
- Machine lifting equipment, if applicable

Some Safety Issues to Consider

- Diagnosis of this fault may require test driving the machine on the school grounds or on a hoist, both of which carry severe risks. Attempt this task only with full permission from your supervisor/instructor and follow all the guidelines exactly.

- Caution: If you are working in an area where there could be "brake dust" present (may contain asbestos, which has been determined to cause cancer when inhaled or ingested), ensure you wear and use all OSHA-approved asbestos protective/removal equipment.
- Lifting equipment such as machine jacks and stands, machine hoists, and engine hoists are important tools that increase productivity and make the job easier. However, they can also cause severe injury or death if used improperly. Make sure you follow the manufacturer's operation procedures. Also make sure you have your supervisor/instructor's permission to use any particular type of lifting equipment.
- Comply with personal and environmental safety practices associated with clothing; eye protection; hand tools; power equipment; proper ventilation; and the handling, storage, and disposal of chemicals/materials in accordance with federal, state, and local regulations.
- Always wear the correct protective eyewear and clothing and use the appropriate safety equipment, as well as fender covers, seat protectors, and floor mat protectors.
- Make sure you understand and observe all legislative and personal safety procedures when carrying out practical assignments. If you are unsure of what these are, ask your supervisor/instructor.

Performance Standard

0–No exposure: No information or practice provided during the program; complete training required

1–Exposure only: General information provided with no practice time; close supervision needed; additional training required

2–Limited practice: Has practiced job during training program; additional training required to develop skill

3–Moderately skilled: Has performed job independently during training program; limited additional training may be required

4–Skilled: Can perform job independently with no additional training

Student/Intern information:

Name _______________________________ Date ___________ Class _______________________

Vehicle used for this activity:

Year _______________ Make _____________________________ Model _____________________

Odometer ____________ Hour meter ___________ VIN _________________________________

▶ TASK Test engine oil pressure and check operation of pressure sensor, gauge, and/or sending unit; test engine oil temperature and check operation of temperature sensor; determine needed action.

AED 5.8

Time off____________

Time on____________

Total time____________

CDX Tasksheet Number: EO186

1. **Referring to the manufacturer's workshop manual, test the engine oil pressure and check operation of the pressure sensor, gauge, and/or sending unit.**

   a. **Meets the manufacturer's specifications:**
      Yes: _________ No: _________

   b. **If no, list your recommendations for any rectifications:**

      ____
      ____
      ____
      ____
      ____

2. **Referring to the manufacturer's workshop manual, test the engine oil temperature and check operation of the temperature sensor.**

   a. **Meets the manufacturer's specifications:**
      Yes: _________ No: _________

   b. **If no, list your recommendations for any rectifications:**

      ____
      ____
      ____
      ____
      ____

3. **Discuss the findings with your instructor.**

Performance Rating

CDX Tasksheet Number: EO186

0	1	2	3	4

Supervisor/instructor signature ____ Date _____________

Student/Intern information:

Name _________________________ Date _________ Class _________________________

Vehicle used for this activity:

Year _____________ Make _________________________ Model _________________________

Odometer ___________ Hour meter ___________ VIN _________________________

▶ TASK Check engine oil level, condition, and consumption; determine needed action.

AED
5.8

Time off_________________

Time on_________________

CDX Tasksheet Number: E0187

1. **Refer to the manufacturer's workshop manual and check the engine oil level, condition, and consumption.**

Total time_________________

   a. **Meets the manufacturer's specifications:**
      Yes: _________ **No:** _________

   b. **If no, list your recommendations for any rectifications:**

      ____

      ____

      ____

      ____

      ____

2. **Discuss the findings with your instructor.**

Performance Rating

CDX Tasksheet Number: E0187

0	**1**	**2**	**3**	**4**

Supervisor/instructor signature _________________________________ Date _____________

Student/Intern information:

Name _______________________________ Date _____________ Class _____________________

Vehicle used for this activity:

Year _______________ Make _______________________ Model _____________________

Odometer _____________ Hour meter _____________ VIN _________________________

▶ **TASK** Inspect and measure oil pump, drives, inlet pipes, and pick-up
screens; check drive gear clearances; determine needed action.

AED
5.5

Time off____________

Time on____________

CDX Tasksheet Number: E0188

Total time____________

1. **Refer to the manufacturer's workshop manual and inspect and measure the oil pump, drives, inlet pipes, and pick-up screens.**

   a. **Meets the manufacturer's specifications:**
   Yes: _________ **No:** _________

   b. **If no, list your recommendations for any rectifications:**

   ____

   ____

   ____

   ____

   ____

2. **Refer to the manufacturer's workshop manual and check the drive gear clearances.**

   a. **Meets the manufacturer's specifications:**
   Yes: _________ **No:** _________

   b. **If no, list your recommendations for any rectifications:**

   ____

   ____

   ____

   ____

   ____

3. **Discuss the findings with your instructor.**

Performance Rating   **CDX Tasksheet Number: E0188**

0	**1**	**2**	**3**	**4**

Supervisor/instructor signature _________________________________ Date _____________

Student/Intern information:

Name _____________________________ Date _____________ Class _____________________

Vehicle used for this activity:

Year _______________ Make _______________________ Model _____________________

Odometer _____________ Hour meter _____________ VIN _______________________

▶ TASK Inspect oil pressure regulator valve(s), bypass and pressure relief valve(s), oil thermostat, and filters; determine needed action.

AED 5.5

CDX Tasksheet Number: E0189

1. **Refer to the manufacturer's workshop manual and inspect the oil pressure regulator valve(s), bypass and pressure relief valve(s), oil thermostat, and filters.**

   a. **Meets the manufacturer's specifications:**
      Yes: _________ No: _________

   b. **If no, list your recommendations for any rectifications:**

      ____
      ____
      ____
      ____
      ____
      ____

2. **Discuss the findings with your instructor.**

Performance Rating

CDX Tasksheet Number: E0189

0	1	2	3	4

Supervisor/instructor signature _________________________________ Date _____________

▶ TASK Inspect, clean, and test oil cooler and components; determine needed action.

AED 5.5

Time off_________________

Time on_________________

Total time_________________

CDX Tasksheet Number: E0190

1. **Refer to the manufacturer's workshop manual and inspect, clean, and test the oil cooler and components.**

   Note: Oil cooler should be placed at time of overhaul.

   a. **Meets the manufacturer's specifications:**
      Yes: _________ **No:** _________

   b. **If no, list your recommendations for any rectifications:**

   ____
   ____
   ____
   ____
   ____

2. **Discuss the findings with your instructor.**

Performance Rating

CDX Tasksheet Number: E0190

0	1	2	3	4

Supervisor/instructor signature _________________________________ Date ___________

Student/Intern information:

Name _______________________________ Date ____________ Class _______________________________

Vehicle used for this activity:

Year _______________ Make _________________________________ Model _______________________________

Odometer _____________ Hour meter _____________ VIN _______________________________________

▶ **TASK** Inspect turbocharger lubrication systems; determine needed action.

AED
5.5

CDX Tasksheet Number: E0191

1. **Refer to the manufacturer's workshop manual and inspect the turbocharger lubrication systems.**

    a. **Meets the manufacturer's specifications:**
    Yes: _________ No: _________

    b. **If no, list your recommendations for any rectifications:**

    ____

    ____

    ____

    ____

    ____

2. **Discuss the findings with your instructor.**

Performance Rating

CDX Tasksheet Number: E0191

0	1	2	3	4

Supervisor/instructor signature ____ Date _____________

Student/Intern information:

Name ___________________________ Date ___________ Class ___________________________

Vehicle used for this activity:

Year ___________ Make ___________________________ Model ___________________________

Odometer ___________ Hour meter ___________ VIN ___________________________

▶ TASK Determine proper lubricant and perform oil and filter change.

AED 5.5

CDX Tasksheet Number: E0192

1. **Referring to the manufacturer's workshop manual, determine the proper lubricant and perform an oil and filter change.**

   a. **Meets the manufacturer's specifications:**
      Yes: ___________ No: ___________

   b. **If no, list your recommendations for any rectifications:**

      ____

      ____

      ____

      ____

      ____

2. **Discuss the findings with your instructor.**

Performance Rating

CDX Tasksheet Number: E0192

0	1	2	3	4

Supervisor/instructor signature ___________________________ Date ___________

Component Repair/Diagnostics: Cooling System I

Learning Objective/Task	CDX Tasksheet Number	2014 Edition Rev2/12/16 AED Standard
• Check engine coolant type, level, condition, and consumption; test coolant for freeze protection and additive package concentration; determine needed action.	E0193	5.6
• Test coolant temperature and check operation of temperature and level sensors, gauge, and/or sending unit; determine needed action.	E0194	5.6
• Inspect and reinstall/replace pulleys, tensioners, and drive belts; adjust drive belts and check alignment.	E0195	5.6
• Inspect thermostat(s), bypasses, housing(s), and seals; replace as needed.	E0196	5.6
• Recover coolant, flush, and refill with recommended coolant/additive package; bleed cooling system.	E0197	5.6

Time off__________

Time on__________

Total time__________

Materials Required

- Machine with possible engine concern
- Machine manufacturer's workshop manual
- Manufacturer-specific tools depending on the concern
- Machine lifting equipment, if applicable

Some Safety Issues to Consider

- Diagnosis of this fault may require test driving the machine on the school grounds or on a hoist, both of which carry severe risks. Attempt this task only with full permission from your supervisor/instructor and follow all the guidelines exactly.
- Caution: If you are working in an area where there could be "brake dust" present (may contain asbestos, which has been determined to cause cancer when inhaled or ingested), ensure you wear and use all OSHA-approved asbestos protective/removal equipment.
- Lifting equipment such as machine jacks and stands, machine hoists, and engine hoists are important tools that increase productivity and make the job easier. However, they can also cause severe injury or death if used improperly. Make sure you follow the manufacturer's operation procedures. Also make sure you have your supervisor/instructor's permission to use any particular type of lifting equipment.

- Comply with personal and environmental safety practices associated with clothing; eye protection; hand tools; power equipment; proper ventilation; and the handling, storage, and disposal of chemicals/materials in accordance with federal, state, and local regulations.
- Always wear the correct protective eyewear and clothing and use the appropriate safety equipment, as well as fender covers, seat protectors, and floor mat protectors.
- Make sure you understand and observe all legislative and personal safety procedures when carrying out practical assignments. If you are unsure of what these are, ask your supervisor/ instructor.

Performance Standard

0–No exposure: No information or practice provided during the program; complete training required

1–Exposure only: General information provided with no practice time; close supervision needed; additional training required

2–Limited practice: Has practiced job during training program; additional training required to develop skill

3–Moderately skilled: Has performed job independently during training program; limited additional training may be required

4–Skilled: Can perform job independently with no additional training

Vehicle used for this activity:

Year _______________ Make _______________________________ Model ___________________________

Odometer _____________ Hour meter ___________ VIN _______________________________________

▶ **TASK** Check engine coolant type, level, condition, and consumption; test coolant for freeze protection and additive package concentration; determine needed action.

AED 5.6

Time off_______________

Time on_______________

Total time_______________

CDX Tasksheet Number: E0193

1. **Refer to the manufacturer's workshop manual and check engine coolant type, level, condition, and consumption.**

   a. **Manufacturer's specifications:**

      i. **Coolant type:** _______________________________

      ii. **Coolant quantity:** _______________ **pts/liters**

      iii. **Cooling system pressure cap:** _______________ **psi/kPa**

2. **Check the coolant level.**

3. **Pressure test the cooling system pressure cap:**
   Actual _______________ **psi/kPa**

4. **Pressure test the cooling system for leaks.**

5. **Draw off a sample of the coolant; have it tested for its condition. Note: If bad coolant caused catastrophic failure, send a sample out to an independent lab for analysis.**

   a. **Meets the manufacturer's specifications:**
      Yes: ________ **No:** ________

   b. **If no, list your recommendations for any rectifications:**

   ____
   ____
   ____
   ____
   ____

6. **Refer to the manufacturer's workshop manual and test the coolant for freeze protection and additive package concentration.**

    a. **Meets the manufacturer's specifications:**
        Yes: _______ No: _______

    b. **If no, list your recommendations for any rectifications:**

        ____
        ____
        ____
        ____
        ____

7. **Discuss the findings with your instructor.**

Performance Rating

CDX Tasksheet Number: EO193

0	1	2	3	4

Supervisor/instructor signature ____ Date _______________

Student/Intern information:

Name _________________________________ Date ___________ Class _____________________

Vehicle used for this activity:

Year _____________ Make _______________________ Model ___________________

Odometer ___________ Hour meter ___________ VIN _________________________

▶ TASK Test coolant temperature and check operation of temperature and level sensors, gauge, and/or sending unit; determine needed action.

AED
5.6

Time off_____________

Time on_____________

Total time_____________

CDX Tasksheet Number: E0194

1. **Refer to the manufacturer's workshop manual and test the coolant temperature.**

   a. **Manufacturer's specifications:**

      i. **Engine coolant operating temperature:** _____________

2. **Check the coolant temperature cold:** _____________

3. **Bring the engine to operating temperature; check and record the coolant temperature:** _____________

   a. **Meets the manufacturer's specifications:**
      Yes: _________ **No:** _________

   b. **If no, list your recommendations for any rectifications:**

      ____
      ____
      ____
      ____
      ____

4. **Refer to the manufacturer's workshop manual and check the operation of temperature and level sensors, gauge, and/or sending unit.**

   a. **Meets the manufacturer's specifications:**
      Yes: _________ **No:** _________

   b. **If no, list your recommendations for any rectifications:**

      ____
      ____
      ____
      ____
      ____

5. **Discuss the findings with your instructor.**

Student/Intern information:

Name _____________________________ Date ___________ Class _____________________

Vehicle used for this activity:

Year _____________ Make _____________________ Model _____________________

Odometer ___________ Hour meter ___________ VIN _____________________

▶ TASK Inspect and reinstall/replace pulleys, tensioners, and drive belts; adjust drive belts and check alignment.

AED 5.6

Time off________________

Time on________________

Total time________________

CDX Tasksheet Number: EO195

1. **Refer to the manufacturer's workshop manual and inspect and reinstall/replace the pulleys, tensioners, and drive belts; adjust the drive belts and check alignment.**

    a. **Manufacturer's specifications:**

        i. **Number of grooves in the pulley:** _____________

        ii. **Type of tensioner:** _____________

        iii. **Engine coolant drive belt tension:** _____________ **ft-lb (N·m)**

2. **Inspect the pulleys.**

    a. **Require replacement: Yes: _______ No: _______**

    b. **Reinstall, if necessary.**

3. **Inspect the belt tensioner(s).**

    a. **Require replacement: Yes: _______ No: _______**

4. **Inspect the drive belt(s).**

    a. **Require replacement: Yes: _______ No: _______**

    b. **Adjust to the manufacturer's specifications.**

5. **Check pulley alignment.**

    a. **Require realignment: Yes: _______ No: _______**

    b. **If directed by your instructor, follow the manufacturer's workshop manual instructions and re-align the pulleys.**

6. **Overall assessment**

    a. **Meets the manufacturer's specifications:**
       Yes: _______ No: _______

    b. **If no, list your recommendations for any rectifications:**

    ____
    ____
    ____
    ____
    ____

7. Discuss the findings with your instructor.

Performance Rating   **CDX Tasksheet Number: E0195**

0	1	2	3	4

Supervisor/instructor signature ______________________________________ Date ______________

Student/Intern information:

Name _______________________________ Date ____________ Class _______________________

Vehicle used for this activity:

Year _____________ Make _________________________ Model ___________________________

Odometer ____________ Hour meter ____________ VIN _______________________________

▶ TASK Inspect thermostat(s), bypasses, housing(s), and seals; replace
as needed.

AED
5.6

Time off________________

Time on_________________

CDX Tasksheet Number: EO196

Total time______________

1. **Refer to the manufacturer's workshop manual and inspect the thermostat(s), bypasses, housing(s), and seals; replace as needed.**

    a. **Manufacturer's specifications:**

        i. **Engine thermostat setting:** _______________

2. **Inspect the thermostat(s). In some cases, this may require the removal of the thermostat. It is a recommended replacement upon engine overhaul.**

    a. **Require replacement: Yes:** _______ **No:** _______

    b. **Reinstall with a new gasket, if necessary.**

3. **Inspect the bypasses.**

    a. **Require replacement: Yes:** _______ **No:** _______

4. **Inspect the housing(s).**

    a. **Require replacement: Yes:** _______ **No:** _______

5. **Check the seals.**

    a. **Require replacement: Yes:** _______ **No:** _______

    b. **Replace the seals, if necessary.**

6. **Overall assessment**

    a. **Meets the manufacturer's specifications:
    Yes:** _______ **No:** _______

    b. **If no, list your recommendations for any rectifications:**

    ____
    ____
    ____
    ____
    ____

7. **Discuss the findings with your instructor.**

© 2019 Jones & Bartlett Learning, LLC, an Ascend Learning Company

▶ **TASK** Recover coolant, flush, and refill with recommended coolant/additive package; bleed cooling system.

AED
5.6

CDX Tasksheet Number: E0197

1. **Refer to the manufacturer's workshop manual and recover, flush, and refill with the recommended coolant/additive package.**

   a. **Manufacturer's specifications:**

   i. **Coolant type:** _________________________

   ii. **Coolant quantity:** ______________ **pts/liters**

   iii. **Cooling system pressure cap:** ______________ **psi/kPa**

2. **Using a collection container adequate for coolant to be drained, drain the coolant.**

3. **Ensure that the discarded coolant is disposed of in accordance with all environmental legislative and government regulations.**

4. **Referring to the manufacturer's workshop manual, flush the cooling system. Use a coolant flush machine to make sure adequate flush of entire system.**

5. **Refill the cooling system with the manufacturer's recommended coolant and with the recommended quantity.**

6. **Run the engine.**

7. **Shut down the engine and recheck the coolant level.**

8. **Referring to the manufacturer's workshop manual, bleed the cooling system.**

9. **Cooling system checks, and inspections/flushing/refilling meets the manufacturer's specifications: Yes: _______ No: _______**

   a. **If no, list your recommendations for any rectifications:**

   ____
   ____
   ____
   ____
   ____

10. **Discuss the findings with your instructor.**

Performance Rating

CDX Tasksheet Number: EO197

| 0 | 1 | 2 | 3 | 4 |

Supervisor/instructor signature ____ Date _____________

Component Repair/Diagnostics: Cooling System II

Student/Intern information:

Name _________________________________ Date ___________ Class _________________________

Vehicle used for this activity:

Year _______________ Make _____________________________ Model _________________________

Odometer ____________ Hour meter ____________ VIN _________________________________

Learning Objective/Task	CDX Tasksheet Number	2014 Edition Rev2/12/16 AED Standard
• Inspect coolant conditioner/filter assembly for leaks; inspect valves, lines, and fittings; replace as needed.	E0198	5.6
• Inspect water pump and hoses; replace as needed.	E0199	5.6
• Inspect, clean, and pressure test the radiator, pressure cap, tank(s), and recovery systems; determine needed action.	E0200	5.6
• Inspect thermostatic cooling fan system (hydraulic, pneumatic, and electronic) and fan shroud; replace as needed.	E0201	5.6
• Inspect turbocharger cooling systems; determine needed action.	E0202	5.6

Time off_________________

Time on_________________

Total time_________________

Materials Required

- Machine with possible engine concern
- Machine manufacturer's workshop manual
- Manufacturer-specific tools depending on the concern
- Machine lifting equipment, if applicable

Some Safety Issues to Consider

- Diagnosis of this fault may require test driving the machine on the school grounds or on a hoist, both of which carry severe risks. Attempt this task only with full permission from your supervisor/instructor and follow all the guidelines exactly.
- Caution: If you are working in an area where there could be "brake dust" present (may contain asbestos, which has been determined to cause cancer when inhaled or ingested), ensure you wear and use all OSHA-approved asbestos protective/removal equipment.
- Lifting equipment such as machine jacks and stands, machine hoists, and engine hoists are important tools that increase productivity and make the job easier. However, they can also cause severe injury or death if used improperly. Make sure you follow the manufacturer's operation procedures. Also make sure you have your supervisor/instructor's permission to use any particular type of lifting equipment.
- Comply with personal and environmental safety practices associated with clothing; eye protection; hand tools; power equipment; proper ventilation; and the handling, storage, and disposal of chemicals/materials in accordance with federal, state, and local regulations.

- Always wear the correct protective eyewear and clothing and use the appropriate safety equipment, as well as fender covers, seat protectors, and floor mat protectors.
- Make sure you understand and observe all legislative and personal safety procedures when carrying out practical assignments. If you are unsure of what these are, ask your supervisor/instructor.

Performance Standard

0–No exposure: No information or practice provided during the program; complete training required

1–Exposure only: General information provided with no practice time; close supervision needed; additional training required

2–Limited practice: Has practiced job during training program; additional training required to develop skill

3–Moderately skilled: Has performed job independently during training program; limited additional training may be required

4–Skilled: Can perform job independently with no additional training

Student/Intern information:

Name _________________________________ Date ____________ Class _____________________

Vehicle used for this activity:

Year _______________ Make ___________________________ Model ___________________________

Odometer ____________ Hour meter ___________ VIN ___________________________________

▶ TASK Inspect coolant conditioner/filter assembly for leaks; inspect valves, lines, and fittings; replace as needed.

AED 5.6

CDX Tasksheet Number: E0198

1. **Refer to the manufacturer's workshop manual and inspect the coolant conditioner/filter assembly for leaks; inspect the valves, lines, and fittings.**

    a. **Manufacturer's specifications:**

        i. **Coolant conditioner/filter type:** ___________________________

2. **Check and inspect the coolant conditioner/filter assembly for leaks.**

3. **Pressurize the cooling system and re-check and inspect the coolant conditioner/filter assembly for leaks.**

4. **Inspect the valves.**

5. **Inspect the lines and fittings.**

6. **Draw off a sample of the coolant; have it tested for its condition.**

    a. **Meets the manufacturer's specifications:**
        Yes: ________ No: ________

    b. **If no, list your recommendations for any rectifications:**

        ____
        ____
        ____
        ____
        ____

7. **Refer to the manufacturer's workshop manual; after inspecting the coolant conditioner/filter assembly for leaks, etc., replace as needed.**

    a. **Meets the manufacturer's specifications:**
        Yes: ________ No: ________

    b. **If no, list your recommendations for any rectifications:**

        ____
        ____
        ____
        ____
        ____

8. Discuss the findings with your instructor.

Student/Intern information:

Name _________________________________ Date ___________ Class _________________________

Vehicle used for this activity:

Year _____________ Make _________________________ Model _________________________

Odometer ___________ Hour meter ___________ VIN _________________________

▶ TASK Inspect water pump and hoses; replace as needed.

AED 5.6

Time off_____________

Time on_____________

CDX Tasksheet Number: E0199

Total time_____________

1. **Refer to the manufacturer's workshop manual and check and inspect the water pump and hoses. Look closely at the weep hole in the bottom of the pump for dried coolant residue. If present, replace the pump. Always replace upon engine overhaul.**

    a. **Meets the manufacturer's specifications:**
       Yes: _________ No: _________

    b. **If no, list your recommendations for any rectifications:**

    ____
    ____
    ____
    ____
    ____

2. **Refer to the manufacturer's workshop manual; after inspection of the water pump and hoses, replace as needed.**

    a. **Meets the manufacturer's specifications:**
       Yes: _________ No: _________

    b. **If no, list your recommendations for any rectifications:**

    ____
    ____
    ____
    ____
    ____

3. **Discuss the findings with your instructor.**

Performance Rating

CDX Tasksheet Number: E0199

0	1	2	3	4

Supervisor/instructor signature _________________________________ Date ___________

Student/Intern information:

Name ______________________________ Date ____________ Class ____________________________

Vehicle used for this activity:

Year ________________ Make ______________________________ Model ______________________________

Odometer ____________ Hour meter ____________ VIN ______________________________

▶ TASK Inspect, clean, and pressure test the radiator, pressure cap, tank(s), and recovery systems; determine needed action.

AED 5.6

Time off____________

Time on____________

CDX Tasksheet Number: E0200

Total time____________

1. **Refer to the manufacturer's workshop manual and inspect, clean, and pressure test the radiator, pressure cap, tank(s), and recovery systems.**

2. **Inspect the radiator cap and pressure test cap.**

3. **Inspect the radiator and pressure test system.**

4. **Inspect the recovery tanks for leaks, corrosion, or splits.**

   a. **Meets the manufacturer's specifications:**
      Yes: __________ No: __________

   b. **If no, list your recommendations for any rectifications:**

   ____
   ____
   ____
   ____
   ____
   ____

5. **Discuss the findings with your instructor.**

Performance Rating

CDX Tasksheet Number: E0200

0	1	2	3	4

Supervisor/instructor signature ____ Date ____________

▶ TASK Inspect thermostatic cooling fan system (hydraulic, pneumatic, and electronic) and fan shroud; replace as needed.

AED 5.6

Time off_____________

Time on_____________

Total time_____________

CDX Tasksheet Number: E0201

1. **Refer to the manufacturer's workshop manual and inspect the thermostatic cooling fan system (hydraulic, pneumatic, and electronic) and fan shroud.**

   a. **List the manufacturer's preferred thermostatic cooling fan system:**

   ____

2. **Check and inspect the thermostatic cooling fan system for security.**

3. **Check and inspect the thermostatic cooling fans for correct operation. Inspect fan fins for malformation or missing or broken condition.**

4. **Check and inspect the thermostatic cooling fans' electrical harness for correct operation and security.**

5. **Check and inspect the cooling fan shroud for security.**

6. **Check and inspect the cooling fan shroud for serviceability.**

   a. **Meets the manufacturer's specifications:**
      Yes: _______ No: _______

   b. **If no, list your recommendations for any rectifications:**

   ____
   ____
   ____
   ____
   ____

7. **Refer to the manufacturer's workshop manual; after inspection of the thermostatic cooling fan system and fan shroud, replace as needed.**

    a. **Meets the manufacturer's specifications:**
       Yes: _________ No: _________

    b. **If no, list your recommendations for any rectifications:**

    ____
    ____
    ____
    ____
    ____

8. **Discuss the findings with your instructor.**

Performance Rating

CDX Tasksheet Number: E0201

0	1	2	3	4
☐	☐	☐	☐	☐

Supervisor/instructor signature ____ Date _____________

Name _________________________________ Date _____________ Class _________________________________

Vehicle used for this activity:

Year _______________ Make _____________________________ Model _________________________________

Odometer ______________ Hour meter ____________ VIN _________________________________

▶ TASK Inspect turbocharger cooling systems; determine needed action.

AED 5.6

Time off_________________

Time on_________________

CDX Tasksheet Number: E0202

Total time_________________

1. **Inspect supply line from the primary oil gallery on side of the block for leakage due to a possible loose connection and or faulty fittings.**

    a. **Condition of supply line:**

    ____
    ____
    ____
    ____

2. **Inspect return line at the bottom of the turbocharger for leakage at the connections.**

    a. **Condition of return line:**

    ____
    ____
    ____
    ____

3. **If necessary, remove and inspect both lines for blockage in case of a catastrophic turbocharger failure.**

4. **Take a supply line oil pressure reading to ensure the turbocharger is receiving the correct pressure to maintain proper operation at all times.**

    a. **Reading of supply line oil pressure:**

    ____
    ____
    ____
    ____

5. **Inspect hood openings and clearance between the hood, turbocharger, and the engine to make sure of proper airflow over turbocharger housing.**

    a. **Condition of hood openings and clearance:**

    ____
    ____
    ____
    ____

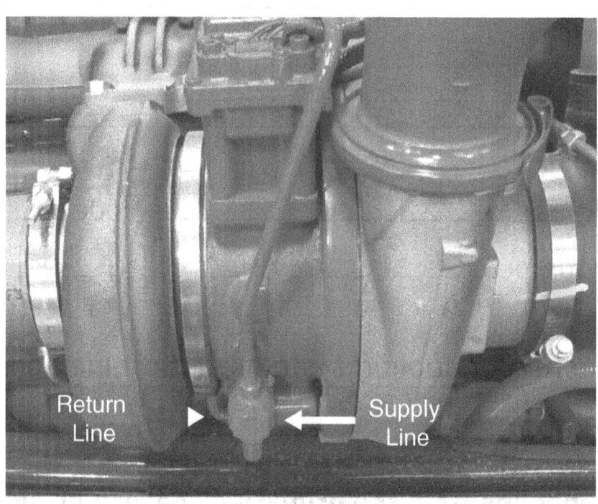

6. **Discuss the findings with your instructor.**

Performance Rating

CDX Tasksheet Number: E0202

| 0 | 1 | 2 | 3 | 4 |

Supervisor/instructor signature ____ Date _____________

Component Repair/Diagnostics: Intake and Exhaust System I

Student/Intern information:

Name _______________________________ Date ____________ Class _______________________

Vehicle used for this activity:

Year ______________ Make ____________________________ Model ______________________

Odometer ____________ Hour meter ____________ VIN _________________________________

Learning Objective/Task	CDX Tasksheet Number	2014 Edition Rev2/12/16 AED Standard
• Perform air intake system restriction and leakage tests; determine needed action.	E0203	5.8
• Perform intake manifold pressure (boost) test; determine needed action.	E0204	5.8
• Check exhaust back pressure; determine needed action.	E0205	5.8
• Inspect turbocharger(s), wastegate, and piping systems; determine needed action.	E0206	5.8
• Inspect turbocharger(s) (variable ratio/geometry VGT), pneumatic, hydraulic, electronic controls, and actuators.	E0207	5.8
• Check air induction system: piping, hoses, clamps, and mounting; service or replace air filter as needed.	E0208	5.8
• Remove and reinstall turbocharger/wastegate assembly.	E0209	5.6

Time off____________

Time on____________

Total time____________

Materials Required

- Machine with possible engine concern
- Machine manufacturer's workshop manual
- Manufacturer-specific tools depending on the concern
- Machine lifting equipment, if applicable

Some Safety Issues to Consider

- Diagnosis of this fault may require test driving the machine on the school grounds or on a hoist, both of which carry severe risks. Attempt this task only with full permission from your supervisor/instructor and follow all the guidelines exactly.
- Caution: If you are working in an area where there could be "brake dust" present (may contain asbestos, which has been determined to cause cancer when inhaled or ingested), ensure you wear and use all OSHA-approved asbestos protective/removal equipment.
- Lifting equipment such as machine jacks and stands, machine hoists, and engine hoists are important tools that increase productivity and make the job easier. However, they can also cause severe injury or death if used improperly. Make sure you follow the manufacturer's operation procedures. Also make sure you have your supervisor/instructor's permission to use any particular type of lifting equipment.

- Comply with personal and environmental safety practices associated with clothing; eye protection; hand tools; power equipment; proper ventilation; and the handling, storage, and disposal of chemicals/materials in accordance with federal, state, and local regulations.
- Always wear the correct protective eyewear and clothing and use the appropriate safety equipment, as well as fender covers, seat protectors, and floor mat protectors.
- Make sure you understand and observe all legislative and personal safety procedures when carrying out practical assignments. If you are unsure of what these are, ask your supervisor/instructor.

Performance Standard

0–No exposure: No information or practice provided during the program; complete training required

1–Exposure only: General information provided with no practice time; close supervision needed; additional training required

2–Limited practice: Has practiced job during training program; additional training required to develop skill

3–Moderately skilled: Has performed job independently during training program; limited additional training may be required

4–Skilled: Can perform job independently with no additional training

Student/Intern information:

Name _________________________________ Date ___________ Class _____________________

Vehicle used for this activity:

Year _______________ Make _______________________________ Model _____________________

Odometer _____________ Hour meter ____________ VIN _______________________________

▶ TASK Perform air intake system restriction and leakage tests; determine needed action.

AED 5.8

Time off______________

Time on_______________

CDX Tasksheet Number: E0203

Total time____________

1. **Refer to the manufacturer's workshop manual and perform air intake system restriction tests using the manufacturer's recommended tooling.**

   a. **Manufacturer's specifications:**

      i. **Maximum restriction/leakage permissible:**

      ____

2. **Perform an air intake system restriction test. Check filter minder for operation. Or, if necessary, use a manometer for flow restriction. Check all plumbing for kinking or obstruction.**

   a. **Meets the manufacturer's specifications:**
      Yes: _________ No: _________

   b. **If no, list your recommendations for any rectifications:**

      ____
      ____
      ____
      ____
      ____

3. **Refer to the manufacturer's workshop manual; perform air intake system leakage tests using the manufacturer's recommended tooling.**

4. **Perform an air intake system leakage test.**

    a. **Meets the manufacturer's specifications:**
      Yes: _________ **No:** _________

    b. **If no, list your recommendations for any rectifications:**

      ____
      ____
      ____
      ____
      ____
      ____

5. **Discuss the findings with your instructor.**

Performance Rating   **CDX Tasksheet Number: EO203**

0	1	2	3	4

Supervisor/instructor signature _________________________________ Date _____________

▶**TASK** Perform intake manifold pressure (boost) test; determine needed action.

AED 5.8

Time off_____________

Time on_____________

Total time_____________

CDX Tasksheet Number: E0204

1. **Refer to the manufacturer's workshop manual and perform the intake manifold pressure (boost) test.**

    a. **Meets the manufacturer's specifications:**
    Yes: _________ No: _________

    b. **If no, list your recommendations for any rectifications (Note: May need to install inline pressure boost gauge to check for restrictions.):**

    ____
    ____
    ____
    ____
    ____

2. **Discuss the findings with your instructor.**

Performance Rating

CDX Tasksheet Number: E0204

0	1	2	3	4

Supervisor/instructor signature ____ Date _____________

▶ **TASK** Check exhaust back pressure; determine needed action.

AED 5.8

Time off______________

Time on______________

CDX Tasksheet Number: E0205

1. **Refer to the manufacturer's workshop manual and perform the exhaust back pressure test; determine needed action.**

    a. **Meets the manufacturer's specifications:**
       Yes: _________ No: _________

       Total time______________

    b. **If no, list your recommendations for any rectifications:**

       ____
       ____
       ____
       ____
       ____

2. **Discuss the findings with your instructor.**

Performance Rating

CDX Tasksheet Number: E0205

0	1	2	3	4

Supervisor/instructor signature _________________________________ Date ___________

Name ________________________ Date ___________ Class ________________________

Vehicle used for this activity:

Year _____________ Make __________________________ Model ________________________

Odometer ___________ Hour meter ___________ VIN ________________________

▶ **TASK** Inspect turbocharger(s), wastegate, and piping systems; determine needed action.

AED 5.8

Time off___________

Time on___________

Total time___________

CDX Tasksheet Number: E0206

1. **Refer to the manufacturer's workshop manual and inspect the turbocharger(s), wastegate, and piping systems; determine needed action.**

2. **Check and inspect the turbocharger for security.**

3. **Check and inspect the turbocharger for correct operation.**

4. **Check and inspect the wastegate for security.**

5. **Check and inspect the wastegate for correct operation.**

6. **Check and inspect the turbocharger piping for security.**

   a. **Meets the manufacturer's specifications:**
      Yes: _______ No: _______

   b. **If no, list your recommendations for any rectifications:**

      ____
      ____
      ____
      ____
      ____

7. **Discuss the findings with your instructor.**

Performance Rating

CDX Tasksheet Number: E0206

0	1	2	3	4

Supervisor/instructor signature ________________________________ Date ___________

Student/Intern information:

Name _________________________________ Date _____________ Class _________________________

Vehicle used for this activity:

Year _______________ Make _________________________________ Model _________________________

Odometer _____________ Hour meter _____________ VIN _________________________________

▶ TASK Inspect turbocharger(s) (variable ratio/geometry VGT), pneumatic, hydraulic, electronic controls, and actuators.

AED 5.8

Time off_________________

Time on_________________

CDX Tasksheet Number: E0207

Total time_________________

1. **Refer to the machine's specifications and identify the type of turbocharger fitted as standard equipment.**

    a. **Type of turbocharger:** _________________________________

2. **Identify and name the type of turbocharger fitted to this machine:**

    a. **Type of turbocharger:** _________________________________

    b. **Type of controls:** _________________________________

> **Note:** The most common cause of turbocharger failure can be caused by hot shut down. Hot shutdown is when the machine is driven hard, exhaust temperature is too hot, and the engine is shut down before the turbocharger has had idle time to cool off.

3. **Refer to the manufacturer's workshop manual and inspect the turbocharger(s) (variable ratio/geometry VGT); pneumatic, hydraulic, and electronic controls; and actuators.**

    a. **Meets the manufacturer's specifications:**
    Yes: _________ **No:** _________

    b. **If no, list your recommendations for any rectifications:**

    ____
    ____
    ____
    ____
    ____

4. Discuss the findings with your instructor.

Student/Intern information:

Name _________________________________ Date _____________ Class _________________________________

Vehicle used for this activity:

Year _______________ Make _________________________________ Model _________________________________

Odometer _____________ Hour meter _____________ VIN _________________________________

▶ TASK Check air induction system: piping, hoses, clamps, and mounting; service or replace air filter as needed.

AED 5.8

Time off______________

Time on______________

Total time______________

CDX Tasksheet Number: E0208

1. **Refer to manufacturer's workshop manual and check the air induction system: piping, hoses, clamps, and mounting.**

2. **Check and inspect the air induction system for security.**

3. **Check and inspect the air induction system housing.**

4. **Check and inspect the air induction piping for security.**

5. **Check and inspect the air induction piping for correct sizing.**

6. **Check and inspect the air induction piping hoses and clamps for serviceability and mounting security.**

   a. **Meets the manufacturer's specifications:**
      Yes: _________ No: _________

   b. **If no, list your recommendations for any rectifications:**

   ____
   ____
   ____
   ____
   ____

7. **Refer to the manufacturer's workshop manual and service or replace the air filter as needed.**

8. **Clean and inspect the air filter in accordance with the manufacturer's specifications.**

9. Inspect the air filter for serviceability.

    a. **Meets the manufacturer's specifications:**
        Yes: _________ No: _________

    b. **If no, list your recommendations for any rectifications:**

       ____
       ____
       ____
       ____
       ____
       ____

10. **If directed by your instructor, replace the air filter if required.**

11. **Discuss the findings with your instructor.**

Performance Rating　　　　**CDX Tasksheet Number: E0208**

0	1	2	3	4

Supervisor/instructor signature ____ Date _______________

Name _________________________ Date _____________ Class _________________________

Vehicle used for this activity:

Year _______________ Make _________________________ Model _________________________

Odometer _____________ Hour meter ____________ VIN _________________________

▶ TASK Remove and reinstall turbocharger/wastegate assembly.

AED 5.6

Time off_________________

Time on_________________

Total time_________________

CDX Tasksheet Number: E0209

1. **Refer to the manufacturer's workshop manual and list the procedure and all safety precautions that must be observed when you remove and reinstall the turbocharger/wastegate assembly.**

   a. **List the steps involved when you remove and reinstall the turbocharger/wastegate assembly:**

   ____
   ____
   ____
   ____
   ____

   b. **Determine what safety precautions must be observed when you remove and reinstall the turbocharger/wastegate assembly:**

   ____
   ____
   ____
   ____
   ____

2. **Following the procedures listed above, and while referencing the manufacturer's workshop manual, remove and reinstall the turbocharger/wastegate assembly.**

   a. **Meets the manufacturer's specifications:**
   Yes: _______ No: _______

   b. **If no, list your recommendations for any rectifications:**

   ____
   ____
   ____
   ____
   ____

3. **Discuss the findings with your instructor.**

CDX Tasksheet Number: E0209

0	1	2	3	4

Supervisor/instructor signature ____ Date _______________

Component Repair/Diagnostics: Intake and Exhaust System II

Student/Intern information:

Name _______________________________ Date _____________ Class _______________________

Vehicle used for this activity:

Year _______________ Make _______________________ Model _______________________

Odometer _____________ Hour meter _____________ VIN _______________________

Learning Objective/Task	CDX Tasksheet Number	2014 Edition Rev2/12/16 AED Standard
• Inspect intake manifold, gaskets, and connections; replace as needed.	EO210	5.8
• Inspect, clean, and test charge air cooler assemblies; replace as needed.	EO211	5.8
• Inspect exhaust manifold, piping, mufflers, and mounting hardware; repair or replace as needed.	EO212	5.8
• Inspect exhaust after treatment devices; determine necessary action.	EO213	5.8
• Inspect and test preheater/inlet air heater, or glow plug system and controls; perform needed action.	EO214	5.8
• Inspect exhaust gas recirculation (EGR) system including EGR valve, cooler, piping, filter, electronic sensors, controls, and wiring; determine needed action.	EO215	5.8

Time off _______________

Time on _______________

Total time _______________

Materials Required

- Machine with possible engine concern
- Machine manufacturer's workshop manual
- Manufacturer-specific tools depending on the concern
- Machine lifting equipment, if applicable

Some Safety Issues to Consider

- Diagnosis of this fault may require test driving the machine on the school grounds or on a hoist, both of which carry severe risks. Attempt this task only with full permission from your supervisor/instructor and follow all the guidelines exactly.
- Caution: If you are working in an area where there could be "brake dust" present (may contain asbestos, which has been determined to cause cancer when inhaled or ingested), ensure you wear and use all OSHA-approved asbestos protective/removal equipment.
- Lifting equipment such as machine jacks and stands, machine hoists, and engine hoists are important tools that increase productivity and make the job easier. However, they can also cause severe injury or death if used improperly. Make sure you follow the manufacturer's operation procedures. Also make sure you have your supervisor/instructor's permission to use any particular type of lifting equipment.

- Comply with personal and environmental safety practices associated with clothing; eye protection; hand tools; power equipment; proper ventilation; and the handling, storage, and disposal of chemicals/materials in accordance with federal, state, and local regulations.
- Always wear the correct protective eyewear and clothing and use the appropriate safety equipment, as well as fender covers, seat protectors, and floor mat protectors.
- Make sure you understand and observe all legislative and personal safety procedures when carrying out practical assignments. If you are unsure of what these are, ask your supervisor/instructor.

Performance Standard

0–No exposure: No information or practice provided during the program; complete training required

1–Exposure only: General information provided with no practice time; close supervision needed; additional training required

2–Limited practice: Has practiced job during training program; additional training required to develop skill

3–Moderately skilled: Has performed job independently during training program; limited additional training may be required

4–Skilled: Can perform job independently with no additional training

Student/Intern information:

Name _________________________ Date _____________ Class _________________________

Vehicle used for this activity:

Year _______________ Make _________________________ Model _________________________

Odometer _____________ Hour meter _____________ VIN _________________________

▶ TASK Inspect intake manifold, gaskets, and connections; replace as needed.

AED
5.8

Time off_________________

Time on_________________

Total time_________________

CDX Tasksheet Number: EO210

1. **Refer to the manufacturer's workshop manual and inspect the intake manifold, gaskets, and connections.**

   a. **Meets the manufacturer's specifications:**
      Yes: _________ No: _________

   b. **If no, list your recommendations for any rectifications:**

   ____
   ____
   ____
   ____
   ____

2. **Discuss the findings with your instructor.**

Performance Rating

CDX Tasksheet Number: EO210

0	1	2	3	4

Supervisor/instructor signature _________________________________ Date _____________

Student/Intern information:

Name _________________________________ Date _____________ Class _________________________

Vehicle used for this activity:

Year _______________ Make _____________________________ Model _________________________

Odometer _____________ Hour meter _____________ VIN _________________________________

▶ **TASK** Inspect, clean, and test charge air cooler assemblies; replace as needed.

AED 5.8

Time off _________________

Time on _________________

Total time _________________

CDX Tasksheet Number: E0211

1. **Refer to the manufacturer's workshop manual and inspect, clean, and test charge the air cooler assemblies.**

2. **Inspect the air cooler assemblies.**

3. **Clean the air cooler assemblies.**

4. **Test charge the air cooler assemblies.**

   a. **Meets the manufacturer's specifications:**
      Yes: _________ No: _________

   b. **If no, list your recommendations for any rectifications:**

      ____
      ____
      ____
      ____
      ____

5. **If directed by your instructor, replace the components as necessary.**

6. **Discuss the findings with your instructor.**

Performance Rating

CDX Tasksheet Number: E0211

0	1	2	3	4

Supervisor/instructor signature ____ Date _____________

Student/Intern information:

Name ________________________________ Date ____________ Class ______________________

Vehicle used for this activity:

Year ______________ Make ________________________ Model ______________________

Odometer ____________ Hour meter ____________ VIN ______________________________

▶ TASK Inspect exhaust manifold, piping, mufflers, and mounting hardware; repair or replace as needed.

AED 5.8

Time off________________

Time on________________

CDX Tasksheet Number: E0212

1. **Refer to the manufacturer's workshop manual and inspect the exhaust manifold, piping, mufflers, and mounting hardware.**

   a. **Inspect the exhaust manifold.**

      i. **Require replacement: Yes: ________ No: ________**

   b. **Inspect the exhaust piping.**

      i. **Require replacement: Yes: ________ No: ________**

   c. **Inspect the muffler.**

      i. **Require replacement: Yes: ________ No: ________**

   d. **Inspect the exhaust pipe mounting hardware.**

      i. **Require replacement: Yes: ________ No: ________**

2. **Overall assessment**

   a. **Meets the manufacturer's specifications:
      Yes: ________ No: ________**

   b. **If no, list your recommendations for any rectifications:**

   ____

   ____

   ____

   ____

   ____

Total time________________

3. **If directed by your instructor, follow manufacturer's workshop manual instructions and repair or replace the components as needed.**

4. **Discuss the findings with your instructor.**

Student/Intern information:

Name _________________________________ Date _____________ Class _________________________

Vehicle used for this activity:

Year _______________ Make _____________________________ Model _____________________________

Odometer _____________ Hour meter _____________ VIN _____________________________________

▶ TASK Inspect exhaust after treatment devices; determine necessary action.

AED 5.8

Time off_____________

Time on_____________

Total time_____________

CDX Tasksheet Number: EO213

1. **Identify the type of treatment device fitted to your machine:**

   ____

2. **Refer to the manufacturer's workshop manual and inspect the exhaust after treatment devices.**

   a. **Carry out a visual inspection of the treatment device.**

      i. **Require replacement: Yes: ________ No: ________**

   b. **If the treatment device is a "diesel particulate filter," attach the diagnostic code reader to the machine.**

   c. **Are there any codes present or stored in the reader?**

      i. **If yes, list the codes:**

      ____

      ____

      ____

      ____

      ____

   d. **Carry out a forced regeneration of the particulate filter.**

3. **If your machine is fitted with another type of exhaust treatment device, name the device:**

   ____

   ____

   ____

   ____

   ____

4. **Briefly describe the testing procedure for this unit as outlined in the appropriate manufacturer's workshop manual:**

   ____

   ____

   ____

   ____

   ____

5. As outlined above, carry out a visual inspection of the treatment device.

   a. Require replacement: Yes: _________ No: _________

   b. If yes, list your recommendations for any rectifications:

   ____
   ____
   ____
   ____
   ____

6. Discuss the findings with your instructor.

CDX Tasksheet Number: EO213

| 0 | 1 | 2 | 3 | 4 |

Supervisor/instructor signature ____ Date _____________

Student/Intern information:

Name _________________________________ Date _____________ Class _____________________

Vehicle used for this activity:

Year _______________ Make _________________________________ Model _____________________

Odometer _____________ Hour meter _____________ VIN _________________________________

▶ TASK Inspect and test preheater/inlet air heater, or glow plug system and controls; perform needed action.

AED 5.8

Time off________________

Time on________________

Total time________________

CDX Tasksheet Number: EO214

1. **Identify the type of air induction system heating device fitted to your engine:**

   ____

2. **Refer to the manufacturer's workshop manual and inspect and test the preheater/inlet air heater.**

   a. **Carry out a visual inspection of the heating device.**

      i. **Require replacement: Yes: _________ No: _________**

   b. **Carry out a visual inspection of the glow plug system.**

      i. **Require replacement: Yes: _________ No: _________**

      ii. **If yes, list your recommendations for any rectifications:**

      ____
      ____
      ____
      ____
      ____

3. **Discuss the findings with your instructor.**

Performance Rating

CDX Tasksheet Number: EO214

0	1	2	3	4

Supervisor/instructor signature ____ Date _____________

Student/Intern information:

Name _________________________________ Date ____________ Class _________________________

Vehicle used for this activity:

Year _______________ Make ___________________________________ Model _________________________

Odometer ____________ Hour meter ____________ VIN ____

▶ TASK Inspect exhaust gas recirculation (EGR) system including
EGR valve, cooler, piping, filter, electronic sensors, controls,
and wiring; determine needed action.

AED
5.8

Time off____________

Time on____________

Total time____________

CDX Tasksheet Number: E0215

1. **Refer to the manufacturer's workshop manual and inspect the exhaust gas recirculation (EGR) system including the EGR valve, cooler, piping, filter, electronic sensors, controls, and wiring. This may require use of OBD II scan tool.**

   a. **Inspect the exhaust gas recirculation (EGR) system including the EGR valve.**

      i. **Require replacement: Yes: _________ No: _________**

   b. **Inspect the cooler and piping.**

      i. **Require replacement: Yes: _________ No: _________**

   c. **Inspect the filter.**

      i. **Require replacement: Yes: _________ No: _________**

   d. **Inspect the electronic sensors.**

      i. **Require replacement: Yes: _________ No: _________**

   e. **Inspect the controls and wiring.**

      i. **Require replacement: Yes: _________ No: _________**

2. **Overall assessment**

   a. **Meets the manufacturer's specifications:
   Yes: _________ No: _________**

   b. **If no, list your recommendations for any rectifications:**

   ____
   ____
   ____
   ____
   ____

3. Discuss the findings with your instructor.

Component Repair/Diagnostics: Fuel System

Student/Intern information:

Name ___________________________ Date ___________ Class ___________________

Vehicle used for this activity:

Year ______________ Make ____________________ Model ____________________

Odometer ___________ Hour meter ___________ VIN ____________________

Learning Objective/Task	CDX Tasksheet Number	2014 Edition Rev2/12/16 AED Standard
• Check fuel level and condition; determine needed action.	E0216	5.8
• Perform fuel supply and return system tests; determine needed action.	E0217	5.8
• Inspect fuel tanks, vents, caps, mounts, valves, screens, crossover system, supply and return lines, and fittings; determine needed action.	E0218	5.8
• Inspect, clean, and test fuel transfer (lift) pump, pump drives, screens, fuel/water separators/indicators, filters, heaters, coolers, ECM cooling plates, and mounting hardware; determine needed action.	E0219	5.8
• Inspect and test pressure regulator systems (check valves, pressure regulator valves, and restrictive fittings); determine needed action.	E0220	5.8
• Check fuel system for air; determine needed action; prime and bleed fuel system; check primer pump.	E0221	5.8

Time off____________

Time on____________

Total time____________

Materials Required

- Machine with possible engine concern
- Machine manufacturer's workshop manual
- Manufacturer-specific tools depending on the concern
- Machine lifting equipment, if applicable

Some Safety Issues to Consider

- Diagnosis of this fault may require test driving the machine on the school grounds or on a hoist, both of which carry severe risks. Attempt this task only with full permission from your supervisor/instructor and follow all the guidelines exactly.
- Caution: If you are working in an area where there could be "brake dust" present (may contain asbestos, which has been determined to cause cancer when inhaled or ingested), ensure you wear and use all OSHA-approved asbestos protective/removal equipment.

- Lifting equipment such as machine jacks and stands, machine hoists, and engine hoists are important tools that increase productivity and make the job easier. However, they can also cause severe injury or death if used improperly. Make sure you follow the manufacturer's operation procedures. Also make sure you have your supervisor/instructor's permission to use any particular type of lifting equipment.
- Comply with personal and environmental safety practices associated with clothing; eye protection; hand tools; power equipment; proper ventilation; and the handling, storage, and disposal of chemicals/materials in accordance with federal, state, and local regulations.
- Always wear the correct protective eyewear and clothing and use the appropriate safety equipment, as well as fender covers, seat protectors, and floor mat protectors.
- Make sure you understand and observe all legislative and personal safety procedures when carrying out practical assignments. If you are unsure of what these are, ask your supervisor/instructor.

Performance Standard

0–No exposure: No information or practice provided during the program; complete training required

1–Exposure only: General information provided with no practice time; close supervision needed; additional training required

2–Limited practice: Has practiced job during training program; additional training required to develop skill

3–Moderately skilled: Has performed job independently during training program; limited additional training may be required

4–Skilled: Can perform job independently with no additional training

____

Student/Intern information:

Name _________________________________ Date _____________ Class _______________________

Vehicle used for this activity:

Year _______________ Make _________________________ Model ____________________________

Odometer _____________ Hour meter ___________ VIN __________________________________

▶ TASK Check fuel level and condition; determine needed action.

AED 5.8

Time off________________

CDX Tasksheet Number: EO216

Time on________________

1. **Check the fuel level(s).**

    a. **Meets the manufacturer's specifications:**
       Yes: __________ **No:** __________

Total time________________

2. **Draw off a sample of the fuel and have it tested for its condition.**

    a. **Meets the manufacturer's specifications:**
       Yes: __________ **No:** __________

    b. **If no, list your recommendations for any rectifications:**

       ____
       ____
       ____
       ____
       ____

3. **Discuss the findings with your instructor.**

Performance Rating   **CDX Tasksheet Number: EO216**

0	1	2	3	4

Supervisor/instructor signature ___________________________________ Date _____________

▶ TASK Perform fuel supply and return system tests; determine needed action.

AED 5.8

CDX Tasksheet Number: EO217

1. **Refer to the manufacturer's workshop manual and perform fuel supply and return system tests; determine needed action.**

2. **Using the manufacturer's recommended tools and gauges, perform fuel supply and return system tests.**

   a. **Meets the manufacturer's specifications:**
      Yes: _________ No: _________

   b. **If no, list your recommendations for any rectifications:**

      ____
      ____
      ____
      ____
      ____

3. **Check the coolant temperature cold: _________________________________**

4. **Discuss the findings with your instructor.**

Performance Rating

CDX Tasksheet Number: EO217

0	1	2	3	4

Supervisor/instructor signature _________________________________ Date ____________

Student/Intern information:

Name _________________________________ Date ____________ Class _______________________

Vehicle used for this activity:

Year _______________ Make _______________________________ Model _______________________

Odometer ____________ Hour meter ____________ VIN _________________________________

▶ **TASK** Inspect fuel tanks, vents, caps, mounts, valves, screens, crossover system, supply and return lines, and fittings; determine needed action.

AED
5.8

Time off________________

Time on________________

Total time________________

CDX Tasksheet Number: EO218

1. **Inspect the surface of the fuel tank for any signs of structural damage or leakage.**

    a. **Any signs of structural damage: Yes: ____________ No: ____________**

    b. **If yes, list your recommendations for any rectifications:**

    ____
    ____
    ____
    ____
    ____

2. **Inspect the fuel tank cap/vents for any signs of damage or blockage.**

    a. **Any signs of damage/blockage: Yes: ____________ No: ____________**

    b. **If yes, list your recommendations for any rectifications:**

    ____
    ____
    ____
    ____
    ____

3. **Inspect the fuel tank mounts for any signs of structural damage or loose bolts.**

    a. **Any signs of structural damage: Yes: ____________ No: ____________**

    b. **If yes, list your recommendations for any rectifications:**

    ____
    ____
    ____
    ____
    ____

4. **Are fuel tank mounting bolts correctly torqued to the manufacturer's specifications or leaking from any of the structurally damaged areas?**

    a. **Manufacturer's recommended torque setting:** __________ ft-lb (N·m)

    b. **Have the mounting bolts been torqued to specifications?**
    Yes: __________ **No:** __________

    c. **If no, detail why these mounts have not been torqued to specifications:**

____

____

____

____

____

5. **Inspect the fuel tank valves for any signs of malfunction.**

    a. **Any signs of malfunction: Yes:** __________ **No:** __________

    b. **If yes, list your recommendations for any rectifications:**

____

____

____

____

____

6. **Inspect the fuel tank screens for any signs of damage.**

    a. **Any signs of damage: Yes:** __________ **No:** __________

    b. **If yes, list your recommendations for any rectifications:**

____

____

____

____

____

7. **Inspect the fuel tank crossover pipe system for any signs of damage.**

    a. **Any signs of damage: Yes:** __________ **No:** __________

    b. **If yes, list your recommendations for any rectifications:**

____

____

____

____

____

8. **Inspect the fuel supply and return lines and fittings for any signs of damage or deterioration.**

    a. **Any signs of damage/deterioration: Yes:** _____________ **No:** _____________

    b. **If yes, list your recommendations for any rectifications:**

        ____

        ____

        ____

        ____

        ____

9. **Discuss the findings with your instructor.**

Performance Rating

CDX Tasksheet Number: E0218

0	1	2	3	4

Supervisor/instructor signature ____ Date _____________

Student/Intern information:

Name ___________________________________ Date ___________ Class ___________________________

Vehicle used for this activity:

Year ______________ Make ________________________ Model ___________________________

Odometer ____________ Hour meter ____________ VIN ___________________________

▶ TASK Inspect, clean, and test fuel transfer (lift) pump, pump drives, screens, fuel/water separators/indicators, filters, heaters, coolers, ECM cooling plates, and mounting hardware; determine needed action.

AED 5.8

Time off___________

Time on___________

Total time___________

CDX Tasksheet Number: E0219

1. **Refer to the manufacturer's workshop manual and inspect, clean, and test the fuel transfer (lift) pump; determine needed action.**

2. **Following the instructions in the manufacturer's workshop manual, inspect, clean, and test the fuel transfer (lift) pump.**

    a. **Meets the manufacturer's specifications:**
       Yes: ___________ **No:** ___________

    b. **If no, list your recommendations for any rectifications:**

    ____
    ____
    ____
    ____
    ____

3. **Refer to the manufacturer's workshop manual and inspect and clean the pump drives and screens; determine needed action.**

    a. **Meets the manufacturer's specifications:**
       Yes: ___________ **No:** ___________

    b. **If no, list your recommendations for any rectifications:**

    ____
    ____
    ____
    ____
    ____

4. **Following the instructions in the manufacturer's workshop manual, inspect and clean the fuel/water separators/indicators.**

    a. **Locate the fuel filter water trap, if fitted.**

       i. **Is there water visible in the water trap fuel bowl?**
          Yes: ___________ **No:** ___________

       ii. **If yes, place a drain container under the sediment bowl of the filter and remove.**

       iii. **Clean the sediment bowl and dry.**

5. **Refer to the manufacturer's workshop manual and inspect and clean the filters, heaters, coolers, ECM cooling plates, and mounting hardware.**

    a. **Meets the manufacturer's specifications:**
       Yes: _____________ No: _____________

    b. **If no, list your recommendations for any rectifications:**

       ____

       ____

       ____

       ____

       ____

6. **Discuss the findings with your instructor.**

Performance Rating

CDX Tasksheet Number: EO219

| 0 | 1 | 2 | 3 | 4 |

Supervisor/instructor signature _______________________________________ Date _____________

▶ TASK Inspect and test pressure regulator systems (check valves, pressure regulator valves, and restrictive fittings); determine needed action.

AED 5.8

Time off___________

Time on___________

Total time___________

CDX Tasksheet Number: E0220

1. **Refer to the manufacturer's workshop manual and inspect and test the pressure regulator systems (check valves, pressure regulator valves, and restrictive   fittings).**

   a. **Use the manufacturer's specified tooling and gauges.**

   b. **Carry out the inspection and testing.**

      i. **Meets the manufacturer's specifications:**
         Yes: ___________ No: ___________

      ii. **If no, list your recommendations for any rectifications:**

      ____

      ____

      ____

      ____

      ____

2. **Discuss the findings with your instructor**

Performance Rating

CDX Tasksheet Number: E0220

| 0 | 1 | 2 | 3 | 4 |

Supervisor/instructor signature _________________________________ Date ___________

▶ TASK Check fuel system for air; determine needed action; prime and bleed fuel system; check primer pump.

AED
5.8

Time off________

Time on________

Total time________

CDX Tasksheet Number: EO221

1. **Refer to the manufacturer's workshop manual and check the fuel system for air. In some cases, a sight glass may be needed to determine if air is present.**

    a. **Meets the manufacturer's specifications:**
       Yes: ___________ No: ___________

    b. **If no, list your recommendations for any rectifications:**

    ____
    ____
    ____
    ____
    ____
    ____

2. **Refer to the manufacturer's workshop manual and prime and bleed the fuel system; check the primer pump.**

    a. **Meets the manufacturer's specifications:**
       Yes: ___________ No: ___________

    b. **If no, list your recommendations for any rectifications:**

    ____
    ____
    ____
    ____
    ____

3. **Discuss the findings with your instructor.**

Performance Rating

CDX Tasksheet Number: EO221

0	1	2	3	4

Supervisor/instructor signature _______________________________________ Date ___________

Diagnostics: Electronic Fuel Management System I

Student/Intern information:

Name _________________________________ Date ___________ Class _________________________

Vehicle used for this activity:

Year _______________ Make _______________________ Model _____________________________

Odometer _____________ Hour meter ___________ VIN ______________________________________

Learning Objective/Task	CDX Tasksheet Number	2014 Edition Rev2/12/16 AED Standard
• Inspect and test power and ground circuits and connections; measure and interpret voltage, voltage drop, amperage, and resistance readings using a digital multimeter (DMM); determine needed action.	E0222	5.8
• Interface with machine's on-board computer; perform diagnostic procedures using electronic service tools (to include PC-based software and/or data scan tools); determine needed action.	E0223	5.8
• Check and record electronic diagnostic codes and trip/operational data; monitor electronic data; clear codes; determine further diagnosis.	E0224	5.8
• Locate and use relevant service information (to include diagnostic procedures, flow charts, and wiring diagrams).	E0225	5.8
• Inspect and replace electrical connector terminals, seals, and locks.	E0226	5.8
• Inspect and test switches, sensors, controls, actuator components, and circuits; adjust or replace as needed.	E0227	5.8
• Using electronic service tools, access and interpret customer programmable parameters.	E0228	5.8

Time off________________

Time on________________

Total time________________

Materials Required

- Machine with possible engine concern
- Machine manufacturer's workshop manual
- Manufacturer-specific tools depending on the concern
- Machine lifting equipment, if applicable

Some Safety Issues to Consider

- Diagnosis of this fault may require test driving the machine on the school grounds or on a hoist, both of which carry severe risks. Attempt this task only with full permission from your supervisor/instructor and follow all the guidelines exactly.

- Caution: If you are working in an area where there could be "brake dust" present (may contain asbestos, which has been determined to cause cancer when inhaled or ingested), ensure you wear and use all OSHA-approved asbestos protective/removal equipment.
- Lifting equipment such as machine jacks and stands, machine hoists, and engine hoists are important tools that increase productivity and make the job easier. However, they can also cause severe injury or death if used improperly. Make sure you follow the manufacturer's operation procedures. Also make sure you have your supervisor/instructor's permission to use any particular type of lifting equipment.
- Comply with personal and environmental safety practices associated with clothing; eye protection; hand tools; power equipment; proper ventilation; and the handling, storage, and disposal of chemicals/materials in accordance with federal, state, and local regulations.
- Always wear the correct protective eyewear and clothing and use the appropriate safety equipment, as well as fender covers, seat protectors, and floor mat protectors.
- Make sure you understand and observe all legislative and personal safety procedures when carrying out practical assignments. If you are unsure of what these are, ask your supervisor/instructor.

Performance Standard

0—No exposure: No information or practice provided during the program; complete training required

1—Exposure only: General information provided with no practice time; close supervision needed; additional training required

2—Limited practice: Has practiced job during training program; additional training required to develop skill

3—Moderately skilled: Has performed job independently during training program; limited additional training may be required

4—Skilled: Can perform job independently with no additional training

Student/Intern information:

Student/Intern information:

Name ________________________________ Date ____________ Class ________________________

Vehicle used for this activity:

Year ______________ Make ____________________________ Model ________________________

Odometer ____________ Hour meter ____________ VIN ________________________________

▶ **TASK** Inspect and test power and ground circuits and connections; measure and interpret voltage, voltage drop, amperage, and resistance readings using a digital multimeter (DMM); determine needed action.

AED 5.8

Time off ________________

Time on ________________

Total time ________________

CDX Tasksheet Number: E0222

1. **Inspect and test power and ground circuits and connections.**

    a. **Meets the manufacturer's specifications:**
       Yes: ________ **No:** ________

    b. **If no, list your recommendations for any rectifications:**

    ____

    ____

    ____

    ____

    ____

2. **Refer to the manufacturer's workshop manual for the appropriate specifications and measure and interpret voltage, voltage drop, amperage, and resistance readings using a digital multimeter (DMM).**

    a. **Meets the manufacturer's specifications:**
       Yes: ________ **No:** ________

    b. **If no, list your recommendations for any rectifications:**

    ____

    ____

    ____

    ____

    ____

3. **Discuss the findings with your instructor.**

Performance Rating

CDX Tasksheet Number: E0222

0	1	2	3	4

Supervisor/instructor signature ________________________________ Date ____________

▶ **TASK** Interface with machine's on-board computer; perform diagnostic procedures using electronic service tools (to include PC-based software and/or data scan tools); determine needed action.

AED 5.8

Time off_______________

Time on_______________

Total time_______________

CDX Tasksheet Number: EO223

1. **Identify the type of electronic service tool available in your workshop.**

   a. **PC-based unit: Yes: _________ No: _________**

   b. **If yes, list the make and model:**

      ____
      ____
      ____
      ____
      ____

   c. **Hand-held data scanner: Yes: _________ No: _________**

   d. **If yes, list the make and model:**

      ____
      ____
      ____
      ____

2. **Refer to the manufacturer's workshop manual and the service tool manual(s) and connect the unit to the machine's interface.**

   a. **Perform diagnostic procedures as outlined by the manufacturer.**

      i. **Meets the manufacturer's specifications:**
         Yes: _________ No: _________

         ____
         ____
         ____
         ____
         ____

      ii. **If no, list your recommendations for any rectifications:**

         ____
         ____
         ____
         ____
         ____

3. **Discuss the findings with your instructor.**

Student/Intern information:

Name _______________________________ Date _____________ Class _______________________

Vehicle used for this activity:

Year _______________ Make _________________________________ Model _______________________

Odometer _____________ Hour meter _____________ VIN _________________________________

▶ TASK Check and record electronic diagnostic codes and trip/operational data; monitor electronic data; clear codes; determine further diagnosis.

AED 5.8

Time off_______________

Time on_______________

Total time_______________

CDX Tasksheet Number: E0224

1. **Record all diagnostic codes and trip/operational data stored in the system:**

   ____

   ____

   ____

   ____

   ____

   ____

2. **Refer to the manufacturer's manual and identify each stored code below.**

   a. **Code:** _____________ **Interpretation:** _______________________________

   b. **Code:** _____________ **Interpretation:** _______________________________

   c. **Code:** _____________ **Interpretation:** _______________________________

   d. **Code:** _____________ **Interpretation:** _______________________________

   e. **Code:** _____________ **Interpretation:** _______________________________

   f. **Code:** _____________ **Interpretation:** _______________________________

   g. **Code:** _____________ **Interpretation:** _______________________________

   h. **Code:** _____________ **Interpretation:** _______________________________

   i. **Code:** _____________ **Interpretation:** _______________________________

   j. **Code:** _____________ **Interpretation:** _______________________________

      i. **Are any of these codes detrimental to the machine's operation?**
         Yes: _____________ **No:** _____________

      ii. **If yes, list your recommendations for any rectifications:**

          ____

          ____

          ____

          ____

          ____

3. Clear all stored codes.

4. Discuss the findings with your instructor.

Student/Intern information:

Name _______________________________ Date ____________ Class _______________________

Vehicle used for this activity:

Year _______________ Make _______________________________ Model _______________________

Odometer ____________ Hour meter ____________ VIN _______________________________

▶ TASK Locate and use relevant service information (to include diagnostic procedures, flow charts, and wiring diagrams).

AED 5.8

Time off_______________

Time on_______________

Total time_______________

CDX Tasksheet Number: E0225

1. **Refer to the manufacturer's workshop manual and locate and use relevant service information (to include diagnostic procedures, flow charts, and wiring diagrams) as required throughout the servicing procedures.**

   a. **Meets the manufacturer's specifications:**
      Yes: ____________ No: ____________

   b. **If no, list your recommendations for any rectifications:**

   ____
   ____
   ____
   ____
   ____
   ____

2. **Discuss the findings with your instructor.**

Performance Rating

CDX Tasksheet Number: E0225

| 0 | 1 | 2 | 3 | 4 |

Supervisor/instructor signature ____ Date ____________

Student/Intern information:

Name _______________________ Date ___________ Class _______________________

Vehicle used for this activity:

Year ______________ Make _______________________ Model _______________________

Odometer ___________ Hour meter ___________ VIN _______________________

▶ TASK Inspect and replace electrical connector terminals, seals, and locks.

AED 5.8

Time off_______________

Time on_______________

Total time_______________

CDX Tasksheet Number: E0226

1. **Refer to the manufacturer's workshop manual to inspect and replace electrical connector terminals, seals, and locks.**

   a. **Meets the manufacturer's specifications:**
      Yes: ___________ **No:** ___________

   b. **If no, list your recommendations for any rectifications:**

      ____
      ____
      ____
      ____
      ____

2. **Discuss the findings with your instructor.**

Performance Rating

CDX Tasksheet Number: E0226

0	1	2	3	4

Supervisor/instructor signature ___________________________________ Date ___________

Name _________________________________ Date ___________ Class _________________________

Vehicle used for this activity:

Year _______________ Make _____________________________ Model _________________________

Odometer ____________ Hour meter ___________ VIN _________________________________

▶ **TASK** Inspect and test switches, sensors, controls, actuator components, and circuits; adjust or replace as needed.

AED 5.8

Time off________________

Time on________________

CDX Tasksheet Number: E0227

Total time________________

1. **Refer to the manufacturer's workshop manual to inspect and test switches, sensors, controls, actuator components, and circuits.**

    a. **Inspect and test the switches.**

       i. **Meets the manufacturer's specifications:**
          Yes: ___________ **No:** ___________

    b. **Inspect and test the sensors.**

       i. **Meets the manufacturer's specifications:**
          Yes: ___________ **No:** ___________

    c. **Inspect and test the controls.**

       i. **Meets the manufacturer's specifications:**
          Yes: ___________ **No:** ___________

    d. **Inspect and test the actuator components.**

       i. **Meets the manufacturer's specifications:**
          Yes: ___________ **No:** ___________

    e. **Inspect and test the circuits.**

       i. **Meets the manufacturer's specifications:**
          Yes: ___________ **No:** ___________

    f. **If no, list your recommendations for any rectifications:**

       ____
       ____
       ____
       ____
       ____

2. **Discuss the findings with your instructor.**

Student/Intern information:

Name _________________________________ Date ___________ Class _____________________

Vehicle used for this activity:

Year _______________ Make ___________________________ Model ___________________________

Odometer ____________ Hour meter ___________ VIN ___________________________________

▶ TASK Using electronic service tools, access and interpret customer programmable parameters.

AED 5.8

Time off_______________

Time on_______________

CDX Tasksheet Number: E0228

Total time_______________

1. **With the diagnostic tool and interpretation tool connected in accordance with the manufacturer's specifications, fill out the table below.**

Programmable Parameter	Default (D) / Programmed (P)	Programmed Parameter Setting
Example: Engine RPM	P	2200 RPM

   a. **Meets the manufacturer's specifications**
      Yes: ____________ No: ____________

   b. **If no, list your recommendations for any rectifications:**

      ____

      ____

      ____

      ____

      ____

2. **Discuss the findings with your instructor.**

Performance Rating

CDX Tasksheet Number: E0228

0	1	2	3	4

Supervisor/instructor signature ____ Date ___________

Diagnostics: Electronic Fuel Management System II

Student/Intern information:

Name _________________________________ Date _____________ Class _________________________

Vehicle used for this activity:

Year _______________ Make _________________________ Model _________________________

Odometer _____________ Hour meter _____________ VIN _________________________

Learning Objective/Task	CDX Tasksheet Number	2014 Edition Rev2/12/16 AED Standard
• Perform on-engine inspections, tests, and adjustments on electronic unit injectors (EUI); determine needed action.	E0229	5.8
• Remove and install electronic unit injectors (EUI) and related components; recalibrate ECM (if applicable).	E0230	5.8
• Perform cylinder contribution test utilizing electronic service tool(s).	E0231	5.8
• Perform on-engine inspections and tests on hydraulic electronic unit injectors (HEUI) and system electronic controls; determine needed action.	E0232	5.8
• Perform on-engine inspections and tests on hydraulic electronic unit injector (HEUI) high-pressure oil supply and control systems; determine needed action.	E0233	5.8
• Perform on-engine inspections and tests on high-pressure common rail (HPCR) type injection systems; determine needed action.	E0234	5.8
• Inspect high-pressure injection lines, hold downs, fittings, and seals; determine needed action.	E0235	5.8

Time off_____________

Time on_____________

Total time_____________

Materials Required

- Machine with possible engine concern
- Machine manufacturer's workshop manual
- Manufacturer-specific tools depending on the concern
- Machine lifting equipment, if applicable

Some Safety Issues to Consider

- Diagnosis of this fault may require test driving the machine on the school grounds or on a hoist, both of which carry severe risks. Attempt this task only with full permission from your supervisor/instructor and follow all the guidelines exactly.
- Caution: If you are working in an area where there could be "brake dust" present (may contain asbestos, which has been determined to cause cancer when inhaled or ingested), ensure you wear and use all OSHA-approved asbestos protective/removal equipment.

- Lifting equipment such as machine jacks and stands, machine hoists, and engine hoists are important tools that increase productivity and make the job easier. However, they can also cause severe injury or death if used improperly. Make sure you follow the manufacturer's operation procedures. Also make sure you have your supervisor/instructor's permission to use any particular type of lifting equipment.
- Comply with personal and environmental safety practices associated with clothing; eye protection; hand tools; power equipment; proper ventilation; and the handling, storage, and disposal of chemicals/materials in accordance with federal, state, and local regulations.
- Always wear the correct protective eyewear and clothing and use the appropriate safety equipment, as well as fender covers, seat protectors, and floor mat protectors.
- Make sure you understand and observe all legislative and personal safety procedures when carrying out practical assignments. If you are unsure of what these are, ask your supervisor/instructor.

Performance Standard

0–No exposure: No information or practice provided during the program; complete training required

1–Exposure only: General information provided with no practice time; close supervision needed; additional training required

2–Limited practice: Has practiced job during training program; additional training required to develop skill

3–Moderately skilled: Has performed job independently during training program; limited additional training may be required

4–Skilled: Can perform job independently with no additional training

Student/Intern information:

Name _________________________________ Date _____________ Class _________________________

Vehicle used for this activity:

Year _______________ Make _________________________________ Model _________________________

Odometer _____________ Hour meter _____________ VIN _________________________________

▶ TASK Perform on-engine inspections, tests, and adjustments on electronic unit injectors (EUI); determine needed action.

AED 5.8

Time off_____________

Time on_____________

CDX Tasksheet Number: E0229

1. **Refer to the manufacturer's workshop manual for the correct procedure and the appropriate specifications to inspect, test, and adjust the EUI.**

   Total time_____________

   a. **Meets the manufacturer's specifications:**
      Yes: _________ No: _________

   b. **If no, list your recommendations for any rectifications:**
      ____
      ____
      ____
      ____
      ____

2. **Discuss the findings with your instructor.**

Performance Rating

CDX Tasksheet Number: E0229

0	1	2	3	4

Supervisor/instructor signature ____ Date _____________

▶ **TASK** Remove and install EUI and related components; recalibrate ECM
(if applicable).

**AED
5.8**

Time off______________

Time on______________

Total time______________

CDX Tasksheet Number: E0230

1. **Refer to the manufacturer's workshop manual and appropriate tool(s) manual(s) and remove the EUI and related components.**

   a. **Remove and number the EUI and related components.**

      i. **Keep all components together for each cylinder in an appropriate parts tray.**

      ii. **Inspect each injector unit and component (including any wiring harness if applicable).**

      ____

      ____

      ____

      ____

      ____

   b. **Meets the manufacturer's specifications:
      Yes: _________ No: _________**

   c. **If no, list your recommendations for any rectifications:**

      ____

      ____

      ____

      ____

      ____

2. **Discuss the findings with your instructor.**

   a. **Perform diagnostic procedures as outlined by the manufacturer.**

      i. **Meets the manufacturer's specifications:
         Yes: _________ No: _________**

      ii. **If no, list your recommendations for any rectifications:**

      ____

      ____

      ____

      ____

      ____

3. **Refer to the manufacturer's workshop manual and appropriate tool(s) manual(s) and reinstall the EUI and related components.**

    a. **Torque all the components to their specified torque and record the manufacturer's torque specifications below.**

        i. **Unit injector holding clamp:** _________________ **ft-lb (N·m)**

        ii. **Fuel connecting pipes (if applicable):** _____________ **ft-lb (N·m)**

    b. **Record the actual torque readings below.**

        i. **Unit injector holding clamp:** _________________ **ft-lb (N·m)**

        ii. **Fuel connecting pipes (if applicable):** _____________ **ft-lb (N·m)**

    c. **Start and run the engine.**

    d. **Check the injectors and components for any leakage.**

    e. **Shut down the engine.**

        i. **Meets the manufacturer's specifications:**
        Yes: _________ **No:** _________

        ii. **If no, list your recommendations for any rectifications:**

        ____

        ____

        ____

        ____

        ____

4. **Discuss the findings with your instructor.**

Performance Rating

CDX Tasksheet Number: EO230

0	1	2	3	4
☐	☐	☐	☐	☐

Supervisor/instructor signature ____ Date _______________

▶ **TASK** Perform cylinder contribution test utilizing electronic service.

AED 5.8

Time off________________

Time on________________

Total time________________

CDX Tasksheet Number: EO231

1. **Research the procedures to be carried out when you perform the cylinder contribution test. Utilize the manufacturer's workshop manual and the tools manual.**

   a. **Referring to the appropriate manual(s), list the procedures involved in carrying out the cylinder contribution test:**

   ____
   ____
   ____
   ____
   ____

2. **Discuss these procedures with your instructor. When your instructor is satisfied with your proposed testing procedures, start the test.**

3. **While following the procedures involved in carrying out the cylinder contribution test, perform the test.**

   a. **Meets the manufacturer's specifications:**
      Yes: _________ No: _________

   b. **If no, list your recommendations for any rectifications:**

   ____
   ____
   ____
   ____
   ____

4. **Discuss the findings with your instructor.**

Performance Rating

CDX Tasksheet Number: EO231

| 0 | 1 | 2 | 3 | 4 |

Supervisor/instructor signature _________________________________ Date __________

Student/Intern information:

Name _________________________________ Date ____________ Class _____________________

Vehicle used for this activity:

Year _______________ Make ___________________________ Model ____________________________

Odometer ____________ Hour meter ____________ VIN _________________________________

▶ TASK Perform on-engine inspections and tests on hydraulic electronic unit injectors (HEUI) and system electronic controls; determine needed action.

AED
5.8

Time off________________

Time on________________

Total time________________

CDX Tasksheet Number: EO232

1. **Research the procedures to perform on-engine inspections and tests on HEUI and system electronic controls. Utilize the manufacturer's workshop manual and the tools manual.**

   a. **Referring to the appropriate manual(s), list the procedures involved in carrying out the inspections and test:**

   ____
   ____
   ____
   ____
   ____
   ____

2. **Discuss these procedures with your instructor. When your instructor is satisfied with your proposed testing procedures, start the test.**

3. **Perform on-engine inspections and tests on HEUI and system electronic controls.**

   a. **Meets the manufacturer's specifications:**
   Yes: ____________ No: ____________

   b. **If no, list your recommendations for any rectifications:**

   ____
   ____
   ____
   ____
   ____

4. Discuss the findings with your instructor.

Student/Intern information:

Name _________________________________ Date _____________ Class _________________________

Vehicle used for this activity:

Year _______________ Make _________________________________ Model _________________________

Odometer _____________ Hour meter ___________ VIN _________________________________

▶ TASK Perform on-engine inspections and tests on hydraulic electronic unit injector (HEUI) high-pressure oil supply and control systems; determine needed action.

AED 5.8

Time off________________

Time on________________

Total time________________

CDX Tasksheet Number: EO233

1. **Research the procedures to perform on-engine inspections and tests on hydraulic electronic unit injector high-pressure oil supply and control systems. Utilize the manufacturer's workshop manual and the tools manual.**

    a. **Referring to the appropriate manual(s), list the procedures involved in carrying out the inspections and test.**

    ____

    ____

    ____

    ____

    ____

2. **Discuss these procedures with your instructor. When your instructor is satisfied with your proposed testing procedures, start the test.**

3. **Perform on-engine inspections and tests on hydraulic electronic unit injector high-pressure oil supply and control systems.**

    a. **Meets the manufacturer's specifications:**
       Yes: ___________ No: ___________

    b. **If no, list your recommendations for any rectifications:**

    ____

    ____

    ____

    ____

    ____

4. **Discuss the findings with your instructor.**

Name _________________________________ Date _____________ Class ___________________

Vehicle used for this activity:

Year _______________ Make _______________________________ Model ___________________

Odometer _____________ Hour meter _____________ VIN _____________________________

▶ **TASK** Perform on-engine inspections and tests on high-pressure common rail (HPCR) type injection systems; determine needed action. **AED 5.8**

Time off_______________

Time on_______________

CDX Tasksheet Number: E0234

Total time_______________

1. **Research the procedures to perform on-engine inspections and tests on common rail-type injection systems. Utilize the manufacturer's workshop manual and the tools manual.**

    a. **Referring to the appropriate manual(s), list the procedures involved in carrying out the inspections and test.**

    ____
    ____
    ____
    ____
    ____

2. **Discuss these procedures with your instructor. When your instructor is satisfied with your proposed testing procedures, start the test.**

3. **Perform on-engine inspections and tests on common rail type injection systems.**

    a. **Meets the manufacturer's specifications:**
    Yes: _____________ No: _____________

    b. **If no, list your recommendations for any rectifications:**

    ____
    ____
    ____
    ____
    ____

4. **Discuss the findings with your instructor.**

Performance Rating

CDX Tasksheet Number: E0234

0	1	2	3	4

Supervisor/instructor signature ____ Date _____________

Name _________________________________ Date _____________ Class _________________________

Vehicle used for this activity:

Year _______________ Make _________________________________ Model _________________________

Odometer _____________ Hour meter _____________ VIN _________________________________

▶ TASK Inspect high-pressure injection lines, hold downs, fittings, and seals; determine needed action.

AED 5.8

Time off_________________

Time on_________________

Total time_________________

CDX Tasksheet Number: EO235

1. **Refer to the manufacturer's workshop manual to locate and use relevant service information; inspect high-pressure injection lines, hold downs, fittings, and seals.**

   a. **Inspect the high-pressure injection lines.**

      i. **Meets the manufacturer's specifications:**
         Yes: _____________ **No:** _____________

   b. **Inspect the hold downs.**

      i. **Meets the manufacturer's specifications:**
         Yes: _____________ **No:** _____________

   c. **Inspect the fittings and seals.**

      i. **Meets the manufacturer's specifications:**
         Yes: _____________ **No:** _____________

   d. **If no to any of the above, list your recommendations for any rectifications:**

      ____

      ____

      ____

      ____

      ____

2. **Discuss the findings with your instructor.**

Performance Rating　　　　　　　　　**CDX Tasksheet Number: EO235**

0	1	2	3	4

Supervisor/instructor signature ____ Date _____________

Diagnostics: Engine Brakes

Learning Objective/Task	CDX Tasksheet Number	2014 Edition Rev2/12/16 AED Standard
• Inspect and adjust engine compression/exhaust brakes; determine needed action.	EO236	5.8
• Inspect, test, and adjust engine compression/exhaust brake control circuits, switches, and solenoids; determine necessary action.	EO237	5.8

Time off ________________

Time on ________________

Total time ______________

Materials Required

- Machine with possible engine concern
- Machine manufacturer's workshop manual
- Manufacturer-specific tools depending on the concern
- Machine lifting equipment, if applicable

Some Safety Issues to Consider

- Diagnosis of this fault may require test driving the machine on the school grounds or on a hoist, both of which carry severe risks. Attempt this task only with full permission from your supervisor/instructor and follow all the guidelines exactly.
- Caution: If you are working in an area where there could be "brake dust" present (may contain asbestos, which has been determined to cause cancer when inhaled or ingested), ensure you wear and use all OSHA-approved asbestos protective/removal equipment.
- Lifting equipment such as machine jacks and stands, machine hoists, and engine hoists are important tools that increase productivity and make the job easier. However, they can also cause severe injury or death if used improperly. Make sure you follow the manufacturer's operation procedures. Also make sure you have your supervisor/instructor's permission to use any particular type of lifting equipment.
- Comply with personal and environmental safety practices associated with clothing; eye protection; hand tools; power equipment; proper ventilation; and the handling, storage, and disposal of chemicals/materials in accordance with federal, state, and local regulations.
- Always wear the correct protective eyewear and clothing and use the appropriate safety equipment, as well as fender covers, seat protectors, and floor mat protectors.
- Make sure you understand and observe all legislative and personal safety procedures when carrying out practical assignments. If you are unsure of what these are, ask your supervisor/instructor.

Performance Standard

0–No exposure: No information or practice provided during the program; complete training required

1–Exposure only: General information provided with no practice time; close supervision needed; additional training required

2–Limited practice: Has practiced job during training program; additional training required to develop skill

3–Moderately skilled: Has performed job independently during training program; limited additional training may be required

4–Skilled: Can perform job independently with no additional training

▶ TASK Inspect and adjust engine compression/exhaust brakes; determine needed action.

AED 5.8

Time off________________

Time on________________

Total time________________

CDX Tasksheet Number: E0236

1. **Refer to the manufacturer's workshop manual and inspect and adjust the engine compression/exhaust brakes.**

    a. **Meets the manufacturer's specifications:**
       Yes: ________ No: ________

    b. **If no, list your recommendations for any rectifications:**

    ____
    ____
    ____
    ____
    ____

2. **Discuss the findings with your instructor.**

Performance Rating

CDX Tasksheet Number: E0236

0	1	2	3	4

Supervisor/instructor signature ____ Date ____________

Student/Intern information:

Name _________________________________ Date _____________ Class _________________________

Vehicle used for this activity:

Year _______________ Make _________________________________ Model _________________________

Odometer _____________ Hour meter _____________ VIN _________________________________

▶ **TASK** Inspect, test, and adjust engine compression/exhaust brake control circuits, switches, and solenoids; determine necessary action.

AED 5.8

Time off_________________

Time on_________________

CDX Tasksheet Number: EO237

1. **Refer to the manufacturer's workshop manual and inspect, test, and adjust engine compression/exhaust brake control circuits, switches, and solenoids.**

Total time_________________

   a. **Meets the manufacturer's specifications:**
      Yes: _________ No: _________

   b. **If no, list your recommendations for any rectifications:**

   ____
   ____
   ____
   ____
   ____

2. **Discuss the findings with your instructor.**

Performance Rating

CDX Tasksheet Number: EO237

0	1	2	3	4

Supervisor/instructor signature ____ Date _____________

Section A6: Air Conditioning and Heating

CONTENTS

Servicing AC Systems

Student/Intern information:

Name _________________________________ Date ___________ Class _________________________

Vehicle used for this activity:

Year _______________ Make _______________________ Model _______________________

Odometer ___________ Hour meter ___________ VIN _________________________________

Learning Objective/Task	CDX Tasksheet Number	2014 Edition Rev2/12/16 AED Standard
• Identify causes of temperature control problems in the A/C system; determine needed action.	E0238	6.3
• Identify refrigerant and lubricant types; check for contamination; determine needed action.	E0239	6.3
• Identify A/C system problems indicated by pressure gauge and temperature readings; determine needed action.	E0240	6.3
• Identify A/C system problems indicated by visual, audible, smell, and touch procedures; determine needed action.	E0241	6.3
• Perform A/C system leak test; determine needed action.	E0242	6.3
• Recover, evacuate, and recharge A/C system using appropriate equipment.	E0243	6.3
• Identify contamination in the A/C system components; determine needed action.	E0244	6.3

Time off_________________

Time on_________________

Total time_________________

Materials Required

- Machines or simulators with HVAC faults
- PC-based software and/or data scan tools
- Machine manufacturer's workshop manual including schematic wiring diagrams
- HVAC systems with orifice tubes and expansion valves
- Specialist refrigeration tools, thermometers, manifold set, refrigerant recycler, and vacuum pump
- HVAC parts and refrigerant
- Manufacturer-specific tools depending on the concern

Some Safety Issues to Consider

- Activities may require test driving the equipment on the school grounds, which carry severe risks. Attempt this task only with full permission from your supervisor/instructor, and follow all the guidelines exactly.
- All practices and procedures must be performed according to current mandates, standards and regulations.

- You may be required to handle refrigerant. Use extreme caution: refrigerant is pressurized and very cold. Always wear eye protection and appropriate clothing when working with refrigerant. Never inhale refrigerant.
- Do not release refrigerant to the atmosphere; always use a recycling system to reclaim refrigerant. Check for local environmental laws or regulations in relation to the management of refrigerant gases.
- Comply with personal and environmental safety practices associated with clothing; eye protection; hand tools; power equipment; proper ventilation; and the handling, storage, and disposal of chemicals/materials in accordance with federal, state, and local regulations.
- Always wear the correct protective eyewear and clothing and use the appropriate safety equipment, as well as fender covers, seat protectors, and floor mat protectors.
- Make sure you understand and observe all legislative and personal safety procedures when carrying out practical assignments. If you are unsure of what these are, ask your supervisor/instructor.

Performance Standard

0—No exposure: No information or practice provided during the program; complete training required

1—Exposure only: General information provided with no practice time; close supervision needed; additional training required

2—Limited practice: Has practiced job during training program; additional training required to develop skill

3—Moderately skilled: Has performed job independently during training program; limited additional training may be required

4—Skilled: Can perform job independently with no additional training

Student/Intern information:

Name _________________________ Date ___________ Class _________________

Vehicle used for this activity:

Year _____________ Make _______________________ Model _________________

Odometer ___________ Hour meter __________ VIN _____________________

▶ **TASK** Identify causes of temperature control problems in the A/C system; determine needed action.

AED
6.3

Time off_______________

Time on_______________

CDX Tasksheet Number: E0238

1. **Research causes of temperature control problems in the A/C system and list them below:**

   ____
   ____
   ____

Total time_______________

2. **Ask your supervisor/instructor for a machine or simulator to check.**

3. **Examine the machine/simulator and determine any faults with temperature control in the A/C system.**

   List the customer concern:

   ____
   ____
   ____

   Using the appropriate service information; determine the steps necessary for testing the components:

   ____
   ____
   ____

4. **Write a short description of the purpose and operation of the suspected components.**

   ____
   ____
   ____

5. **Determine and list any necessary action(s):**

   ____
   ____
   ____

6. **Return the machine to beginning condition and return any tools that you may have used to their proper locations.**

7. **Discuss the findings with the instructor.**

____

____

____

▶ **TASK** Identify refrigerant and lubricant types; check for contamination; determine needed action.

**AED
6.3**

Time off________________

Time on________________

Total time________________

CDX Tasksheet Number: E0239

1. **Research the types of refrigerants and lubricants and list them below. Types of refrigerants and lubricants can be identified by the factory sticker or placard on the radiator support or underside of the hood.**

    a. **Refrigerant types:**

    ____

    ____

    ____

    b. **Lubricant types:**

    ____

    ____

    ____

2. **Research the causes of refrigerant and lubricant contamination and list them below:**

    ____

    ____

    ____

3. **Ask your supervisor/instructor for a machine or simulator to check.**

4. **Examine the machine/simulator and determine the causes of refrigerant and lubrication contamination.**

5. **List the nature of contamination:**

    ____

    ____

    ____

6. **Determine and list any necessary action(s):**

    ____

    ____

    ____

7. **Return the machine to beginning condition and return any tools that you may have used to their proper locations.**

8. **Discuss the findings with the instructor.**

____

____

____

____

Performance Rating

CDX Tasksheet Number: EO239

| 0 | 1 | 2 | 3 | 4 |

Supervisor/instructor signature ____ Date _______________

Student/Intern information:

Name _________________________________ Date ____________ Class _________________________

Vehicle used for this activity:

Year ______________ Make _________________________ Model _________________________

Odometer ___________ Hour meter __________ VIN _________________________________

▶ **TASK** Identify A/C system problems indicated by pressure gauge and temperature readings; determine needed action.

AED 6.3

Time off_________________

Time on_________________

Total time_________________

CDX Tasksheet Number: E0240

1. **Research how to identify problems on A/C systems by pressure gauge and temperature readings and list the findings below:**

   a. **Pressure gauge.**

      i. **List correct system operation gauge readings:**

      High Pressure: _________________ **Low Pressure:** _________________

      ii. **List common problems and expected gauge readings:**

Problem	High Pressure	Low Pressure
_________	_________	_________
_________	_________	_________
_________	_________	_________
_________	_________	_________
_________	_________	_________
_________	_________	_________
_________	_________	_________

   b. **Temperature readings.**

   List correct system operation temperature readings:

2. **Have your supervisor/instructor verify your research.**

   Supervisor/instructor's initials: _________________________

3. **Ask your supervisor for a machine to check. Install the necessary test equipment, set the parking brake, and lift the cabin as necessary.**

4. **Identify A/C system problems indicated by pressure gauge and temperature readings. List your observations here:**

5. **Compare your results to the manufacturer's specifications. List your observations:**

6. **Determine and list any necessary action(s):**

   ____

   ____

   ____

7. Return the machine to beginning condition and return any tools that you may have used to their proper locations.

8. Discuss the findings with the instructor.

____
____
____

▶ TASK Identify A/C system problems indicated by visual, audio, smell, and touch procedures.

**AED
6.3**

Time off________________

Time on________________

Total time________________

CDX Tasksheet Number: E0241

1. **Research how to identify problems on A/C systems by visual, audio, smell, and touch procedures and list them below:**

    a. **Problems identified by looking—describe the signs and the problem:**

    ____

    ____

    ____

    b. **Problems identified by listening—describe the sound and the problem:**

    ____

    ____

    ____

    c. **Problems identified by smell—describe smell and problem. This usually requires the machine air-conditioning system in the running position. Pay attention to the air ducts for any unusual smells. Also check cabin filter for moisture or mold. Check condensation drain for clogging if smells are present.**

    ____

    ____

    ____

    d. **Problems identified by touch—describe the feeling and the problem:**

    ____

    ____

    ____

2. **Have your supervisor/instructor verify your research.**

    Supervisor/instructor's initials: ____________________

3. **Ask your supervisor/instructor for a machine or simulator to check.**

4. **Using visual, audio, smell, and touch procedures, test the system to determine system performance. List your observations here:**

    ____

    ____

    ____

5. Compare your results to the manufacturer's specifications. List your observations:

____

____

____

6. Determine and list any necessary action(s):

____

____

____

7. Return the machine to beginning condition and return any tools that you may have used to their proper locations.

8. Discuss the findings with the instructor.

____

____

____

Performance Rating

CDX Tasksheet Number: EO241

0	1	2	3	4

Supervisor/instructor signature _________________________________ Date _____________

Student/Intern information:

Name _________________________________ Date ___________ Class _________________________

Vehicle used for this activity:

Year _____________ Make _________________________ Model ___________________________

Odometer ___________ Hour meter ___________ VIN ________________________________

▶ **TASK** Perform A/C system leak test; determine needed action. **AED 6.3**

CDX Tasksheet Number: E0242

1. **Research the different methods of refrigerant leak testing and list them below:**

    ____

    ____

    ____

2. **Describe each method of refrigerant leak testing below:**

    ____

    ____

    ____

3. **Ask your supervisor/instructor for a machine or simulator to check.**

4. **Examine the machine /simulator and test for refrigerant leaks.**

5. **List the location(s) and cause(s) of refrigerant leaks:**

    ____

    ____

    ____

6. **Determine and list any necessary action(s):**

    ____

    ____

    ____

7. **Return the machine to beginning condition and return any tools that you may have used to their proper locations.**

____

____

____

Performance Rating

CDX Tasksheet Number: E0242

<table>
<tr><td>☐</td><td>☐</td><td>☐</td><td>☐</td><td>☐</td></tr>
<tr><td>0</td><td>1</td><td>2</td><td>3</td><td>4</td></tr>
</table>

Supervisor/instructor signature ______________________________________ Date ______________

Student/Intern information:

Name _________________________________ Date ____________ Class ________________________

Vehicle used for this activity:

Year _______________ Make _________________________ Model ___________________________

Odometer ____________ Hour meter ___________ VIN _____________________________________

▶ **TASK** Recover, evacuate, and recharge A/C system using appropriate equipment.

AED
6.3

Time off________________

Time on________________

Total time________________

CDX Tasksheet Number: E0243

1. **Research the recovery, evacuation, and recharge of an A/C system using the appropriate equipment.**

    a. **List the A/C system recovery process and tools required.**
    b. **List the A/C system evacuation procedure and tools required.**
    c. **List the A/C system recharge procedure and tools required.**
    d. **Have your supervisor/instructor verify this information and ask for a machine or simulator to undertake the activities.**

2. **Recover refrigerant from an A/C system. Ensure manufacturer's recommendations are followed.**

    a. **Record amount of refrigerant recovered: _________________ lb or oz**

    b. **Have your supervisor/instructor verify refrigerant recovery:**

    Supervisor/instructor's initials: _________________

3. **Evacuate an A/C system. Ensure manufacturer's recommendations are followed.**

    a. **Record vacuum in inches achieved: _________________ in of vacuum**

    b. **Have your supervisor/instructor verify.**

    Supervisor/instructor's initials: _________________

4. **Recharge an A/C system with refrigerant. Ensure manufacturer's recommendations are followed.**

    a. **Record amount of refrigerant used: _________________ lb or oz**

5. **Return the machine to beginning condition and return any tools that you may have used to their proper locations.**

6. **Discuss the findings with the instructor.**

____
____
____

Student/Intern information:

Name ________________________________ Date ____________ Class ________________________

Vehicle used for this activity:

Year ______________ Make ________________________ Model ________________________

Odometer ____________ Hour meter ____________ VIN ________________________________

▶ TASK Identify contamination in the A/C system components; determine needed action.

AED
6.3

CDX Tasksheet Number: E0244

1. **Research how to identify contaminated A/C system components and list findings below:**

   a. **How to identify contaminated compressors:**

   ____
   ____
   ____
   ____

   b. **How to identify a contaminated evaporator:**

   ____
   ____
   ____
   ____

   c. **How to identify contaminated condensers:**

   ____
   ____
   ____
   ____

   d. **How to identify contaminated expansion valves or fixed orifices:**

   ____
   ____
   ____
   ____

   e. **How to identify contaminated receiver driers/filters:**

   ____
   ____
   ____
   ____

   f. **How to identify contaminated hoses:**

   ____
   ____
   ____
   ____

2. **Have your supervisor/instructor verify your research.**

   Supervisor/instructor's initials: ____________________

3. Ask your supervisor for a machine or simulator to check, set the parking brake, and lift the cabin as necessary.

4. Identify A/C system contaminated components. List your observations here:

5. Determine and list any necessary action(s):

____

____

____

6. Return the machine to beginning condition and return any tools that you may have used to their proper locations.

7. Discuss the findings with the instructor.

____

____

____

Performance Rating

CDX Tasksheet Number: E0244

0	1	2	3	4

Supervisor/instructor signature _______________________________________ Date _____________

Testing, Troubleshooting, Diagnosing, and Repairing AC Systems: General System

Student/Intern information:

Name _______________________________ Date _____________ Class _____________________

Vehicle used for this activity:

Year _______________ Make _______________________ Model _____________________

Odometer _____________ Hour meter _____________ VIN _____________________________

<table>
<tr><td rowspan="2">Learning Objective/Task</td><td>CDX Tasksheet Number</td><td>2014 Edition Rev2/12/16 AED Standard</td></tr>
<tr></tr>
<tr><td>• Verify the need for service or repair of HVAC systems based on unusual operating noises; determine needed action.</td><td>E0245</td><td>6.4</td></tr>
<tr><td>• Verify the need for service or repair of HVAC systems based on unusual visual, smell, and touch conditions; determine needed action.</td><td>E0246</td><td>6.4</td></tr>
<tr><td>• Identify system type and components (cycling clutch orifice tube-CCOT, expansion valve) and conduct performance test(s) on HVAC systems; determine needed action.</td><td>E0247</td><td>6.4</td></tr>
<tr><td>• Retrieve diagnostic codes; determine needed action.</td><td>E0248</td><td>6.4</td></tr>
</table>

Time off________________

Time on________________

Total time________________

Materials Required

- Machines or simulators with HVAC faults
- PC-based software and/or data scan tools
- Machine manufacturer's workshop manual including schematic wiring diagrams
- HVAC systems with orifice tubes and expansion valves
- Specialist refrigeration tools, thermometers, manifold set, refrigerant recycler, and vacuum pump
- HVAC parts and refrigerant
- Manufacturer-specific tools depending on the concern

Some Safety Issues to Consider

- Activities may require test driving the equipment on the school grounds, which carry severe risks. Attempt this task only with full permission from your supervisor/instructor, and follow all the guidelines exactly.
- All practices and procedures must be performed according to current mandates, standards and regulations.
- You may be required to handle refrigerant. Use extreme caution: refrigerant is pressurized and very cold. Always wear eye protection and appropriate clothing when working with refrigerant. Never inhale refrigerant.
- Do not release refrigerant to the atmosphere; always use a recycling system to reclaim refrigerant. Check for local environmental laws or regulations in relation to the management of refrigerant gases.

- Comply with personal and environmental safety practices associated with clothing; eye protection; hand tools; power equipment; proper ventilation; and the handling, storage, and disposal of chemicals/materials in accordance with federal, state, and local regulations.
- Make sure you understand and observe all legislative and personal safety procedures when carrying out practical assignments. If you are unsure of what these are, ask your supervisor/ instructor.

Performance Standard

0–No exposure: No information or practice provided during the program; complete training required

1–Exposure only: General information provided with no practice time; close supervision needed; additional training required

2–Limited practice: Has practiced job during training program; additional training required to develop skill

3–Moderately skilled: Has performed job independently during training program; limited additional training may be required

4–Skilled: Can perform job independently with no additional training

Student/Intern information:

Name _________________________________ Date ____________ Class _________________________

Vehicle used for this activity:

Year _______________ Make _________________________________ Model _________________________

Odometer ______________ Hour meter ______________ VIN _________________________________

▶ **TASK** Research causes of unusual operating noises with HVAC systems and list them below. This will require starting the machine, operating the air-conditioning system, and listening for any unusual noises that may be present while in operation.

AED
6.4

Time off_________________

Time on_________________

Total time_________________

CDX Tasksheet Number: EO245

1. **Research causes of unusual operating noises with HVAC systems and list them below. This will require starting the machine, operating the air-conditioning system, and listening for any unusual noises that may be present while in operation.**

2. **List the customer concern:**

   ____
   ____
   ____

3. **Examine the machine/simulator and determine the need for service or repair based on unusual operating noises. List the unusual operating noises and possible faults:**

   ____
   ____
   ____

4. **Write a short description of the purpose and operation of the suspected component(s):**

   ____
   ____
   ____

5. **Determine and list any necessary action(s):**

   ____
   ____
   ____

6. **Return the machine to beginning condition and return any tools that you may have used to their proper locations.**

7. **Discuss the findings with the instructor.**

____

____

____

Student/Intern information:

Name ________________________________ Date _____________ Class _____________________

Vehicle used for this activity:

Year _______________ Make _____________________________ Model ____________________

Odometer ____________ Hour meter ____________ VIN ________________________________

▶ **TASK** Verify the need for service or repair of HVAC systems based on unusual visual, smell, and touch conditions; determine needed action.

AED 6.4

Time off_______________

Time on_______________

Total time_______________

CDX Tasksheet Number: E0246

1. **Research causes of unusual visual, smell, and touch conditions with HVAC systems and list them below:**

   ____
   ____
   ____

2. **List the customer concern:**

   ____
   ____
   ____

3. **Examine the machine/simulator and determine the need for service or repair based on unusual visual, smell, and touch conditions. List the unusual visual, smell, and touch conditions and possible faults.**

   ____
   ____
   ____

4. **Check cab vents for any foul odors or clouding coming from them.**

   a. **Odors or clouding present? Yes: _____________ No: _____________**

5. **Check HVAC system for any components that may look damaged or show signs of leakage.**

6. **Write a short description of the purpose and operation of the suspected component(s):**

   ____
   ____
   ____

7. **Determine and list any necessary action(s):**

   ____
   ____
   ____

8. Return the machine to beginning condition and return any tools that you may have used to their proper locations.

9. Discuss the findings with the instructor.

____
____
____

Performance Rating

CDX Tasksheet Number: EO246

| 0 | 1 | 2 | 3 | 4 |

Supervisor/instructor signature _______________________________________ Date _______________

Student/Intern information:

Name _________________________________ Date ____________ Class _____________________

Vehicle used for this activity:

Year ______________ Make ________________________ Model _____________________

Odometer ____________ Hour meter ___________ VIN _________________________________

▶ TASK Identify system type and components (cycling clutch orifice tube (CCOT), expansion valve) and conduct performance test(s) on HVAC systems.

AED 6.4

Time off________

Time on________

Total time________

CDX Tasksheet Number: E0247

1. **Research and identify system-type components.**

    a. **Cycling clutch orifice tube CCOT systems:**

    ____
    ____
    ____
    ____

    b. **Expansion valve systems:**

    ____
    ____
    ____
    ____

2. **Have your supervisor/instructor verify your research.**

    Supervisor/instructor's initials: __________________________

3. **Ask your supervisor for a machine to check. Install the necessary test equipment, set the parking brake, and lift the cabin as necessary.**

4. **Test the system to determine system performance. When doing a system performance test, always wear eye protection to avoid any damage to eyes. List your observations here:**

    ____
    ____
    ____

5. **Compare your results to the manufacturer's specifications. List your observations:**

    ____
    ____
    ____

6. **Determine and list any necessary action(s):**

    ____
    ____
    ____

7. **Return the machine to beginning condition and return any tools that you may have used to their proper locations.**

8. **Discuss the findings with the instructor.**

____

____

____

Performance Rating

CDX Tasksheet Number: EO247

| 0 | 1 | 2 | 3 | 4 |

Supervisor/instructor signature _____________________________________ Date ____________

Student/Intern information:

Name _________________________________ Date _____________ Class _______________________

Vehicle used for this activity:

Year _______________ Make _________________________________ Model _______________________

Odometer _____________ Hour meter _____________ VIN _________________________________

▶ TASK Retrieve diagnostic codes; determine needed action.

AED 6.4

Time off _______________

Time on _______________

Total time _______________

CDX Tasksheet Number: E0248

1. **Research how to retrieve diagnostic codes. List below the procedure and type of software and/or data scan tools that will be used for the machine:**

   ____
   ____
   ____

2. **Connect the software and/or data scan tool to the machine's Data Link Connector (DLC), and retrieve any Diagnostic Trouble Codes (DTC). Using the process described in the flow chart as a guide, undertake the test and list any DTCs here:**

   ____
   ____
   ____

3. **Research the DTCs for this machine in the appropriate service manual.**

   a. **List the DTC code descriptions for each code stored:**

      ____
      ____
      ____

   b. **Print out the procedure for diagnosing each code and attach a copy to this sheet.**

4. **Have your supervisor/instructor verify satisfactory completion of this section of the procedure.**

   Supervisor/instructor's initials: _______________

5. **Follow the machine's service manual procedure to diagnose the specific cause of the DTC and use the PC-based software and/or data scan tool to assist in the diagnosis of the problem. List the data related to this DTC here:**

   ____
   ____
   ____

6. **Compare this data to the service manual specifications and list your interpretations:**

   ____
   ____
   ____

7. **Determine and list any necessary action(s):**

____

____

____

8. **Return the machine to beginning condition and return any tools that you may have used to their proper locations.**

9. **Discuss the findings with the instructor.**

____

____

____

Performance Rating

CDX Tasksheet Number: E0248

0	1	2	3	4
☐	☐	☐	☐	☐

Supervisor/instructor signature _______________________________ Date _____________

Testing, Troubleshooting, Diagnosing, and Repairing AC Systems: Compressor and Protection

Student/Intern information:

Name _________________________ Date _________ Class _________________

Vehicle used for this activity:

Year _____________ Make _________________ Model _________________

Odometer __________ Hour meter __________ VIN _________________

Learning Objective/Task	CDX Tasksheet Number	2014 Edition Rev2/12/16 AED Standard
• Identify A/C system problems that cause protection devices (pressure, thermal, and electronic) to interrupt system operation; determine needed action.	E0249	6.4
• Inspect, test, and replace A/C system pressure, thermal, and electronic protection devices.	E0250	6.4
• Inspect and replace A/C compressor drive belts, pulleys, and tensioners; adjust belt tension and check alignment.	E0251	6.4
• Inspect, test, adjust, service, or replace A/C compressor clutch components or assembly.	E0252	6.4
• Inspect and correct A/C compressor lubricant level (if applicable).	E0253	6.4
• Inspect, test, or replace A/C compressor.	E0254	6.4
• Inspect, repair, or replace A/C compressor mountings and hardware.	E0255	6.4

Time off_____________

Time on_____________

Total time_____________

Materials Required

- Machines or simulators with HVAC faults
- PC-based software and/or data scan tools
- Machine manufacturer's workshop manual including schematic wiring diagrams
- HVAC systems with orifice tubes and expansion valves
- Specialist refrigeration tools, thermometers, manifold set, refrigerant recycler, and vacuum pump
- HVAC parts and refrigerant
- Manufacturer-specific tools depending on the concern

Some Safety Issues to Consider

- Activities may require test driving the equipment on the school grounds, which carry severe risks. Attempt this task only with full permission from your supervisor/instructor, and follow all the guidelines exactly.
- All practices and procedures must be performed according to current mandates, standards and regulations.

- You may be required to handle refrigerant. Use extreme caution: refrigerant is pressurized and very cold. Always wear eye protection and appropriate clothing when working with refrigerant. Never inhale refrigerant.
- Do not release refrigerant to the atmosphere; always use a recycling system to reclaim refrigerant. Check for local environmental laws or regulations in relation to the management of refrigerant gases.
- Comply with personal and environmental safety practices associated with clothing; eye protection; hand tools; power equipment; proper ventilation; and the handling, storage, and disposal of chemicals/materials in accordance with federal, state, and local regulations.
- Always wear the correct protective eyewear and clothing and use the appropriate safety equipment, as well as fender covers, seat protectors, and floor mat protectors.
- Make sure you understand and observe all legislative and personal safety procedures when carrying out practical assignments. If you are unsure of what these are, ask your supervisor/instructor.

Performance Standard

0–No exposure: No information or practice provided during the program; complete training required

1–Exposure only: General information provided with no practice time; close supervision needed; additional training required

2–Limited practice: Has practiced job during training program; additional training required to develop skill

3–Moderately skilled: Has performed job independently during training program; limited additional training may be required

4–Skilled: Can perform job independently with no additional training

▶ TASK Identify A/C system problems that cause protection devices (pressure, thermal, and electronic) to interrupt system operation; determine needed action.

AED 6.4

Time off_______________

Time on_______________

Total time_______________

CDX Tasksheet Number: E0249

1. **Research what causes protection devices (pressure, thermal, and electronic) to interrupt system operation; use the appropriate service information for the machine you are working on and list the causes below:**

    a. **Causes of pressure protection devices interrupting system operation:**

    ____
    ____
    ____

    b. **Causes of thermal protection devices interrupting system operation:**

    ____
    ____
    ____

    c. **Causes of electronic protection devices interrupting system operation:**

    ____
    ____
    ____

2. **Ask your supervisor/instructor for a machine or simulator to check.**

3. **Examine the machine/simulator and identify A/C system problems that cause protection devices (pressure, thermal, and electronic) to interrupt system operation.**

4. **List the problems identified:**

    ____
    ____
    ____

5. **Determine and list any necessary action(s):**

    ____
    ____
    ____

6. **Return the machine to beginning condition and return any tools that you may have used to their proper locations.**

7. **Discuss the findings with the instructor.**

____

____

____

Student/Intern information:

Name _________________________________ Date ____________ Class _________________________

Vehicle used for this activity:

Year _______________ Make ________________________________ Model ________________________

Odometer ______________ Hour meter ______________ VIN _________________________________

▶ TASK Inspect, test, and replace A/C system pressure, thermal, and electronic protection.

AED 6.4

Time off_______________

Time on_______________

Total time_______________

CDX Tasksheet Number: E0250

1. **Research how to inspect, test, and replace A/C system pressure, thermal, and electronic protection devices in the appropriate service information for the machine you are working on and list them below:**

   a. **Pressure protection:**

   ____

   ____

   ____

   b. **Thermal protection:**

   ____

   ____

   ____

   c. **Electronic protection:**

   ____

   ____

   ____

2. **Have your supervisor/instructor verify your research.**

   Supervisor/instructor's initials: _____________________________

3. **Ask your supervisor/instructor for a machine or simulator to check.**

4. **Examine the machine/simulator and inspect, test, and replace A/C system pressure, thermal, and electronic protection devices.**

5. **Using the appropriate service information, determine the steps necessary for testing the components:**

   ____

   ____

   ____

6. **Write a short description of the purpose and operation of the suspected component(s):**

   ____

   ____

   ____

7. Determine and list any necessary action(s):

____

____

____

8. **Return the machine to beginning condition and return any tools that you may have used to their proper locations.**

9. **Discuss the findings with the instructor.**

____

____

____

Performance Rating

CDX Tasksheet Number: E0250

☐	☐	☐	☐	☐
0	1	2	3	4

Supervisor/instructor signature _______________________________ Date ___________

▶ TASK Inspect and replace A/C compressor drive belts, pulleys, and tensioners; adjust belt.

AED
6.4

Time off________________

Time on________________

Total time________________

CDX Tasksheet Number: EO251

1. **Research how to inspect and replace A/C compressor drive belts, pulleys, and tensioners. Use the appropriate service information for the machine you are working on.**

    a. **List the steps outlined in the service information:**

    ____
    ____
    ____
    ____

    b. **Write down the specified tension of the compressor drive belt:**

    ____
    ____
    ____
    ____

    c. **List faults to look for when inspecting drive belts, pulleys, and tensioners:**

    ____
    ____
    ____
    ____

    d. **Describe how to check correct pulley and belt alignment:**

    ____
    ____
    ____
    ____

2. **Remove machine A/C compressor drive belts.**

    a. **Inspect machine drive belts, pulleys, tensioners, and mounting brackets for faults.**

    b. **List any faults identified on the machine A/C compressor drive belts, pulleys, and tensioners:**

    ____
    ____
    ____

    c. **Refit machine A/C compressor drive belts using appropriate service information.**

d. Adjust the retention of A/C compressor drive belts using appropriate service information.

e. Check for correct pulley, tensioner, and A/C compressor drive belt alignment.

3. Return the machine to beginning condition and return any tools that you may have used to their proper locations.

4. Discuss the findings with the instructor.

____

____

____

Performance Rating

CDX Tasksheet Number: EO251

0	1	2	3	4

Supervisor/instructor signature _________________________________ Date _____________

Vehicle used for this activity:

Year ______________ Make ____________________________ Model ____________________________

Odometer ____________ Hour meter ____________ VIN ________________________________

▶ **TASK** Inspect, test, adjust, service, or replace A/C compressor clutch components or assembly.

AED 6.4

Time off________________

Time on________________

Total time________________

CDX Tasksheet Number: EO252

1. **Research how to inspect, test, adjust, service, or replace A/C compressor clutch components or assembly in the appropriate service information for the machine you are working on and list them below:**

   ____

   ____

   ____

2. **Ask your supervisor for a machine or simulator to check the A/C compressor clutch or components.**

3. **Inspect, test, adjust, service, or replace the A/C compressor clutch components or assembly. Check and record for resistance and voltages necessary to operate the clutch properly. List your observations here:**

   ____

   ____

   ____

4. **Compare your results to the manufacturer's specifications. List your observations:**

   ____

   ____

   ____

5. **Return the machine to beginning condition and return any tools that you may have used to their proper locations.**

6. **Discuss the findings with the instructor.**

____
____
____

Performance Rating

CDX Tasksheet Number: E0252

☐	☐	☐	☐	☐
0	1	2	3	4

Supervisor/instructor signature _________________________________ Date ___________

▶ TASK Inspect and correct A/C compressor lubricant.

AED 6.4

Time off_________________

Time on_________________

CDX Tasksheet Number: E0253

Total time_________________

1. **Research how to inspect and correct A/C compressor lubricant levels in the appropriate service information for the machine you are working on and list the results below:**

   ____

   ____

   ____

2. **Ask your supervisor/instructor for a machine or simulator to check A/C compressor lubricant level.**

3. **Examine the machine/simulator A/C compressor.**

   a. **Record the amount of oil necessary to bring the system to manufacturer's specification.**

   ____

   ____

   ____

4. **List the results of your inspection:**

   ____

   ____

   ____

5. **Return the machine to beginning condition and return any tools that you may have used to their proper locations.**

6. **Discuss the findings with the instructor.**

____
____
____
____

Performance Rating

CDX Tasksheet Number: EO253

| 0 | 1 | 2 | 3 | 4 |

Supervisor/instructor signature ____ Date ______________

Student/Intern information:

Name _________________________ Date __________ Class _________________

Vehicle used for this activity:

Year _____________ Make __________________________ Model _________________

Odometer ____________ Hour meter ____________ VIN _________________________

▶ TASK Inspect, test, or replace A/C compressor.

AED 6.4

Time off________________

Time on________________

CDX Tasksheet Number: E0254

Total time________________

1. **Research how to inspect, test, or replace the A/C compressor in the appropriate service information for the machine you are working on and list them below:**

   a. **Inspection of A/C compressor:**

   ____
   ____
   ____

   b. **Testing of A/C compressor:**

   ____
   ____
   ____

   c. **Replacing A/C compressor (remember to vacuum the system for the recommended time to remove moisture and contaminants):**

   ____
   ____
   ____

2. **Have your supervisor/instructor verify your research.**

   Supervisor/instructor's initials: _________________________

3. **Ask your supervisor/instructor for a machine or simulator to check.**

4. **Examine the machine/simulator A/C compressor and inspect and test the A/C compressor.**

5. **List the results of your inspection:**

   ____
   ____
   ____

6. **Return the machine to beginning condition and return any tools that you may have used to their proper locations.**

7. **Discuss the findings with the instructor.**

____

____

____

Student/Intern information:

Name _________________________________ Date _____________ Class _________________________

Vehicle used for this activity:

Year _______________ Make _____________________________ Model _________________________

Odometer _____________ Hour meter _____________ VIN _________________________________

▶ **TASK** Inspect, repair, or replace A/C compressor mountings and hardware.

AED
6.4

Time off_________________

Time on_________________

Total time_______________

CDX Tasksheet Number: E0255

1. **Research how to inspect, repair, or replace the A/C compressor mounting and hardware in the appropriate service information for the machine you are working on and list them below:**

   a. **Inspection of A/C compressor mountings and hardware.**

   b. **Repair of A/C compressor mountings and hardware.**

   c. **Replacement of A/C compressor mountings and hardware.**

2. **Have your supervisor/instructor verify your research.**

   Supervisor/instructor's initials: _________________________

3. **Ask your supervisor/instructor for a machine or simulator to check.**

4. **Examine the machine/simulator A/C compressor and inspect, repair, and replace the A/C compressor mountings and hardware.**

5. **List the results of your inspection:**

   ____

   ____

   ____

6. **Return the machine to beginning condition and return any tools that you may have used to their proper locations.**

7. **Discuss the findings with the instructor.**

   ____

   ____

   ____

Performance Rating

CDX Tasksheet Number: E0255

0	1	2	3	4

Supervisor/instructor signature ____ Date _____________

Testing, Troubleshooting, Diagnosing, and Repairing AC Systems: Major Components

Student/Intern information:

Name _______________________________ Date _____________ Class _______________________

Vehicle used for this activity:

Year _______________ Make _______________________ Model _____________________

Odometer _____________ Hour meter _____________ VIN _______________________

<table>
<tr><td rowspan="2">Learning Objective/Task</td><td>CDX Tasksheet Number</td><td>2014 Edition Rev2/12/16 AED Standard</td></tr>
<tr><td></td><td></td></tr>
<tr><td>• Correct system lubricant level when replacing the evaporator, condenser, receiver/drier or accumulator/drier, and hoses.</td><td>E0256</td><td>6.4</td></tr>
<tr><td>• Inspect A/C system hoses, lines, filters, fittings, and seals; determine needed action.</td><td>E0257</td><td>6.4</td></tr>
<tr><td>• Inspect and test A/C system condenser. Check for proper airflow and mountings; determine needed action.</td><td>E0258</td><td>6.4</td></tr>
<tr><td>• Inspect and replace receiver/drier or accumulator/drier.</td><td>E0259</td><td>6.4</td></tr>
<tr><td>• Inspect and test refrigerant solenoid, expansion valve(s); check placement of thermal bulb (capillary tube); determine needed action.</td><td>E0260</td><td>6.4</td></tr>
<tr><td>• Remove and replace orifice tube.</td><td>E0261</td><td>6.4</td></tr>
<tr><td>• Inspect and test evaporator core; determine needed action.</td><td>E0262</td><td>6.4</td></tr>
<tr><td>• Inspect, clean, or repair evaporator housing and water drain; inspect and service/replace evaporator air filter.</td><td>E0263</td><td>6.4</td></tr>
<tr><td>• Identify and inspect A/C system service ports (gauge connections); determine needed action.</td><td>E0264</td><td>6.4</td></tr>
<tr><td>• Identify the cause of system failures resulting in refrigerant loss from the A/C system high-pressure relief device; determine needed action.</td><td>E0265</td><td>6.4</td></tr>
</table>

Time off_______________

Time on_______________

Total time_______________

Materials Required

- Machines or simulators with HVAC faults
- PC-based software and/or data scan tools
- Machine manufacturer's workshop manual including schematic wiring diagrams
- HVAC systems with orifice tubes and expansion valves
- Specialist refrigeration tools, thermometers, manifold set, refrigerant recycler, and vacuum pump
- HVAC parts and refrigerant
- Manufacturer-specific tools depending on the concern

Some Safety Issues to Consider

- Activities may require test driving the equipment on the school grounds, which carries severe risks. Attempt this task only with full permission from your supervisor/instructor, and follow all the guidelines exactly.
- All practices and procedures must be performed according to current mandates, standards and regulations.
- You may be required to handle refrigerant. Use extreme caution: refrigerant is pressurized and very cold. Always wear eye protection and appropriate clothing when working with refrigerant. Never inhale refrigerant.
- Do not release refrigerant to the atmosphere; always use a recycling system to reclaim refrigerant. Check for local environmental laws or regulations in relation to the management of refrigerant gases.
- Comply with personal and environmental safety practices associated with clothing; eye protection; hand tools; power equipment; proper ventilation; and the handling, storage, and disposal of chemicals/materials in accordance with federal, state, and local regulations.
- Always wear the correct protective eyewear and clothing and use the appropriate safety equipment, as well as fender covers, seat protectors, and floor mat protectors.
- Make sure you understand and observe all legislative and personal safety procedures when carrying out practical assignments. If you are unsure of what these are, ask your supervisor/instructor.

Performance Standard

0–No exposure: No information or practice provided during the program; complete training required

1–Exposure only: General information provided with no practice time; close supervision needed; additional training required

2–Limited practice: Has practiced job during training program; additional training required to develop skill

3–Moderately skilled: Has performed job independently during training program; limited additional training may be required

4–Skilled: Can perform job independently with no additional training

Name _________________________________ Date ____________ Class ________________________

Vehicle used for this activity:

Year _______________ Make _______________________ Model ____________________________

Odometer ____________ Hour meter ____________ VIN ___________________________________

▶ TASK Correct system lubricant level when replacing the evaporator, condenser, receiver/drier or accumulator/drier, and hoses.

AED 6.4

Time off_______________

Time on_______________

CDX Tasksheet Number: E0256

Total time_______________

1. **Consult the manufacturer's workbook manual for procedures to correct or refill the lubricant level when replacing any major components in the A/C system. Record the procedures below:**

   ____
   ____
   ____

2. **After any component replacements have been performed, consult the manufacturer's workbook manual for the correct type of lubricant and amount to be replaced or refilled.**

   a. **Type of oil used in the system:** _______________

   b. **Amount of oil to be replaced or added:** _______________ **ounces**

   > **NOTE:** If any contamination of the fluid is present, consult local and state laws for disposing the lubricant properly. Also, be sure to properly clean the system to avoid premature failure.

3. **Is contamination present? Yes:** _______________ **No:** _______________

   a. **If yes, record the procedures to correct the contaminated components and system:**

   ____
   ____
   ____
   ____

4. **List the results of your inspection:**

   ____
   ____
   ____

5. **Return the machine to beginning condition and return any tools that you may have used to their proper locations.**

6. Discuss the findings with the instructor.

____

____

____

Name ___________________________________ Date ____________ Class _______________________

Vehicle used for this activity:

Year _______________ Make _________________________________ Model _______________________

Odometer _____________ Hour meter ____________ VIN ___________________________________

▶ **TASK** Inspect A/C system hoses, lines, filters, fittings, and seals;
determine needed action.

AED
6.4

Time off_______________

Time on_______________

CDX Tasksheet Number: E0257

Total time_______________

1. **Research how to inspect A/C system hoses, lines, filters, fittings, and seals and determine needed action for the machine you are working on and list them below:**

   a. **Hoses and lines**

   ____

   ____

   ____

   b. **Filters**

   ____

   ____

   ____

   c. **Fittings and seals**

   ____

   ____

   ____

2. **Have your supervisor/instructor verify your research.**

   Supervisor/instructor's initials: _______________

3. **Ask your supervisor/instructor for a machine or simulator to check**

4. **Examine the machine/simulator and inspect A/C system hoses, lines, filters, fittings, and seals.**

5. **List the results of your inspection:**

   ____

   ____

   ____

6. **Return the machine to beginning condition and return any tools that you may have used to their proper locations.**

7. **Discuss the findings with the instructor.**

____

____

____

▶ TASK Inspect and test A/C system condenser. Check for proper airflow and mountings; determine needed action.

**AED
6.4**

Time off_____________

Time on_____________

Total time_____________

CDX Tasksheet Number: EO258

1. **Research how to inspect and test the A/C system condenser and mountings in the appropriate service information for the machine you are working on and list below:**

   ____

   ____

   ____

   a. **List the possible obstructions and reasons for blockage of air flow:**

      ____

      ____

      ____

2. **Have your supervisor/instructor verify your research.**

   Supervisor/instructor's initials: _______________

3. **Ask your supervisor/instructor for a machine or simulator to check.**

4. **Examine the machine/simulator and inspect and test the A/C system condenser and mountings.**

5. **Compare your results to the manufacturer's specifications. List your observations:**

   ____

   ____

   ____

6. **List the results of your inspection:**

   ____

   ____

   ____

7. **Return the machine to beginning condition and return any tools that you may have used to their proper locations.**

8. Discuss the findings with the instructor.

____

____

____

Student/Intern information:

Name _________________________________ Date _____________ Class _______________________

Vehicle used for this activity:

Year _______________ Make _________________________________ Model _______________________

Odometer _____________ Hour meter _____________ VIN _______________________________

▶ **TASK** Inspect and replace receiver/drier or accumulator/drier.

AED 6.4

Time off__________

Time on__________

Total time__________

CDX Tasksheet Number: EO259

1. **Research how to inspect and replace receiver/drier or accumulator/drier in the appropriate service information for the machine you are working on and list the results below. The receiver/drier or accumulator/ drier is something that should be replaced when replacing the compressor. Some compressor manufacturers will void the warranty if not replaced.**

   ____

   ____

   ____

2. **Have your supervisor/instructor verify your research.**

   Supervisor/instructor's initials: _______________

3. **Ask your supervisor/instructor for a machine or simulator to check.**

4. **Inspect and replace receiver/drier or accumulator/drier.**

5. **Compare your results to the manufacturer's specifications. List your observations:**

   ____

   ____

   ____

6. **List the results of your inspection:**

   ____

   ____

   ____

7. **Return the machine to beginning condition and return any tools that you may have used to their proper locations.**

8. **Discuss the findings with the instructor.**

Name _________________________ Date ___________ Class _______________

Vehicle used for this activity:

Year _____________ Make _________________________ Model _______________

Odometer ____________ Hour meter ___________ VIN _______________

▶ **TASK** Inspect and test refrigerant solenoid, expansion valve(s). **AED 6.4**

Time off_______________

Time on_______________

Total time_______________

CDX Tasksheet Number: E0260

1. **Research how to inspect and test the cab/sleeper refrigerant solenoid, expansion valve(s); check placement of thermal bulb (capillary tube) in the appropriate service information for the machine you are working on and list them below:**

   a. **Inspection and test of refrigerant solenoid.**

   b. **Inspect and replace the expansion valves and thermal bulb (capillary tube).**

2. **Have your supervisor/instructor verify your research.**

   Supervisor/instructor's initials: ______________

3. **Ask your supervisor/instructor for a machine or simulator to check.**

4. **Examine the machine/simulator A/C compressor and inspect and test the cab/sleeper refrigerant solenoid, and expansion valve(s); check placement of thermal bulb (capillary tube).**

5. **Compare your results to the manufacturer's specifications. List your observations:**

   ____
   ____
   ____

6. **List the results of your inspection and test:**

   ____
   ____
   ____

7. **Return the machine to beginning condition and return any tools that you may have used to their proper locations.**

8. **Discuss the findings with the instructor.**

____
____
____

Performance Rating

CDX Tasksheet Number: E0260

0	1	2	3	4

Supervisor/instructor signature _________________________________ Date ____________

▶ TASK Remove and replace orifice tube. **AED 6.4**

CDX Tasksheet Number: E0261

Time off_____________

Time on_____________

Total time_____________

1. **Research the procedure to locate, remove, and replace the orifice tube in the appropriate service information for the machine you are working on.**

    a. **List the steps outlined in the service information to remove the orifice tube:**

    ____

    ____

    ____

    ____

2. **List the steps outlined in the service information to replace the orifice tube. Pay attention to the color of the orifice tube as it designates the amount of restriction needed in the system.**

    ____

    ____

    ____

    ____

3. **Have your supervisor/instructor verify your research.**

    Supervisor/instructor's initials: _______________

4. **Ask your supervisor/instructor for a machine or simulator to check.**

5. **Remove and replace the orifice tube as per the service information.**

6. **Return the machine to beginning condition and return any tools that you may have used to their proper locations.**

7. **Discuss the findings with the instructor.**

    ____

    ____

    ____

Performance Rating

CDX Tasksheet Number: E0261

0	1	2	3	4

Supervisor/instructor signature ____ Date _____________

Student/Intern information:

Name _________________________ Date ___________ Class _______________________

Vehicle used for this activity:

Year _____________ Make _____________________ Model _____________________

Odometer ____________ Hour meter ___________ VIN _________________________

▶ TASK Inspect and test evaporator core; determine needed action. **AED 6.4**

Time off____________

Time on____________

CDX Tasksheet Number: E0262

1. **Research how to inspect and test the evaporator core in the appropriate service information for the machine you are working on and list your findings below:**

   ____

   ____

   ____

   ____

Total time____________

2. **Have your supervisor/instructor verify your research.**

   Supervisor/instructor's initials: _______________

3. **Ask your supervisor/instructor for a machine or simulator to check.**

4. **Compare your results to the manufacturer's specifications. List your observations:**

   ____

   ____

   ____

   ____

5. **Return the machine to beginning condition and return any tools that you may have used to their proper locations.**

6. **Discuss the findings with the instructor.**

   ____

   ____

   ____

Performance Rating

CDX Tasksheet Number: E0262

0	1	2	3	4

Supervisor/instructor signature _____________________________________ Date _____________

▶ TASK Inspect, clean, or repair evaporator housing and water drain;
inspect and service/replace evaporator air filter.

**AED
6.4**

Time off____________________

Time on____________________

Total time__________________

CDX Tasksheet Number: E0263

1. **Research how to inspect, clean, or repair evaporator housing and water drain; inspect and service/replace evaporator air filter in the appropriate service information for the machine you are working on and list the results below:**

    a. **Inspect, clean, or repair evaporator housing and water drain. If condensation is not able to exit the housing, a foul smell may be present. Mold may also grow in the housing which may require housing removal and cleaning.**

    ____

    ____

    ____

    b. **Inspect and service/replace evaporator air filter:**

    ____

    ____

    ____

2. **Have your supervisor/instructor verify your research.**

    Supervisor/instructor's initials: _______________

3. **Ask your supervisor/instructor for a machine or simulator to inspect, clean, or repair evaporator housing and water drain; inspect and service/replace evaporator air filter.**

4. **Inspect, clean, or repair evaporator housing and water drain; inspect and service/replace evaporator air filter; list the actions undertaken below:**

    ____

    ____

    ____

5. **Return the machine to beginning condition and return any tools that you may have used to their proper locations.**

6. **Discuss the findings with the instructor.**

____

____

____

Student/Intern information:

Name _______________________________ Date _____________ Class _______________________________

Vehicle used for this activity:

Year _______________ Make _______________________________ Model _______________________________

Odometer _____________ Hour meter _____________ VIN _______________________________

▶ **TASK** Identify and inspect A/C system service ports (gauge connections); determine needed action.

AED 6.4

Time off_______________

Time on_______________

Total time_______________

CDX Tasksheet Number: E0264

1. **Identify and inspect A/C system service ports (gauge connections).**

2. **Ask your supervisor/instructor for a machine or simulator to identify and inspect A/C system service ports (gauge connections).**

3. **Examine the machine/simulator A/C compressor and identify and inspect A/C system service ports (gauge connections).**

   List the results of your inspection:

   ____
   ____
   ____

   Determine and list any necessary corrective action(s):

   ____
   ____
   ____

4. **Return the machine to beginning condition and return any tools that you may have used to their proper locations.**

5. **Discuss the findings with the instructor.**

   ____
   ____
   ____

Performance Rating

CDX Tasksheet Number: E0264

0	1	2	3	4

Supervisor/instructor signature ____ Date _______________

Student/Intern information:

Name _________________________ Date __________ Class _________________

Vehicle used for this activity:

Year ____________ Make _____________________ Model _________________

Odometer ___________ Hour meter ___________ VIN _________________

▶ TASK Identify the cause of system failures resulting in refrigerant loss from the A/C system high-pressure relief device.

AED 6.4

Time off_______________

Time on_______________

Total time_______________

CDX Tasksheet Number: E0265

1. **Research how to identify causes of system failures that result in refrigerant loss from the A/C system high-pressure relief device. Use the appropriate service information for the machine you are working on and list the findings below:**

   Have your supervisor/instructor verify your research.

   Supervisor/instructor's initials: _______________

2. **Ask your supervisor for a machine to identify the cause of system failures resulting in refrigerant loss from the A/C system high-pressure relief device.**

3. **Test the system to determine the cause of the complaint. List your observations here:**

   ____
   ____
   ____

   Determine and list any necessary corrective action(s):

   ____
   ____
   ____

4. **Return the machine to beginning condition and return any tools that you may have used to their proper locations.**

5. **Discuss the findings with the instructor.**

   ____
   ____
   ____

Performance Rating

CDX Tasksheet Number: E0265

0	1	2	3	4

Supervisor/instructor signature ___________________________ Date ___________

Testing, Troubleshooting, Diagnosing, and Repairing AC Systems: Electrical

Student/Intern information:

Name _________________________________ Date _____________ Class _________________________

Vehicle used for this activity:

Year _______________ Make _____________________________ Model _________________________

Odometer _____________ Hour meter _____________ VIN _________________________________

Learning Objective/Task	CDX Tasksheet Number	2014 Edition Rev2/12/16 AED Standard
• Identify causes of HVAC electrical control system problems; determine needed action.	E0266	6.4
• Inspect and test HVAC blower motors, resistors, switches, relays, modules, wiring, and protection devices; determine needed action.	E0267	6.4
• Inspect and test A/C compressor clutch relays, modules, wiring, sensors, switches, diodes, and protection devices; determine needed action determine needed action.	E0268	6.4
• Inspect and test A/C-related electronic engine control systems; determine needed action.	E0269	6.4
• Inspect and test engine cooling/condenser fan motors, relays, modules, switches, sensors, wiring, and protection devices; determine needed action.	E0270	6.4
• Inspect and test electric actuator motors, relays/modules, switches, sensors, wiring, and protection devices; determine needed action.	E0271	6.4
• Inspect and test HVAC system electrical/electronic control panel assemblies; determine needed action.	E0272	6.4
• Interface with machine's on-board computer; perform diagnostic procedures using recommended electronic service tool(s) (including PC-based software and/or data scan tools); determine needed action.	E0273	6.4

Time off_________________

Time on_________________

Total time_________________

Materials Required

- Machines or simulators with HVAC electrical faults
- Machine manufacturer's workshop manual including schematic wiring diagrams
- DVOM, ammeter, current clamp, PC-based software, and/or data scan tools
- HVAC electrical spare parts including, fuses, circuit breakers, switches, relays, wires, and connectors
- Manufacturer-specific tools depending on the concern
- Coolant recovery system

Some Safety Issues to Consider

- Activities require you to measure electrical values. Always ensure the instructor/supervisor checks test instrument connections prior to connecting power or taking measurements.
- Activities will require you to work with the HVAC system. It will contain refrigerant under pressure. Show caution around high-pressure refrigerant hoses.
- You may be required to handle refrigerant. Use extreme caution: refrigerant is pressurized and very cold. Always wear eye protection and appropriate clothing when working with refrigerant. Never inhale refrigerant.
- Do not release refrigerant to the atmosphere; always use a recycling system to reclaim refrigerant.
- Comply with personal and environmental safety practices associated with clothing; eye protection; hand tools; power equipment; proper ventilation; and the handling, storage, and disposal of chemicals/materials in accordance with federal, state, and local regulations.
- Always wear the correct protective eyewear and clothing and use the appropriate safety equipment, as well as fender covers, seat protectors, and floor mat protectors.
- Make sure you understand and observe all legislative and personal safety procedures when carrying out practical assignments. If you are unsure of what these are, ask your supervisor/instructor.

Performance Standard

0—No exposure: No information or practice provided during the program; complete training required

1—Exposure only: General information provided with no practice time; close supervision needed; additional training required

2—Limited practice: Has practiced job during training program; additional training required to develop skill

3—Moderately skilled: Has performed job independently during training program; limited additional training may be required

4—Skilled: Can perform job independently with no additional training

Student/Intern information:

Name _______________________________________ Date _____________ Class _______________________________

Vehicle used for this activity:

Year _______________ Make _______________________________ Model _______________________

Odometer _______________ Hour meter _______________ VIN _______________________________________

▶ **TASK** Identify causes of HVAC electrical control system problems.

CDX Tasksheet Number: E0266

AED
6.4

Time off_______________

Time on_______________

Total time_______________

1. **Research causes of HVAC electrical control system problems and list them below:**

    ____

    ____

    ____

2. **Ask your supervisor/instructor for a machine or simulator to check.**

3. **Examine the machine/simulator and operate all the electrical controls for the HVAC system to determine fault(s) with HVAC electrical control system in the A/C system.**

    ____

    ____

    ____

4. **Using the appropriate service information, determine the steps necessary for testing the component(s):**

    ____

    ____

    ____

5. **Write a short description of the purpose and operation of the suspected component(s):**

    ____

    ____

    ____

6. **Determine and list any necessary action(s):**

    ____

    ____

    ____

7. **Discuss the findings with your instructor and record the recommendations.**

____

____

____

Performance Rating   CDX Tasksheet Number: E0266

0	1	2	3	4

Supervisor/instructor signature ____ Date ______________

Name _________________________________ Date ___________ Class _________________________

Vehicle used for this activity:

Year _______________ Make _____________________________ Model _________________________

Odometer _____________ Hour meter _____________ VIN _______________________________

▶ TASK Inspect and test HVAC blower motors, resistors, switches, relays, modules, wiring, and protection devices.

AED 6.4

Time off________________

Time on________________

CDX Tasksheet Number: E0267

Total time________________

1. **Research how to inspect and test HVAC blower motors, resistors, switches, relays, modules, wiring, and protection devices and list your findings below. When testing these components, a working knowledge of a DVOM (digital volt-ohmmeter) is necessary for the diagnosis of problems.**

   a. **Inspect and test blower motors:**

   ____
   ____
   ____

   b. **Inspect and test resistors:**

   ____
   ____
   ____

   c. **Inspect and test switches:**

   ____
   ____
   ____

   d. **Inspect and test modules:**

   ____
   ____
   ____

   e. **Inspect and test wiring:**

   ____
   ____
   ____

   f. **Inspect and test protection devices:**

   ____
   ____
   ____

2. **Ask your supervisor/instructor for a machine or simulator to check.**

3. **Using the appropriate service information, inspect and test HVAC blower motors, resistors, switches, relays, modules, wiring, and protection devices.**

4. **List the results of conducting your inspection and tests:**

   ____
   ____
   ____

5. **Write a short description of the purpose and operation of the suspected component(s):**

____

____

____

6. **Determine and list any necessary action(s):**

____

____

____

7. **Return the machine to beginning condition and return any tools that you may have used to their proper locations.**

8. **Discuss the findings with your instructor and record the recommendations.**

____

____

____

▶ TASK Inspect and test A/C compressor clutch relays, modules, wiring, sensors, switches, diodes, and protection devices; determine needed action.

AED 6.4

Time off_______________

Time on_______________

Total time_______________

CDX Tasksheet Number: E0268

1. **Research how to inspect and test A/C compressor clutch relays, modules, wiring, sensors, switches, diodes, and protection devices in the appropriate service information for the machine you are working on and list the findings below. When testing these components, a working knowledge of a DVOM is necessary for the diagnosis of problems.**

   a. **A/C compressor clutch relays:**

   ____

   ____

   ____

   b. **Modules:**

   ____

   ____

   ____

   c. **Wiring:**

   ____

   ____

   ____

   d. **Sensors:**

   ____

   ____

   ____

   e. **Switches:**

   ____

   ____

   ____

   f. **Diodes:**

   ____

   ____

   ____

   g. **Protection devices:**

   ____

   ____

   ____

2. Check your documented procedures with your supervisor/instructor.
    Supervisor/instructor's initials: _______________

3. Using the appropriate service information, inspect and test A/C compressor
    clutch relays, modules, wiring, sensors, switches, diodes, and protection
    devices.

4. List the results of conducting your inspection and tests:

    ____
    ____
    ____

5. Determine and list any necessary action(s):

    ____
    ____
    ____

6. Return the machine to beginning condition and return any tools that you
    may have used to their proper locations.

7. Discuss the findings with your instructor and record the recommendations.

    ____
    ____
    ____

Performance Rating

CDX Tasksheet Number: EO268

0	1	2	3	4

Supervisor/instructor signature _________________________________ Date _____________

Student/Intern information:

Name _________________________________ Date _____________ Class _________________________

Vehicle used for this activity:

Year _______________ Make _________________________ Model _________________________

Odometer _____________ Hour meter _____________ VIN _________________________

▶ TASK Inspect and test A/C-related electronic engine control systems. **AED 6.4**

Time off________________

Time on________________

CDX Tasksheet Number: E0269

1. **Research how to inspect and test A/C-related electronic engine control systems and list the findings below. A working knowledge of a scan tool will aid this task.**

   ____
   ____
   ____

Total time________________

2. **Ask your supervisor/instructor for a machine or simulator to check.**

3. **Examine the machine/simulator; inspect and test A/C-related electronic engine control.**

4. **List identified problems**

   ____
   ____
   ____

5. **Using the appropriate service information, determine the steps necessary for testing the component(s):**

   ____
   ____
   ____

6. **Write a short description of the purpose and operation of the suspected component(s):**

   ____
   ____
   ____

7. **Determine and list any necessary action(s):**

   ____
   ____
   ____

8. **Return the machine to beginning condition and return any tools that you may have used to their proper locations.**

9. **Discuss the findings with your instructor and record the recommendations.**

____

____

____

CDX Tasksheet Number: E0269

| 0 | 1 | 2 | 3 | 4 |

Supervisor/instructor signature ______________________________________ Date ____________

Student/Intern information:

Name _________________________________ Date _____________ Class _________________________

Vehicle used for this activity:

Year _______________ Make _________________________ Model _________________________

Odometer _____________ Hour meter ___________ VIN _________________________

▶ **TASK** Inspect and test engine cooling/condenser fan motors, relays, modules, switches, sensors, wiring, and protection devices. **AED 6.4**

Time off_________________

Time on_________________

Total time_________________

CDX Tasksheet Number: E0270

1. **Research how to inspect and test engine cooling/condenser fan motors, relays, modules, switches, sensors, wiring, and protection devices in the appropriate service information for the machine you are working on and list the results below. When testing these components, a working knowledge of a DVOM (digital volt-ohmmeter) is necessary for diagnosis of problems. A scan tool will also be required for this task.**

   a. **Cooling/condenser fan motors:**

   ____
   ____
   ____

   b. **Relays:**

   ____
   ____
   ____

   c. **Modules:**

   ____
   ____
   ____

   d. **Sensors:**

   ____
   ____
   ____

   e. **Switches:**

   ____
   ____
   ____

   f. **Wiring:**

   ____
   ____
   ____

   g. **Protection devices:**

   ____
   ____
   ____

2. Check your documented procedures with your supervisor/instructor. Supervisor/instructor's initials: ______________

3. Using the appropriate service information, inspect and test engine cooling/condenser fan motors, relays, modules, switches, sensors, wiring, and protection devices.

4. List the results of conducting your inspection and tests:

____
____
____

5. Determine and list any necessary action(s):

____
____
____

6. Return the machine to beginning condition and return any tools that you may have used to their proper locations.

7. Discuss the findings with your instructor and record the recommendations.

____
____
____

Performance Rating　　　　　**CDX Tasksheet Number: EO270**

| 0 | 1 | 2 | 3 | 4 |

Supervisor/instructor signature _______________________________ Date ____________

Student/Intern information:

Name ________________________________ Date ___________ Class ________________________

Vehicle used for this activity:

Year _______________ Make _________________________ Model ________________________

Odometer ____________ Hour meter ____________ VIN ________________________________

▶ **TASK** Inspect and test electric actuator motors, relays/modules, switches, sensors, wiring, and protection devices.

AED 6.4

Time off________________

Time on________________

Total time________________

CDX Tasksheet Number: EO271

1. **Research how to inspect and test electric actuator motors, relays/modules, switches, sensors, wiring, and protection devices in the appropriate service information for the machine you are working on and list the results below. When testing these components, a working knowledge of a DVOM (digital volt-ohmmeter) is necessary for diagnosis of problems. A scan tool will also be required for this task.**

   a. **Electric actuator motors:**

   ____

   ____

   ____

   b. **Relays:**

   ____

   ____

   ____

   c. **Modules:**

   ____

   ____

   ____

   d. **Sensors:**

   ____

   ____

   ____

   e. **Switches:**

   ____

   ____

   ____

   f. **Wiring:**

   ____

   ____

   ____

   g. **Protection devices:**

   ____

   ____

   ____

2. Check your documented procedures with your supervisor/instructor. Supervisor/instructor's initials: _______________

3. Using the appropriate service information, inspect and test electric actuator motors, relays/modules, switches, sensors, wiring, and protection devices.

4. List the results of conducting your inspection and tests:

____

____

____

5. Determine and list any necessary action(s):

____

____

____

6. Return the machine to beginning condition and return any tools that you may have used to their proper locations.

7. Discuss the findings with your instructor and record the recommendations.

____

____

____

Performance Rating

CDX Tasksheet Number: E0271

0	1	2	3	4

Supervisor/instructor signature ____ Date _____________

▶ TASK Inspect and test HVAC system electrical/electronic control panel assemblies.

AED 6.4

Time off_________________

Time on_________________

Total time_________________

CDX Tasksheet Number: EO272

1. **Research how to inspect and test HVAC system electrical/electronic control panel assemblies in the appropriate service information for the machine you are working on and list the results below:**

   ____

   ____

   ____

2. **Check your documented procedures with your supervisor/instructor. Supervisor/instructor's initials:** _______________

3. **Using the appropriate service information, inspect and test HVAC system electrical/electronic control panel assemblies.**

4. **List the results of conducting your inspection and tests:**

   ____

   ____

   ____

5. **Determine and list any necessary action(s):**

   ____

   ____

   ____

6. **Return the machine to beginning condition and return any tools that you may have used to their proper locations.**

7. **Discuss the findings with your instructor and record the recommendations.**

   ____

   ____

   ____

Performance Rating

CDX Tasksheet Number: EO272

0	1	2	3	4

Supervisor/instructor signature ____ Date _______________

Student/Intern information:

Name _________________________________ Date ___________ Class _________________________

Vehicle used for this activity:

Year _______________ Make _____________________________ Model _____________________________

Odometer ______________ Hour meter _____________ VIN ___________________________________

▶ TASK Interface with machine's on-board computer; perform diagnostic procedures using recommended electronic service tool(s) (including PC-based software and/or data scan tools).

AED 6.4

Time off_____________

Time on_____________

Total time_____________

CDX Tasksheet Number: E0273

1. **Consult the manufacturer's workbook manual for the proper procedures to connect an electronic service tool or scan tool to the machine.**

2. **Read the diagnostic codes and record them below.**

   a. **Code # 1** _______________

   b. **Code # 2** _______________

   c. **Code # 3** _______________

   d. **Code # 4** _______________

   e. **Code # 5** _______________

   f. **Code # 6** _______________

3. **Consult the manufacturer's workbook manual and record the code explanations and the procedures to repair codes that are present.**

   a. **Code # 1**

   ____
   ____
   ____
   ____

   b. **Code # 2**

   ____
   ____
   ____
   ____

   c. **Code # 3**

   ____
   ____
   ____
   ____

   d. **Code # 4**

   ____
   ____
   ____
   ____

  e. **Code # 5**

____
____
____
____

  f. **Code # 6**

____
____
____
____

4. **Consult with your instructor for recommendations to repair or replace failed parts. Record the recommended procedure(s) below.**

____
____
____

Performance Rating   **CDX Tasksheet Number: E0273**

0	1	2	3	4

Supervisor/instructor signature ______________________________ Date ___________

Testing, Troubleshooting, Diagnosing, and Repairing AC Systems: Control and Distribution

Student/Intern information:

Name _______________________________ Date _____________ Class _______________________

Vehicle used for this activity:

Year _______________ Make _____________________________ Model _____________________

Odometer _____________ Hour meter _____________ VIN _______________________________

Learning Objective/Task	CDX Tasksheet Number	2014 Edition Rev2/12/16 AED Standard
• Identify the causes of HVAC mechanical control problems; determine needed action.	E0274	6.4
• Inspect, test, and adjust HVAC system air and mechanical control cables and linkages; determine needed action.	E0275	6.4
• Inspect and test HVAC system air and mechanical control panel assemblies.	E0276	6.4
• Inspect and test HVAC system actuators and hoses.	E0277	6.4
• Inspect, test, and adjust HVAC system ducts, doors, and outlets.	E0278	6.4

Time off_______________

Time on_______________

Total time_______________

Materials Required

- Machines or simulators with HVAC faults in operating systems and controls
- Machine manufacturer's workshop manual including HVAC information
- Thermometer, DVOM, PC-based software, and/or data scan tools
- HVAC spare parts
- Manufacturer-specific tools depending on the concern

Some Safety Issues to Consider

- Activities require you to measure electrical values. Always ensure the instructor/supervisor checks test instrument connections prior to connecting power or taking measurements.
- Activities will require you to work with the HVAC system. It will contain refrigerant under pressure. Show caution around high-pressure refrigerant hoses.
- You may be required to handle refrigerant. Use extreme caution: refrigerant is pressurized and very cold. Always wear eye protection and appropriate clothing when working with refrigerant. Never inhale refrigerant.
- Do not release refrigerant to the atmosphere; always use a recycling system to reclaim refrigerant.
- Comply with personal and environmental safety practices associated with clothing; eye protection; hand tools; power equipment; proper ventilation; and the handling, storage, and disposal of chemicals/materials in accordance with federal, state, and local regulations.

- Always wear the correct protective eyewear and clothing and use the appropriate safety equipment, as well as fender covers, seat protectors, and floor mat protectors.
- Make sure you understand and observe all legislative and personal safety procedures when carrying out practical assignments. If you are unsure of what these are, ask your supervisor/instructor.

Performance Standard

0—No exposure: No information or practice provided during the program; complete training required

1—Exposure only: General information provided with no practice time; close supervision needed; additional training required

2—Limited practice: Has practiced job during training program; additional training required to develop skill

3—Moderately skilled: Has performed job independently during training program; limited additional training may be required

4—Skilled: Can perform job independently with no additional training

▶ **TASK** Identify causes of HVAC air and mechanical control problems.

AED 6.4

Time off_________________

Time on_________________

Total time_________________

CDX Tasksheet Number: E0274

1. **Research how to identify causes of HVAC air and mechanical control problems and list the findings below:**

   ____

   ____

   ____

2. **Ask your supervisor/instructor for a machine or simulator to check.**

3. **Using the appropriate service information, identify causes of HVAC air and mechanical control problems in the machine or simulator.**

4. **List the customer's concern:**

   ____

   ____

   ____

5. **Write a short description of the purpose and operation of the suspected component(s):**

   ____

   ____

   ____

6. **Determine and list any necessary action(s):**

   ____

   ____

   ____

7. **Return the machine to beginning condition and return any tools that you may have used to their proper locations.**

8. **Discuss the findings with your instructor and record the recommendations.**

____

____

____

Performance Rating　　　**CDX Tasksheet Number: EO274**

0	1	2	3	4

Supervisor/instructor signature _________________________________ Date ____________

Student/Intern information:

Name ________________________________ Date ____________ Class ________________________

Vehicle used for this activity:

Year ______________ Make ________________________ Model ____________________________

Odometer ____________ Hour meter ____________ VIN ________________________________

▶ **TASK** Inspect and test HVAC system air and mechanical control cables and linkages.

AED 6.4

Time off______________________

Time on______________________

CDX Tasksheet Number: E0275

1. **Research how to inspect and test HVAC system air and mechanical control cables and linkages in the appropriate service information for the machine you are working on and list the procedures below:**

   ____

   ____

   ____

2. **Check your documented procedures with your supervisor/instructor. Supervisor/instructor's initials:** ________________

3. **Ask your supervisor/instructor for a machine or simulator to check.**

4. **Using the appropriate service information, inspect and test HVAC system air and mechanical control cables and linkages.**

5. **Write a short description of the purpose and operation of the suspected component(s):**

   ____

   ____

   ____

6. **Determine and list any necessary action(s):**

   ____

   ____

   ____

7. **Return the machine to beginning condition and return any tools that you may have used to their proper locations.**

Total time______________________

8. Discuss the findings with your instructor and record the recommendations.

____
____
____
____

Performance Rating

CDX Tasksheet Number: E0275

0	1	2	3	4

Supervisor/instructor signature ___________________________________ Date ___________

Student/Intern information:

Name _________________________________ Date ____________ Class _________________________

Vehicle used for this activity:

Year _______________ Make _____________________________ Model _____________________

Odometer _____________ Hour meter ____________ VIN _________________________________

▶ TASK Inspect and test HVAC system air and mechanical control panel assemblies.

AED 6.4

Time off_________________

Time on_________________

CDX Tasksheet Number: E0276

1. **Research how to inspect and test HVAC system air and mechanical control panel assemblies in the appropriate service information for the machine you are working on and list the procedures below:**

   ____

   ____

   ____

Total time_________________

2. **Check your documented procedures with your supervisor/instructor. Supervisor/instructor's initials:** _______________

3. **Ask your supervisor/instructor for a machine or simulator to check.**

4. **Using the appropriate service information, inspect and test HVAC system air and mechanical control panel assemblies.**

5. **Write a short description of the purpose and operation of the suspected component(s):**

   ____

   ____

   ____

6. **Determine and list any necessary action(s):**

   ____

   ____

   ____

7. **Return the machine to beginning condition and return any tools that you may have used to their proper locations.**

8. **Discuss the findings with your instructor and record the recommendations.**

____

____

____

Performance Rating   **CDX Tasksheet Number: E0276**

0	1	2	3	4

Supervisor/instructor signature ____ Date _______________

Student/Intern information:

Name _________________________ Date ___________ Class _________________________

Vehicle used for this activity:

Year _____________ Make _________________________ Model _________________________

Odometer ___________ Hour meter ___________ VIN _________________________

▶ TASK Inspect and test HVAC system actuators and hoses. **_AED_ 6.4**

CDX Tasksheet Number: EO277

Time off_________________

Time on_________________

Total time_________________

1. **Research how to inspect and test HVAC system actuators and hoses in the appropriate service information for the machine you are working on and list the procedures below:**

   ____

   ____

   ____

2. **Check your documented procedures with your supervisor/instructor. Supervisor/instructor's initials: _______________**

3. **Ask your supervisor/instructor for a machine or simulator to check.**

4. **Using the appropriate service information, inspect and test HVAC system actuators and hoses.**

5. **List the results of conducting your inspection and test of HVAC system actuators and hoses. Are all the system actuators and hoses operating in accordance with the manufacturer's specification?**

   ____

   ____

   ____

6. **Determine and list any necessary action(s):**

   ____

   ____

   ____

7. **Return the machine to beginning condition and return any tools that you may have used to their proper locations.**

8. Discuss the findings with your instructor and record the recommendations.

____

____

____

____

Student/Intern information:

Name ________________________________ Date ____________ Class ________________________________

Vehicle used for this activity:

Year ______________ Make ________________________________ Model ________________________________

Odometer ____________ Hour meter ____________ VIN ________________________________

▶ TASK Inspect and test and adjust HVAC system ducts, doors, and outlets.

AED 6.4

Time off________________

Time on________________

Total time________________

CDX Tasksheet Number: EO278

1. **Research how to inspect, test, and adjust HVAC system ducts, doors, and outlets in the appropriate service information for the vehicle you are working on and list the procedures below:**

   a. **Inspect, test, and adjust HVAC system ducts.**

   b. **Inspect, test, and adjust HVAC system doors.**

   c. **Inspect, test, and adjust HVAC system outlets.**

2. **Check your documented procedures with your supervisor/instructor. Supervisor/instructor's initials: ______________**

3. **Ask your supervisor/instructor for a vehicle or simulator to check.**

4. **Using the appropriate service information, inspect, test, and adjust HVAC system ducts, doors, and outlets.**

5. **List the results of conducting your test, and adjustment of HVAC system ducts, doors, and outlets. Are all the system ducts, doors, and outlets operating in accordance with the manufacturer's specification?**

   ____
   ____
   ____

6. **Determine and list any necessary action(s):**

   ____
   ____
   ____

7. **Return the machine to beginning condition and return any tools that you may have used to their proper locations.**

8. Discuss the findings with your instructor and record the recommendations.

____

____

____

Testing, Troubleshooting, Diagnosing, and Repairing AC Systems: Refrigerant Recovery, Recycling, and Handling

Student/Intern information:

Name _________________________________ Date _____________ Class _____________________

Vehicle used for this activity:

Year _______________ Make _________________________________ Model _____________________

Odometer ____________ Hour meter ____________ VIN _________________________________

Learning Objective/Task	CDX Tasksheet Number	2014 Edition Rev2/12/16 AED Standard
• Maintain and verify correct operation of certified equipment.	E0279	6.4
• Identify and recover A/C system refrigerant.	E0280	6.4
• Recycle or properly dispose of refrigerant.	E0281	6.4
• Handle, label, and store refrigerant.	E0282	6.4
• Test recycled refrigerant for non-condensable gases.	E0283	6.4

Time off_____________

Time on_____________

Total time_____________

Materials Required

- Machine manufacturer's workshop manual including HVAC information
- Safety data sheet (SDS) for refrigerants
- Specialized HVAC tools including refrigerant recovery, vacuum pump, recharging station, and thermometer
- Refrigerant
- Manufacturer-specific tools depending on the concern

Some Safety Issues to Consider

- Activities require you to measure electrical values. Always ensure the instructor/supervisor checks test instrument connections prior to connecting power or taking measurements.
- Activities will require you to work with the HVAC system. It will contain refrigerant under pressure. Show caution around high-pressure refrigerant hoses.
- You may be required to handle refrigerant. Use extreme caution: refrigerant is pressurized and very cold. Always wear eye protection and appropriate clothing when working with refrigerant. Never inhale refrigerant.
- Do not release refrigerant to the atmosphere; always use a recycling system to reclaim refrigerant.
- Comply with personal and environmental safety practices associated with clothing; eye protection; hand tools; power equipment; proper ventilation; and the handling, storage, and disposal of chemicals/materials in accordance with federal, state, and local regulations.
- Always wear the correct protective eyewear and clothing and use the appropriate safety equipment, as well as fender covers, seat protectors, and floor mat protectors.

- Make sure you understand and observe all legislative and personal safety procedures when carrying out practical assignments. If you are unsure of what these are, ask your supervisor/instructor.

Performance Standard

0–No exposure: No information or practice provided during the program; complete training required

1–Exposure only: General information provided with no practice time; close supervision needed; additional training required

2–Limited practice: Has practiced job during training program; additional training required to develop skill

3–Moderately skilled: Has performed job independently during training program; limited additional training may be required

4–Skilled: Can perform job independently with no additional training

Student/Intern information:

Name _______________________________ Date ____________ Class _______________________

Vehicle used for this activity:

Year _______________ Make _______________________________ Model _______________________

Odometer ____________ Hour meter ____________ VIN _______________________________

▶ TASK Maintain and verify correct operation of certified equipment.   **AED 6.4**

Time off________________

CDX Tasksheet Number: E0279

Time on________________

1. Consult the certified equipment manufacturer's maintenance manual for procedures to verify correct operation of the equipment being used. Record the procedures.

Total time________________

____

____

____

2. Record service intervals: _______________ months/years

3. Date of last service: _______________ / _______________ / _______________

4. Perform and record any scheduled service procedures required.

____

____

____

5. Consult with the instructor on recommendations for repairing or replacing service equipment. Record the recommendations.

____

____

____

Performance Rating

CDX Tasksheet Number: E0279

☐	☐	☐	☐	☐
0	1	2	3	4

Supervisor/instructor signature ____ Date ____________

Student/Intern information:

Name _________________________________ Date _____________ Class _________________________

Vehicle used for this activity:

Year _______________ Make _________________________ Model _________________________

Odometer ______________ Hour meter _____________ VIN _________________________

▶ TASK Identify and recover A/C system refrigerant. **AED 6.4**

Time off_______________

CDX Tasksheet Number: E0280

Time on_______________

1. **Research how to identify and recover A/C system refrigerant in the appropriate service information for the machine you are working on and list below:**

    a. **How to identify A/C refrigerant type:**

    ____
    ____
    ____
    ____

Total time_______________

    b. **List the procedures for recovering refrigerant:**

    ____
    ____
    ____
    ____

2. **Check your documented procedures with your supervisor/instructor. Supervisor/instructor's initials:** _______________

3. **Ask your supervisor/instructor for a machine or simulator to identify and recover refrigerant.**

4. **Using the appropriate service information, identify and recover refrigerant.**

5. **List the results of identifying and recovering refrigerant:**

    a. **The type of refrigerant:**

    ____
    ____
    ____
    ____

    b. **The amount of refrigerant recovered:**

    ____
    ____
    ____
    ____

    c. **The amount of oil removed from the system:**

    ____
    ____
    ____
    ____

6. Return the machine to beginning condition and return any tools that you may have used to their proper locations.

7. Discuss the findings with the instructor.

____

____

____

Student/Intern information:

Name _________________________ Date __________ Class _________________

Vehicle used for this activity:

Year _____________ Make ___________________ Model ___________________

Odometer __________ Hour meter __________ VIN _______________________

▶ TASK Recycle or properly dispose of refrigerant.

AED 6.4

Time off__________

Time on__________

Total time__________

CDX Tasksheet Number: E0281

1. **Research how to recycle or properly dispose of refrigerant in the appropriate service information for the machine you are working on, and list the procedure below. Check state and federal requirements for this process.**

   ____
   ____
   ____

2. **Check your documented procedures with your supervisor/instructor. Supervisor/instructor's initials:** _____________

3. **Recycle or properly dispose of refrigerant.**

4. **List the steps you undertook to recycle or properly dispose of refrigerant:**

   ____
   ____
   ____

5. **Discuss the findings with the instructor.**

   ____
   ____
   ____

Performance Rating

CDX Tasksheet Number: E0281

0	1	2	3	4

Supervisor/instructor signature _______________________________ Date __________

Student/Intern information:

Name _______________________________ Date ____________ Class _________________________

Vehicle used for this activity:

Year _______________ Make _________________________________ Model _____________________

Odometer ______________ Hour meter ______________ VIN ________________________________

▶ **TASK** Handle, label, and store refrigerant. _______________________________________

AED 6.4

CDX Tasksheet Number: E0282

1. **Research how to handle, label, and store refrigerant in appropriate service information and safety data sheet (SDS) for the types of refrigerant in your workshop and list them below. Check state and federal requirements for this process.**

    a. **Handling requirements:**

    ____

    ____

    ____

    b. **PPE required when handling refrigerant:**

    ____

    ____

    ____

    c. **How is refrigerant labeled:**

    ____

    ____

    ____

    d. **How is refrigerant to be stored:**

    ____

    ____

    ____

2. **Check your documented procedures with your supervisor/instructor. Supervisor/instructor's initials:** ______________

3. **Handle, label, and store refrigerant according to manufacturer's and workshop requirements.**

4. **Ensure refrigerant is returned to storage area.**

5. Discuss the findings with the instructor.

____

____

____

Student/Intern information:

Name _________________________________ Date _____________ Class _________________________

Vehicle used for this activity:

Year _______________ Make _____________________________ Model _________________________

Odometer _______________ Hour meter _______________ VIN _________________________________

▶ **TASK** Test recycled refrigerant for non-condensable gas. **AED 6.4**

Time off_________________

Time on_________________

CDX Tasksheet Number: E0283

1. **Research how to test recycled refrigerant for non-condensable gases in the appropriate service information and list the procedures below:**

   ____

   ____

   ____

   Total time_________________

2. **Check your documented procedures with your supervisor/instructor. Supervisor/instructor's initials:** _______________

3. **Ask your supervisor/instructor for recycled refrigerant to test for non-condensable gases.**

4. **Using the appropriate service information, test recycled refrigerant for non-condensable gases.**

5. **List the results of conducting your test:**

   ____

   ____

   ____

6. **Determine and list any necessary action(s):**

   ____

   ____

   ____

7. **Return the recycled refrigerant and tools to appropriate storage areas.**

8. **Discuss the findings with the instructor.**

   ____

   ____

   ____

Performance Rating

CDX Tasksheet Number: E0283

☐	☐	☐	☐	☐
0	1	2	3	4

Supervisor/instructor signature ____ Date _______________

Servicing Heating Systems

Student/Intern information:

Name _______________________________ Date _____________ Class _____________________________

Vehicle used for this activity:

Year _______________ Make _______________________ Model _____________________________

Odometer _____________ Hour meter _____________ VIN _____________________________

Learning Objective/Task	CDX Tasksheet Number	2014 Edition Rev2/12/16 AED Standard
• Identify the causes of outlet air temperature control problems in the HVAC system.	E0284	6.6
• Identify window fogging problems.	E0285	6.6
• Perform engine cooling system tests for leaks, protection level, contamination, coolant level, coolant type, temperature, and conditioner concentration; determine needed action.	E0286	6.6
• Inspect engine cooling and heating system hoses, lines, and clamps; determine needed action.	E0287	6.6
• Inspect and test radiator, pressure cap, and coolant recovery system (surge tank); determine needed action.	E0288	6.6
• Inspect water pump; determine needed action.	E0289	6.6
• Inspect and test thermostats, bypasses, housings, and seals; determine needed repairs.	E0290	6.6
• Recover, flush, and refill with recommended coolant/additive package; bleed cooling system.	E0291	6.6
• Inspect thermostatic cooling fan system (hydraulic, pneumatic, and electronic) and fan shroud; replace as needed.	E0292	6.6
• Inspect and test heating system coolant control valve(s) and manual shutoff valves.	E0293	6.6
• Inspect and flush heater core; determine needed.	E0294	6.6

Time off_____________

Time on_____________

Total time_____________

Materials Required
- Machines or simulators with cooling system faults
- PC-based software and/or data scan tools
- Machine manufacturer's workshop manual including schematic wiring diagrams
- Cooling system spare parts including coolant, clamps, hoses, water pump, seals, and valves
- Manufacturer-specific tools depending on the concern
- Coolant recovery system

Some Safety Issues to Consider

- Activities may require test driving the equipment on the school grounds, which carry severe risks. Attempt this task only with full permission from your supervisor/instructor, and follow all the guidelines exactly.
- All practices and procedures must be performed according to current mandates, standards and regulations.
- Activities require you work on the machine cooling system. The cooling system operates under pressure and with high temperatures. Never remove or release the pressure cap or hoses while the radiator is hot.
- Comply with personal and environmental safety practices associated with clothing; eye protection; hand tools; power equipment; proper ventilation; and the handling, storage, and disposal of chemicals/materials in accordance with federal, state, and local regulations.
- Always wear the correct protective eyewear and clothing and use the appropriate safety equipment, as well as fender covers, seat protectors, and floor mat protectors.
- Make sure you understand and observe all legislative and personal safety procedures when carrying out practical assignments. If you are unsure of what these are, ask your supervisor/instructor.

Performance Standard

0—No exposure: No information or practice provided during the program; complete training required

1—Exposure only: General information provided with no practice time; close supervision needed; additional training required

2—Limited practice: Has practiced job during training program; additional training required to develop skill

3—Moderately skilled: Has performed job independently during training program; limited additional training may be required

4—Skilled: Can perform job independently with no additional training

Student/Intern information:

Name _________________________________ Date ___________ Class _______________________

Vehicle used for this activity:

Year _______________ Make _______________________ Model ___________________________

Odometer _____________ Hour meter ___________ VIN ____________________________________

▶ **TASK** Identify causes of outlet air temperature control problems in the
HVAC system.

AED
6.6

CDX Tasksheet Number: E0284

1. **Consult the manufacturer's workbook manual for the procedures to identify
   and test outlet air temperature control problems. Record the procedures.**

   ____

   ____

   ____

2. **Utilizing an analog thermometer or infrared thermometer, measure the
   outlet temperature at one of the air ducts on the dash board.**

   a. **Temperature reading:** ______________ **degrees**

   b. **Is the reading acceptable? Yes:** ______________ **No:** ______________

      i. **If no, record the procedure to repair or replace the components to
         correct the situation.**

      ____

      ____

      ____

      ____

   c. **Once the repair has been made, measure the temperature again and
      record your findings:**

      i. **Temperature reading:** ______________ **degrees**

      ii. **Is the reading acceptable? Yes:** ______________ **No:** ______________

3. **Consult with your instructor. Record the recommendations to make
   necessary repairs and bring the machine within specification.**

   ____

   ____

   ____

4. **Return the machine to beginning condition and return any tools that you
   may have used to their proper locations.**

5. **Discuss the findings with the instructor.**

____

____

____

Performance Rating

CDX Tasksheet Number: E0284

0	1	2	3	4

Supervisor/instructor signature _____________________________ Date __________

Student/Intern information:

Name _________________________ Date ___________ Class _________________________

Vehicle used for this activity:

Year _____________ Make _________________________ Model _________________________

Odometer ___________ Hour meter ___________ VIN _________________________

▶ TASK Identify causes of window fogging problems. **AED 6.6**

Time off________

CDX Tasksheet Number: E0285

Time on________

1. **Research causes of window fogging problems in the HVAC system and list them below:**

   ____

   ____

   ____

Total time________

2. **Examine the machine/simulator and identify window fogging problems and list possible causes:**

   ____

   ____

   ____

3. **Write a short description of the purpose and operation of the suspected component(s):**

   ____

   ____

   ____

4. **Return the machine to beginning condition and return any tools that you may have used to their proper locations.**

5. **Discuss the findings with the instructor.**

   ____

   ____

   ____

Performance Rating

CDX Tasksheet Number: E0285

0	1	2	3	4

Supervisor/instructor signature _________________________________ Date ____________

Student/Intern information:

Name _________________________________ Date ____________ Class _________________________

Vehicle used for this activity:

Year ______________ Make _____________________________ Model _________________________

Odometer ____________ Hour meter ____________ VIN _________________________________

▶ TASK Perform engine cooling system tests for leaks, protection level, contamination, coolant level, coolant type, temperature, and conditioner concentration.

AED 6.6

Time off_________________

Time on_________________

Total time_________________

CDX Tasksheet Number: E0286

1. **Research how to perform engine cooling system tests for leaks, protection level, contamination, coolant level, coolant type, temperature, and conditioner concentration and list your findings below.**

   a. **Engine cooling system leak tests:**

   ____
   ____
   ____
   ____

   b. **Protection level:**

   ____
   ____
   ____
   ____

   c. **Contamination:**

   ____
   ____
   ____
   ____

   d. **Coolant level:**

   ____
   ____
   ____
   ____

   e. **Coolant type:**

   ____
   ____
   ____
   ____

   f. **Temperature:**

   ____
   ____
   ____
   ____

    g. Conditioner concentration:

    ____

    ____

    ____

    ____

2. Have your supervisor/instructor verify your research.

    a. Supervisor/instructor's initials: _______________

3. Ask your supervisor/instructor for a machine or simulator to check.

4. Examine the machine/simulator, and perform engine cooling system tests for leaks, protection level, contamination, coolant level, coolant type, temperature, and conditioner concentration.

5. List the issues and faults found:

____

____

____

6. Write a short description of the purpose and operation of the suspected component(s):

____

____

____

____

7. Determine and list any necessary action(s):

____

____

____

____

8. Return the machine to beginning condition and return any tools that you may have used to their proper locations.

9. Discuss the findings with the instructor.

____

____

____

Performance Rating

CDX Tasksheet Number: EO286

0	1	2	3	4

Supervisor/instructor signature _______________________________________ Date _____________

Student/Intern information:

Name ________________________________ Date ____________ Class ________________________

Vehicle used for this activity:

Year ____________ Make ________________________ Model ________________________

Odometer ____________ Hour meter ____________ VIN ________________________

▶ TASK Inspect engine cooling and heating system hoses, lines, and clamps.

AED 6.6

Time off____________

Time on____________

CDX Tasksheet Number: EO287

1. **Consult the manufacturer's workbook manual for the procedures to inspect engine cooling and heating system hoses. Record the procedures below:**

   ____

   ____

   ____

Total time____________

2. **Inspect all hoses for dry rotting and cracking.**

   a. **Condition of hoses: Good: ____________ Bad: ____________**

   b. **If the condition of the hoses is bad, record the procedures to repair or replace the hoses.**

      ____

      ____

      ____

      ____

3. **Inspect all lines for damage and kinking.**

   a. **Condition of lines: Good: ____________ Bad: ____________**

   b. **If the condition of the hoses is bad, record the procedures to repair or replace the hoses.**

      ____

      ____

      ____

      ____

4. **Inspect all clamps for tightness and damage.**

   a. **Condition of clamps: Good: ____________ Bad: ____________**

   b. **If the condition of the clamps is bad, record the procedures to repair or replace the clamps.**

      ____

      ____

      ____

      ____

5. Discuss your findings with your instructor. Record any recommendations to repair or replace the bad components in the system.

____

____

____

Performance Rating

CDX Tasksheet Number: E0287

| 0 | 1 | 2 | 3 | 4 |

Supervisor/instructor signature _______________________________ Date _____________

▶ **TASK** Inspect and test radiator, pressure cap, and coolant recovery system (surge tank).

AED 6.6

Time off_________________

Time on_________________

Total time_________________

CDX Tasksheet Number: E0288

1. **Consult the manufacturer's workbook manual for the procedures to inspect and test the radiator, pressure cap, and coolant recovery system. Record the procedures.**

   ____

   ____

   ____

2. **Using a cooling system pressure tester, pressure test the radiator to check for leaks.**

3. **Attach the pressure tester to the radiator fill neck and pump the tester to the recommended pressure.**

4. **Condition of radiator: Good: ______________ Bad: ______________**

5. **If the condition of the radiator is bad, record the procedure to replace or repair the radiator.**

   ____

   ____

   ____

6. **Pressure test the radiator cap by attaching the pressure tester to the radiator. Pump it up to the recommended pressure on the cap.**

   a. **Recommended pressure on the cap: ______________ psi**

   b. **Condition of the radiator cap:
      Good: ______________ Bad: ______________**

   c. **If the condition of the radiator cap is bad, record the suggested procedure to replace the cap.**

   ____

   ____

   ____

   ____

7. **Inspect the coolant recovery system and all attaching hoses.**

   a. **Inspect the coolant bottle for cracks and leaks.**

   b. **Condition of coolant bottle: Good: ______________ Bad: ______________**

   c. **Inspect all hoses that are connected to the coolant bottle for dry rotting and kinking.**

   d. **Condition of hoses: Good: ______________ Bad: ______________**

e. If any of these components are bad, record the procedure to replace or repair the coolant recovery system below:

____

____

____

____

8. Discuss your findings with your instructor and record the recommendations.

____

____

____

Student/Intern information:

Name _______________________________ Date ____________ Class _______________________

Vehicle used for this activity:

Year _______________ Make _______________________ Model _____________________________

Odometer ____________ Hour meter ___________ VIN _________________________________

▶ **TASK** Inspect water pump; determine needed action.　**AED 6.6**

Time off____________

Time on____________

Total time____________

CDX Tasksheet Number: EO289

1. **Consult the manufacturer's workbook manual for the procedures to inspect the water pump and list below:**

   ____

   ____

   ____

2. **Visually inspect the water pump for leakage at the weep hole and around the pump shaft.**

   a. **Condition of weep hole and shaft:**
      Good: _______________ **Bad:** _______________

   b. **If the condition of the weep hole and/or shaft is bad, record the recommended procedure to replace the pump.**

      ____

      ____

      ____

      ____

3. **Inspect the water pump mountings and gaskets for any leakage that may be present.**

   a. **Condition of mountings and gaskets:**
      Good: _______________ **Bad:** _______________

   b. **If the condition of the mountings and/or gaskets is bad, record the recommended procedure to repair or replace the mountings and/or gaskets.**

      ____

      ____

      ____

      ____

4. **Discuss your findings with your instructor and record the recommendations.**

____
____
____

Performance Rating

CDX Tasksheet Number: E0289

0	1	2	3	4

Supervisor/instructor signature _______________________________ Date _____________

Student/Intern information:

Name _____________________________ Date ____________ Class _____________________

Vehicle used for this activity:

Year _____________ Make _____________________ Model _____________________

Odometer ____________ Hour meter ____________ VIN _____________________

▶ TASK Inspect and test thermostats, bypasses, housings, and seals; determine needed repairs.

AED 6.6

Time off____________

Time on____________

CDX Tasksheet Number: E0290

Total time____________

1. **Consult the manufacturer's workbook manual for the procedures to inspect and test thermostats, bypasses, housings, and seals. Record the procedures.**

   ____

   ____

   ____

2. **Inspect and test the thermostats by utilizing a hot plate and pot of water to test the thermostat.**

   a. **Heat the water to the boiling point.**

   b. **Put the thermostat into the pot of hot water.**

   c. **Observe the thermostat to see if it opens fully.**

   d. **If it does not open fully, the recommended procedure is to replace it.**

   e. **Condition of the thermostat:**
      Good: _____________ Bad: _____________

3. **Inspect all bypass hoses for dry rotting and cracking.**

   a. **Condition of bypass hoses: Good: _____________ Bad: _____________**

   b. **If the condition of the bypass hoses is bad, the recommended procedure is to replace all bypass hoses.**

4. **Inspect all thermostat housings and bypass housings (if present) and seals for leakage or physical damage.**

   a. **Condition of housings and seals:**
      Good: _____________ Bad: _____________

   b. **If the condition of the housings and seals is bad, the recommended procedure is to replace the housings and seals.**

5. Discuss your findings with your instructor and record the recommendations.

____

____

____

<table>
<tr><td colspan="5">Performance Rating   CDX Tasksheet Number: E0290</td></tr>
<tr><td>☐</td><td>☐</td><td>☐</td><td>☐</td><td>☐</td></tr>
<tr><td>0</td><td>1</td><td>2</td><td>3</td><td>4</td></tr>
</table>

Supervisor/instructor signature _______________________________________ Date _____________

Student/Intern information:

Name _________________________________ Date ____________ Class _____________________________

Vehicle used for this activity:

Year ______________ Make ___________________________ Model _________________________________

Odometer ____________ Hour meter ____________ VIN ____

▶ **TASK** Recover, flush, and refill with recommended coolant/additive
package; bleed cooling system.

AED
6.6

Time off______________________

Time on______________________

CDX Tasksheet Number: E0291

Total time____________________

1. **Consult the manufacturer's workbook manual for the procedures to recover, flush, and refill the system with recommended coolant/additive package and bleed the cooling system.**

2. **Recover the coolant from the cooling system and dispose of the fluid according to state and local guidelines.**

3. **Flush the cooling system to remove any leftover coolant and contaminants in the system.**

4. **Note: Utilizing a cooling system flush machine will aid in the removal of all coolant and contaminants in the system. These machines can remove, flush, and refill with the proper amount of coolant automatically.**

5. **Refill the system with the proper coolant and the correct amount of fluid.**

   a. **Manufacturer-specified coolant:** _________________________________

   b. **Recommended coolant amount:** _________________________________

   c. **Actual amount of coolant used:** _________________________________

6. **If additive packages are specified, add as needed.**

> **NOTE:** Additives are used to maintain proper pH levels and to prevent cavitation from eroding internal engine parts due to coolant consistency. As coolant ages, acid levels rise and can damage liners and blocks.

7. **Bleeding the system requires running the engine and opening the bleeder fittings on the water manifold on top of the engine. Some fittings may be found on the thermostat housing also.**

8. **Open the bleeder fittings and let the liquid run until there are no more bubbles present.**

9. **Discuss the findings with your instructor and record the recommendations.**

____

____

____

| 0 | 1 | 2 | 3 | 4 |

Supervisor/instructor signature _______________________________________ Date ___________

Student/Intern information:

Name _________________________________ Date ___________ Class _________________________

Vehicle used for this activity:

Year _______________ Make _________________________ Model _____________________

Odometer ____________ Hour meter ____________ VIN _________________________________

▶ TASK Inspect thermostatic cooling fan system (hydraulic, pneumatic, and electronic) and fan shroud.

AED 6.6

Time off____________________

Time on____________________

CDX Tasksheet Number: E0292

Total time____________________

1. **Consult the manufacturer's workbook manual for the procedures to inspect the thermostatic cooling fan system (hydraulic, pneumatic, and electronic) and fan shroud. Record the procedure.**

   ____

   ____

   ____

2. **Inspect the thermostatic cooling fan components and sensors for proper operation.**

   a. **Inspect fan for cracks or missing blades.**

      i. **Condition of blades: Good: _______________ Bad: _______________**

      ii. **If the condition of the blades is bad, consult the manufacturer's workbook manual for the proper procedure to replace the fan blade. Record the procedure.**

         ____

         ____

         ____

         ____

   b. **Inspect sensors for proper connections (pneumatic or electronic).**

      i. **Condition of connections:**
         Good: _______________ Bad: _______________

      ii. **If the condition of the connections is bad, consult the manufacturer's workbook manual for the proper procedure to replace the sensors and/or repair the connections. Record the procedure.**

         ____

         ____

         ____

         ____

3. **Inspect the hydraulic fan center unit for leakage.**

   a. **Leakage present: Yes: _______________ No: _______________**

   b. **If fluid is present, it will require the replacement of the hydraulic unit portion of the fan.**

4. **Inspect the fan shroud for proper routing of the air over the engine.**

5. Inspect the check for cracks or missing pieces of the fan shroud.

    a. Condition of fan shroud components:
       Good: _______________ Bad: _______________

    b. If the condition of the shroud components is bad, consult the manufacturer's workbook manual for proper procedures to repair or replace the fan shroud components.

> **NOTE:** Some fan shrouds use rubber sections to close off the engine compartment for proper air flow.

6. Discuss the findings with your instructor and record the recommendations.

____
____
____

Performance Rating

CDX Tasksheet Number: E0292

0	1	2	3	4

Supervisor/instructor signature _____________________________________ Date _______________

Student/Intern information:

Name ___________________________ Date ___________ Class ___________________

Vehicle used for this activity:

Year _____________ Make ___________________________ Model ___________________

Odometer ___________ Hour meter ___________ VIN ___________________________

▶ **TASK** Inspect and test heating system coolant control valve(s) and manual shutoff valves.

AED 6.6

CDX Tasksheet Number: E0293

1. **Consult the manufacturer's workbook manual for the procedures to inspect and test the heating system coolant control valve(s) and manual shutoff valves. Record the procedures.**

   ____
   ____
   ____

2. **Inspect the coolant control valves for coolant leakage.**

   a. **Leakage present: Yes: _____________ No: _____________**

   b. **If leakage is present, consult the manufacturer's workbook manual for proper procedures to replace or repair the coolant control valves.**

   ____
   ____
   ____

3. **Check for proper operation of electronically controlled valves.**

   a. **This may require the use of an electronic service tool or scan tool.**

   b. **Record any data the scan tool provides on electronic operation.**

   ____
   ____
   ____

4. **Inspect the manual control valves for leakage.**

   a. **Leakage present: Yes: _____________ No: _____________**

   b. **If leakage is present, consult the manufacturer's workbook manual for proper procedures to replace or repair the manual control valves.**

   ____
   ____
   ____

5. **Inspect manual valve vacuum lines for dry rotting and kinking.**

   a. **Condition of vacuum lines: Good: _____________ Bad: _____________**

   b. **If the condition of the vacuum lines is bad, consult the manufacturer's workbook manual. Record the proper procedures to repair or replace the vacuum lines.**

   ____
   ____
   ____

6. Inspect manual valve cable linkage for proper operation.

   a. Condition of cable linkage: Good: _______________ Bad: _______________

   b. If the condition of the cable linkage is bad, consult the manufacturer's workbook manual. Record the proper procedures to repair or replace the cable linkage.

7. Discuss the findings with the instructor. Record any recommendations for repair or replacement of coolant control valves.

____

____

____

Performance Rating   **CDX Tasksheet Number: E0293**

| 0 | 1 | 2 | 3 | 4 |

Supervisor/instructor signature _________________________________ Date __________

Name _______________________________ Date _____________ Class _______________________

Vehicle used for this activity:

Year _______________ Make _______________________ Model _______________________

Odometer _____________ Hour meter _______________ VIN _______________________________

▶ TASK Inspect and flush heater core.   **AED 6.6**

CDX Tasksheet Number: E0294

Time off_______________

Time on_______________

Total time_______________

1. **Research how to inspect and flush heater core and list your findings below:**

   a. **How to inspect heater core:**

   ____

   ____

   ____

   b. **How to flush heater core:**

   ____

   ____

   ____

2. **Have your supervisor/instructor verify your research.**

   a. **Supervisor/instructor's initials:** _______________

3. **Ask your supervisor/instructor for a machine/simulator to check.**

4. **Inspect and flush heater core.**

5. **List the issues and faults found.**

   ____

   ____

   ____

6. **Write a short description of the purpose and operation of the suspected component(s):**

   ____

   ____

   ____

7. Discuss the findings with the instructor. Record any recommendations for
   repair or replacement of heater core.

   ____

   ____

   ____

▶ **TASK** Inspect and flush heater core.

AED
6.6

Time off________________

Time on________________

CDX Tasksheet Number: E0294

1. **Research how to inspect and flush heater core and list your findings below:**

   a. **How to inspect heater core:**

   ____
   ____
   ____

   Total time________________

   b. **How to flush heater core:**

   ____
   ____
   ____

2. **Have your supervisor/instructor verify your research.**

   a. **Supervisor/instructor's initials:** _______________

3. **Ask your supervisor/instructor for a machine/simulator to check.**

4. **Inspect and flush heater core.**

5. **List the issues and faults found.**

   ____
   ____
   ____

6. **Write a short description of the purpose and operation of the suspected component(s):**

   ____
   ____
   ____

7. Discuss the findings with the instructor. Record any recommendations for repair or replacement of heater core.

____

____

____

Performance Rating

CDX Tasksheet Number: E0294

0	1	2	3	4

Supervisor/instructor signature ____ Date _______________